Handbook of Petrochemicals

Handbook of Petrochemicals

Edited by **Gene Saxe**

NY RESEARCH
P R E S S

New York

Published by NY Research Press,
23 West, 55th Street, Suite 816,
New York, NY 10019, USA
www.nyresearchpress.com

Handbook of Petrochemicals
Edited by Gene Saxe

© 2015 NY Research Press

International Standard Book Number: 978-1-63238-262-7 (Hardback)

Printed in the United States of America.

Contents

Preface

This book has been an outcome of determined endeavour from a group of educationists in the field. The primary objective was to involve a broad spectrum of professionals from diverse cultural background involved in the field for developing new researches. The book not only targets students but also scholars pursuing higher research for further enhancement of the theoretical and practical applications of the subject.

A chemical byproduct obtained from petroleum and natural gas is defined as a petrochemical. The petrochemical industry forms a significant part of our pursuit of economic development, generation of employment and basic needs. It is a vast field that covers several commercial chemicals and polymers. This book is formulated to assist the reader, specifically students and researchers of petroleum science and engineering, in comprehending the mechanics and methodologies. The information provided in this book accompanied by the graphs, tables and examples employed to elucidate them are driven by the fact that this book is targeted mainly at the petroleum science and engineering technologists. It will act as a good source of knowledge for engineers and researchers and even students concerned with vast field of petrochemicals. This book gives valuable and worthwhile insights into the significant chemical reactions and mechanisms. The aim of this book is to serve as an informative, concise and useful source of information.

It was an honour to edit such a profound book and also a challenging task to compile and examine all the relevant data for accuracy and originality. I wish to acknowledge the efforts of the contributors for submitting such brilliant and diverse chapters in the field and for endlessly working for the completion of the book. Last, but not the least; I thank my family for being a constant source of support in all my research endeavours.

Editor

Part 1

Introduction

Chemical Reactor Control

Azzouzi Messaouda[1] and Popescu Dumitru[2]
[1]Ziane Achour University of Djelfa
[2]Politehnica University of Bucharest
[1]Algeria
[2]Romania

1. Introduction

The principal technological element of the industrial plants and the chemical reactions is the chemical reactor. In this chapter, it is considered that the chemical reactor is an apparatus in which the chemical process can be effectuated to obtain certain substances in technological process. The automatic control systems by their dynamic bring those processes to a point where the profile is optimal, a fact which imposes several methods to achieve the desired performances. This chapter proposes a hierarchical configuration of control which treats aspects related to the primary processing of data, processes identification, control and the robustness analysis under some conditions of the operating regime for representative plants in the chemical and petrochemical industry. It is a design of numerical control laws in a pyrolysis reactor in order to achieve an efficient regime of operation that allows as much as possible to optimize the concentration of produced ethylene by the chemical reactions (Popescu et al., 2006a; Landau & Zito, 2006).

2. Technologic overview of the plant

2.1 Technical description and operating conditions

The pyrolysis reactor as shown in figure (1) is planted in petrochemical plants and is intended to obtain ethylene, the combination of a quantity of gas with water under certain operative conditions can produce a chemical reaction, which has the ethylene as a result among the reaction products (Popescu et al., 2006b).

The block diagram shown in figure (1) describes the chemical transformations selected from a petrochemical plant whose pyrolysis took a large space. The kinetic parameters of reactions are determined by the mathematical adjustment to the experimental results obtained in several scientific resources in the combined model in the simplification of the oil and on its derivatives (Mellol, 2004).

2.2 Automation solutions of the reactor

Gas and steam are introduced to the plant of two access roads at a constant flow, but the amount of the necessary heat for the reaction is obtained from a heating system which uses

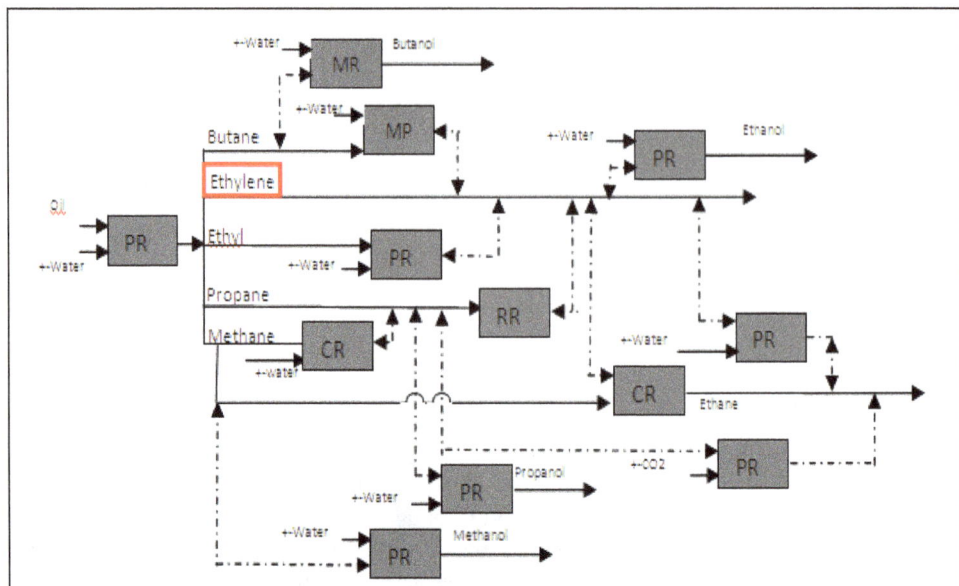

Fig. 1. Integration of Pyrolysis reactor into a petrochemical plant
Where
PR: pyrolysis reactor
MR: Membrane reactor
RR: Reforming reactor
CR: Catalytic reactor

a straight fuel powered methane gas (Harriott, 2002). For the insurance of the desired operation of the reactor, it is necessary to achieve an automation mechanism that provides a measurement system, automatic control and supervision of the parameters I/O, the automation solution is shown in figure (2), which there are implemented the four controllers:

Data were recorded using an acquisition card connected to a computer (Azzouzi, 2008). Variables are limited in technological point of view such as:

$x1$: flow of primary matter (oil) (1000 to 1600 Nm³/h)
$x2$: flow of steam (430-540 Nm³/h)
$x3$: working pressure (3.2-4.5 atm)
$x4$: working temperature (820-860°C)
y (%): the concentration of the useful product (ethylene) at the reactor output

2.2.1 Flow control

Automatic systems of flow control structure are made of simple adjustment depending on the error, such a structure is used only to maintain a flow to a specified value or as a secondary loop in a control structure that changes in cascade the flow with the level or concentration (Borne et al., 1993).

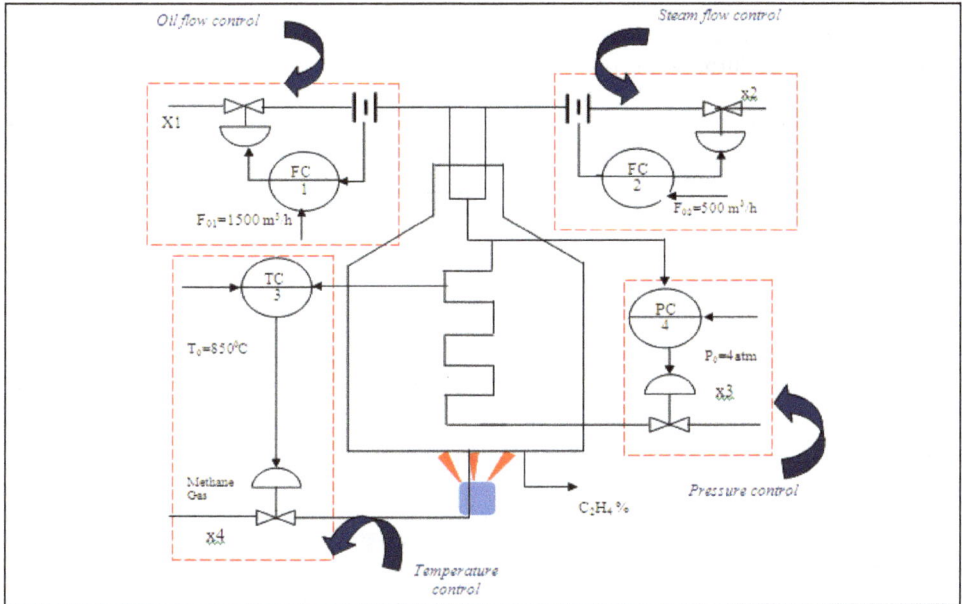

Fig. 2. Controller loops that govern the pyrolysis reactor.

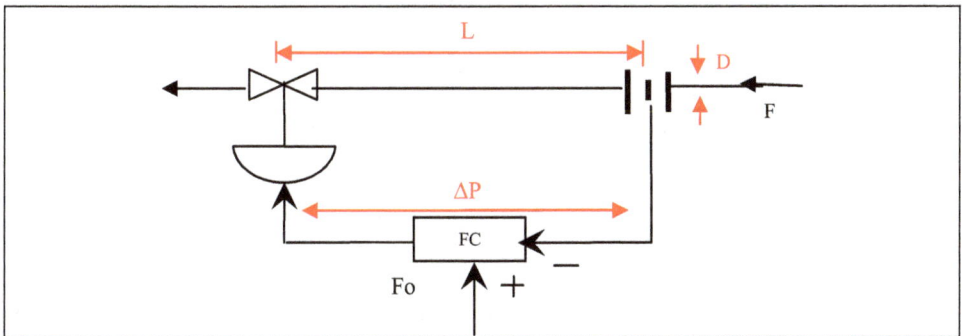

Fig. 3. Flow controller
Where:
R : radius of pipe
L: length de pipe
F : flow of fluid
F0: intial value of flow
ΔP: pressure reduction by restriction
α: flow coefficient
ρ: density of fluid
kp: amplification factor, kp=0.5

τpa: delay constant of channel, $\tau_{pa} = \chi^2 \dfrac{V_0}{F_0}$

V0: volume of fluid in the pipe in case of steady

M: mass of liquid through the pipe

v: flow rate of liquid through the pipe

For the flow control systems, using the theorem of short lines: pipeion is equivalent to a hydraulic resistance defined by the known relationship:

$$F = \alpha S \sqrt{\frac{2\Delta P}{\rho}} \qquad (1)$$

For steady flow, the applied forces into the system are balanced, which implies:

$$\Delta P_0 - \frac{F_0^2 \rho}{2\alpha^2 S} = 0 \qquad (2)$$

where:

ΔP_0: is the active force to push the liquid in the pipe

$\dfrac{F_0^2 \rho}{2\alpha^2 S}$ is the reaction force by the restriction

In dynamic regime, the deference between these two forces is compensated by the rate of change of pulse time of the system.

$$\Delta P(t)S - \frac{F^2(t)\rho}{2\alpha^2 S} = \frac{d}{dt}(Mv) \qquad (3)$$

Which imply that:

$$\Delta P(t)S - \frac{F^2(t)\rho}{2\alpha^2 S} = \rho L S \frac{1}{S}\frac{d}{dt}(F(t)) \qquad (4)$$

Values are obtained that depend on t , if the variations of two arbitrary values of steady state are given as:

$$\Delta P(t) = \Delta P_0 + \Delta(\Delta P(t)) = \Delta P_0 + \Delta p(t)$$
$$F(t) = F_0 + \Delta F(t) \qquad (5)$$

From (4) and (5):

$$(\Delta P_0 + \Delta p(t))S - \frac{\rho(F_0 + \Delta F(t))^2}{2\alpha^2 S^2}S = \rho L \frac{d}{dt}(F_0 + \Delta F(t)) \qquad (6)$$

By extacting (6) and the steady state expressed by (2) and by ignoring the quadratic term $\Delta F^2(t)$, the following equation can be obtained:

$$\Delta p(t)S - \frac{2\rho F_0 \Delta F(t)}{2\alpha^2 s^2} S = \rho L \frac{d}{dt}(\Delta F(t)) \tag{7}$$

By the normativity in the steady state:

$$y(t) = \frac{\Delta F(t)}{F_0} \text{ and } M(t) = \frac{\Delta p(t)}{\Delta P_0}$$

Which results lthe linear model with adimensional variables:

$$\alpha^2 \frac{V_0}{F_0} \frac{dy(t)}{dt} + y(t) = \frac{1}{2} m(t) \tag{8}$$

From equation (6), and by applying the Laplace transform, the transfer function of the execution channel can be got easily.

$$H_{pa}(s) = \frac{k_p}{\tau_{pa} s + 1} \tag{9}$$

Control of oil flow

The oil flow controller is shown in figure (4). In this study case of the pyrolysis reactor technology parameters are given by:

The length of the pipe L = 10 m
The diameter of the pipe is 0.2m which means that R = 0.1m
The flow of rated speed; F_{10}=1500m³/h=0.41m³/sec
The flow coefficient α=0.9
The amplification factor kp=0.9

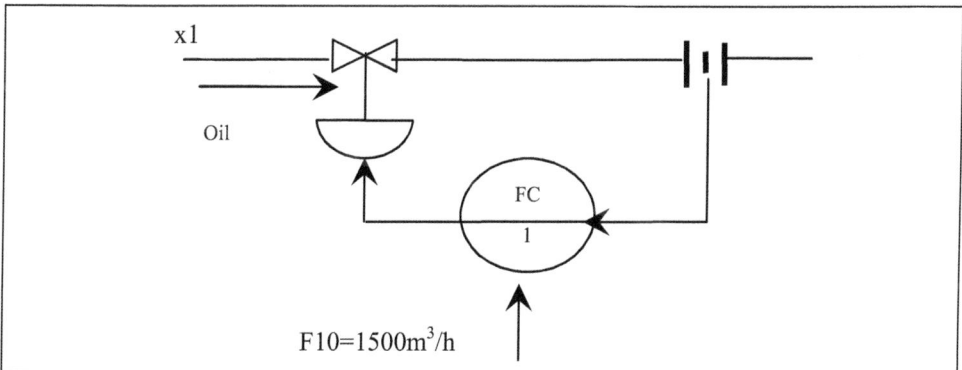

Fig. 4. Oil flow controller.

$$H_p(s) = \frac{0.5}{\alpha^2} \frac{V}{F_{10} s + 1} = \frac{0.5}{0.62s + 1} \tag{10}$$

$$V = \pi * r^2 * L = \pi \left(\frac{0.2}{2}\right)^2 * 10 = 3.14 * (0.01) * (10) = 0.314 m^3 \tag{11}$$

$$a^2 \frac{V}{F_{10}} = 0.9^2 * \frac{0.314}{0.41} = 0.62 \tag{12}$$

$H_E(s) = \dfrac{0.66}{8s+1}$ transfer function of the actuator

$H_p(s) = \dfrac{0.5}{0.62s+1}$ transfer function of the process

$H_T = \dfrac{\Delta I}{\Delta F_1} \cdot \dfrac{F_{10}}{I_0} = \dfrac{16}{2000} * \dfrac{1500}{16} = 0.75$ transfer function of the sensor

$$H_F = H_E * H_p * H_T = \frac{0.66}{8s+1} * \frac{0.5}{0.62s+1} * 0.75$$
$$= \frac{0.24}{(8s+1)(0.62s+1)} \tag{13}$$

The parasite time constant 0.62 can be ignored because is too small compared to the main time constant 8s, so the controller has the following transfer function:

$$H_F(s) = \frac{0.24}{(8s+1)} = \frac{K_p}{T_p s + 1} \tag{14}$$

It is recommended that a PI controller which has the form $H_R(s) = K_R \cdot \left(1 + \dfrac{1}{T_i s}\right)$

If Ti=Tp=8s

$$H_d = H_R(s).H_F(s) = K_R \frac{(1+T_i s)}{T_i s} \frac{K_p}{T_p s + 1} = \frac{K_R.K_p}{T_p s} \tag{15}$$

$$H_0 = \frac{H_d}{H_d + 1} = \frac{K_R K_p / T_p s}{(K_R K_p / T_p s) + 1} = \frac{1}{\dfrac{T_p}{K_R K_p} s + 1} \tag{16}$$

$$\Rightarrow T_0 = \frac{T_p}{K_R.K_p} = \frac{8}{0.24 K_R} \Rightarrow K_R = \frac{33.33}{T_0} \tag{17}$$

To set a time of 8s. It is necessary to choose, $T_0 = 2s \Rightarrow K_R \approx 16$ to facilitate the calculation.

$$H = 16.\frac{8s+1}{8s}.\frac{0.24}{(8s+1)(0.62s+1)} \tag{18}$$

MatLab is used to obtain graphs of the step response of this closed loop system. The transfer function of BF

$$HBF = \frac{30.72s+3.84}{39.68s^3 + 68.96s^2 + 38.72s + 3.84} \tag{19}$$

From the above equation, the oil flow, the time response, the step response and the robustness diagrams can be ploted as shown in figure (5).

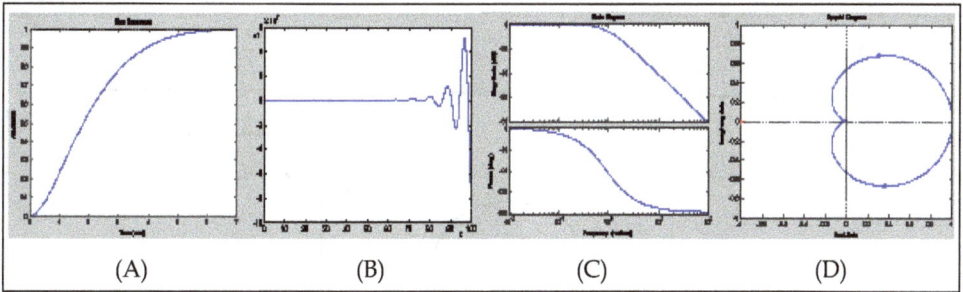

| (A) | (B) | (C) | (D) |

Fig. 5. (A) Step reponse, (B) Time response, (C) Bode diagram and (D) Nyquist diagram.

By the bilinear transformation $s = \frac{2}{T_p}.\frac{1-q^{-1}}{1+q^{-1}}$, the deduced formula of the model and its

discrete PI controller is $H_{PI}(q^{-1}) = \frac{k_p}{2T_i}\left(\frac{(2T_i+T_p)+q^{-1}(T_p-2T_i)}{2T_i(1-q^{-1})}\right)$, which results:

Model

$$H(q^{-1}) = \frac{0.85q^{-2} + 1.4q^{-1} + 3.45}{q^{-2} + 2q^{-1} + 1} \tag{20}$$

Controller

$$R(q^{-1}) = 3 - q^{-1} \tag{21}$$

$$S(q^{-1}) = 2(1-q^{-1}) \tag{22}$$

$$T(q^{-1}) = 2 \tag{23}$$

Steam flow control

The controller in case of steam flow is shown in figure (6). Such as the technological parameters which are given by:

The length of the pipe 5 m
The diameter of the pipe is 0.1m which means that R = 0.05m
The flow of nominal regime $F_{20}=500m^3/h=0.14m^3/sec$
Flow coefficient $\alpha = 0.95$

$$H_p(s) = \frac{0.5}{\alpha^2} \frac{1}{\alpha^2 V/F_{10} s + 1} = \frac{0.5}{0.16s + 1} \tag{24}$$

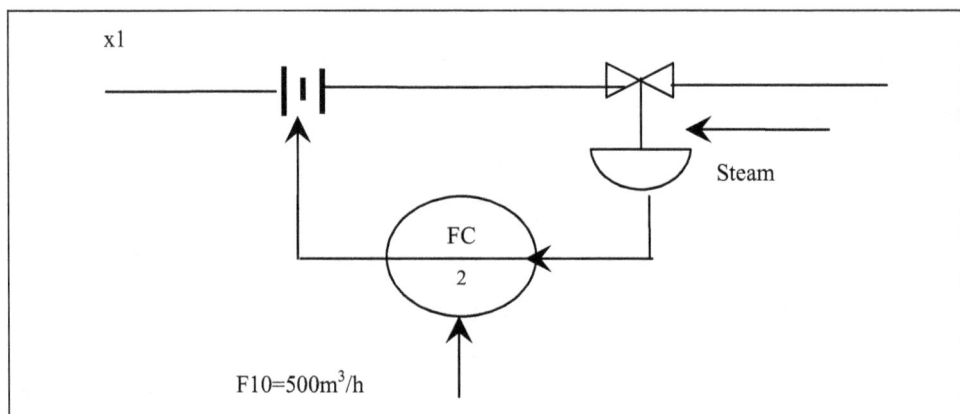

Fig. 6. Control of steam flow.

$$V = \pi * r^2 * L = \pi \left(\frac{0.1}{2}\right)^2 * 5 = 3.14 * (0.05)^2 * (5) = 0.4 \tag{25}$$

$$\alpha^2 \frac{V}{F_{20}} = 0.74^2 * \frac{0.4}{0.14} = 0.16 \tag{26}$$

$H_E(s) = \dfrac{0.66}{4s+1}$ transfer function of the actuator

$H_p(s) = \dfrac{0.5}{0.16s+1}$ transfer function of the process

$H_T = \dfrac{\Delta I}{\Delta F_1} \cdot \dfrac{F_{10}}{I_0} = \dfrac{16}{2000} \cdot \dfrac{500}{12} = 0.66$ transfer function of the sensor

$$H_F = H_E.H_p.H_T = \frac{0.66}{4s+1} \cdot \frac{0.5}{0.16s+1} . 0.66 = \frac{0.21}{(4s+1)(0.16s+1)} \tag{27}$$

The parasite time constant 0.16s because it is too small compared to the main time constant 4s, then the controller transfer function will be: $H_F(s) = \dfrac{0.21}{(4s+1)} = \dfrac{K_p}{T_p s + 1}$

It is recommended the use of PI controller $H_R(s) = K_R \cdot \left(1 + \dfrac{1}{T_i s}\right)$, when Ti=Tp=4s

$$H_d = H_R(s).H_F(s) = K_R \frac{(1+T_i s)}{T_i s} \frac{K_p}{T_p s + 1} = \frac{K_R \cdot K_p}{T_p s} \tag{28}$$

$$H_0 = \frac{H_d}{H_d + 1} = \frac{K_R K_p / T_p s}{(K_R K_p / T_p s) + 1} = \frac{1}{\dfrac{T_p}{K_R K_p} s + 1} \tag{29}$$

$$\Rightarrow T_0 = \frac{T_p}{K_R . K_p} = \frac{4}{0.21 K_R} \Rightarrow K_R = \frac{19}{T_0} \tag{30}$$

To specify a transient time of 8s, it is necessary to choose

$$T_0 = 2s \Rightarrow K_R \approx 9 \Rightarrow H_R(s) = 9\left(1 + \frac{1}{8s}\right)$$

The use of MatLab to check the robustness of the closed loop system gives: The transfer function:

$$HBF = \frac{0.21}{0.84s^2 + 4.16s + 1.21} \tag{31}$$

The time response with the step response and Nyquist and Bode plots are respectively represented in figure (7).

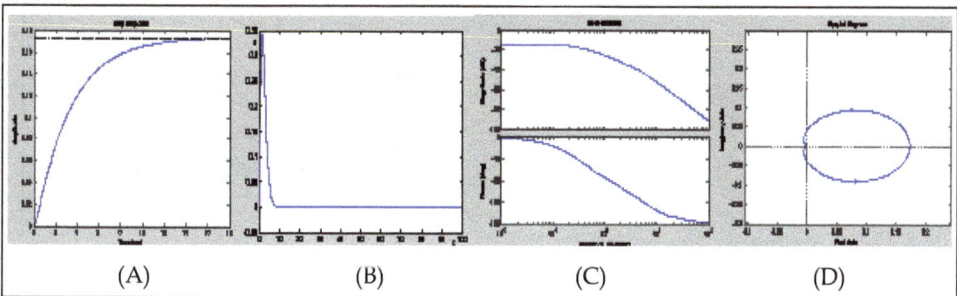

| (A) | (B) | (C) | (D) |

Fig. 7. (A) Step reponse, (B) Time response, (C) Bode diagram and (D) Nyquist diagram.

After the bilinear transformation $s = \dfrac{2}{T_p} \cdot \dfrac{1-q^{-1}}{1+q^{-1}}$, so the discrete formula of the controller is

PI $H_{PI}(q^{-1}) = \dfrac{k_p}{2T_i}\left(\dfrac{\left(2T_i+T_p\right)+q^{-1}\left(T_p-2T_i\right)}{2T_i\left(1-q^{-1}\right)} \right)$ which results:

Model

$$H(q^{-1}) = \frac{-0.92q^{-2}+1.68q^{-1}+3.24}{q^{-2}+2q^{-1}+1} \tag{32}$$

Controller

$$R\left(q^{-1}\right) = 9\left(3-q^{-1}\right) \tag{33}$$

$$S\left(q^{-1}\right) = 16\left(1-q^{-1}\right) \tag{34}$$

$$T\left(q^{-1}\right) = 18 \tag{35}$$

2.2.2 Control of the reaction pressure

In systems of pressure control, It is determined for example, a mathematical model for a pneumatic capacity powered by a fluid (gas phase), the structure of a pressure control system, is given in figure (8) (Bozga & Muntean, 2000).

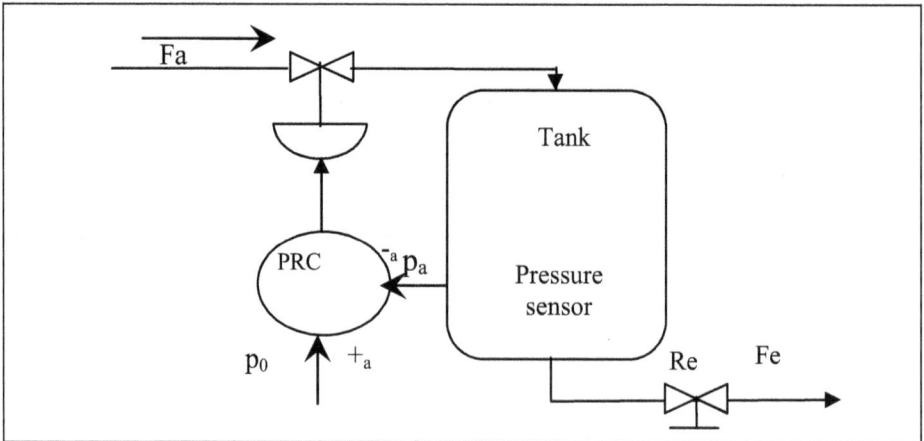

Fig. 8. Pressure control
Where
R: universal gas constant;
F: feed rate;

Fe extraction rate;
p: pressure;
p0: required value for the pressure.

After filtering data recorded previously, and estimated degrees of polynomials by the software WinPim (Azzouzi, 2009b). It was found that the polynomial A is of the second degree, while the polynomial B is of the first degree, the validation test has confirmed the quality of the chosen model, the results of the identification and regulation respectively by using WinPim and WinReg are:

Model

Structure of model of identification system: ARX
Identification method: recursive least squares
Adaptation algorithm parametric decreasing gain
Te=3s
Delay: D=0

$$H\left(q^{-1}\right) = \frac{B\left(q^{-1}\right)}{A\left(q^{-1}\right)} = \frac{0.0471q^{-1}}{1 - 1.614q^{-1} + 0.653q^{-2}} \tag{36}$$

Controller

Method : poles placement

$$R\left(q^{-1}\right) = 43.173 - 45.371q^{-1} + 13.87q^{-2} \tag{37}$$

$$S\left(q^{-1}\right) = 1 - q^{-1} \tag{38}$$

$$T\left(q^{-1}\right) = 21.227 - 12.35q^{-1} + 2.76q^{-2} \tag{39}$$

Refence model

$$Am\left(q^{-1}\right) = 1 - 0.697q^{-1} + 0.151q^{-2} \tag{40}$$

$$Bm\left(q^{-1}\right) = 0.297 + 0.157q^{-1} \tag{41}$$

To maximize the natural frequency and the damping factor, it should be considered initially that in tracking w_0=0.35 rad/s and xi=0.9 then for the regulation w_0=0.4 rad/s and xi=0.85. Figure (9) shows the Nyquist diagram with the shift of the poles from a proposed system P1 to a robust P4 after the change of the tracking and control polynomials.

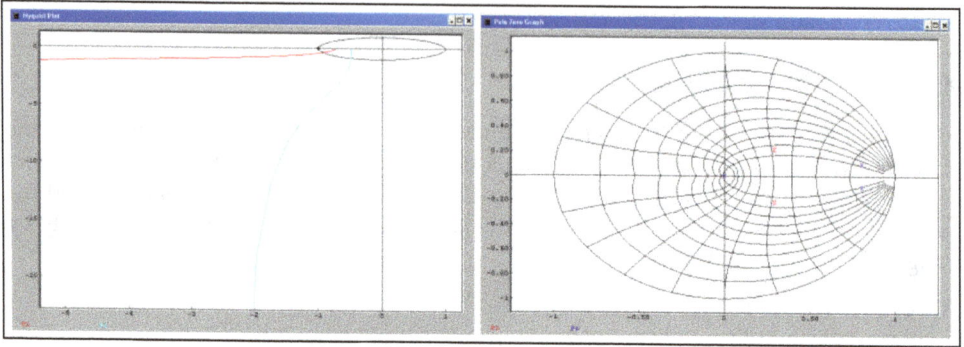

Fig. 9. Nyquist diagram and poles in closed loop.

Robustness margins are respectively shown in table (1), the goal here is to approach the margins form known robustness margins. A slow change in parameter values and further regulation and traking may increase the system robustness in closed loop (Oustaloup, 1994).

Nr	Traking Double		Regulation Double		ΔG (dB)	$\Delta\Phi(°)$	$\Delta\tau(s)$	ΔM(dB)
	w_0	xi	w_0	xi				
1	0.35	0.9	0.4	0.85	2.64	22.8	0.69	11.64
2	0.1	0.65	0.2	0.5	4.94	41.4	1.84	-7.25
3	0.08	0.7	0.1	0.6	6.16	61.0	3.23	-5.88
4	0.075	0.88	0.08	0.87	6.03	59.9	3.02	-6.01

Table 1. Robustness margins in function of tracking and control parameters.

The robustness test is important to identify the operating factors which are not necessarily considered in the development phase of the method, but could influence the results, and therefore to anticipate problems that may occur during the application of the chosen method. A series of curves to the robust sensitivity function analyzed by using WinReg is shown in figure (10). The new polynomials of the controller and the reference model are given as follows:

Controller

$$R\left(q^{-1}\right) = 21.274 - 34.150q^{-1} + 13.87q^{-2} \tag{42}$$

$$S\left(q^{-1}\right) = 1 - q^{-1} \tag{43}$$

$$T\left(q^{-1}\right) = 21.227 - 34.213q^{-1} + 13.98q^{-2} \tag{44}$$

Reference model

$$Am\left(q^{-1}\right) = 1 - 1.631q^{-1} + 0.673q^{-2} \qquad (45)$$

$$Bm\left(q^{-1}\right) = 0.022 + 0.019q^{-1} \qquad (46)$$

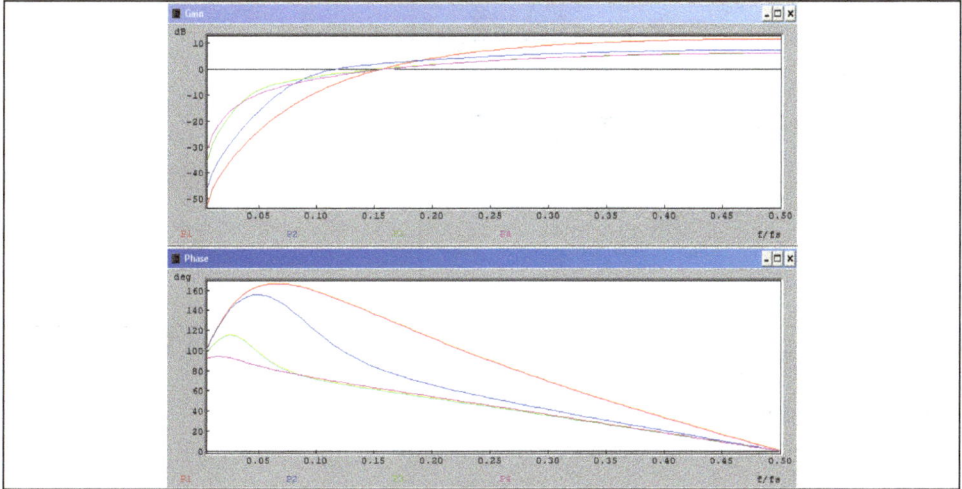

Fig. 10. Sentivity fucntion for the system robustification.

The introduction of an input signal step type in closed-loop system, with a delay of 10s and an amplitude of 1%, by adding a perturbation amplitude of 3.5 * 10-3% applied at time 40s, results a small attenuation of response at the same time of its application, Figure (11), shows the step response and the effect of disturbance on this response, which demonstrates that the tracking and control performances are provided (Azzouzi & Popescu, 2008).

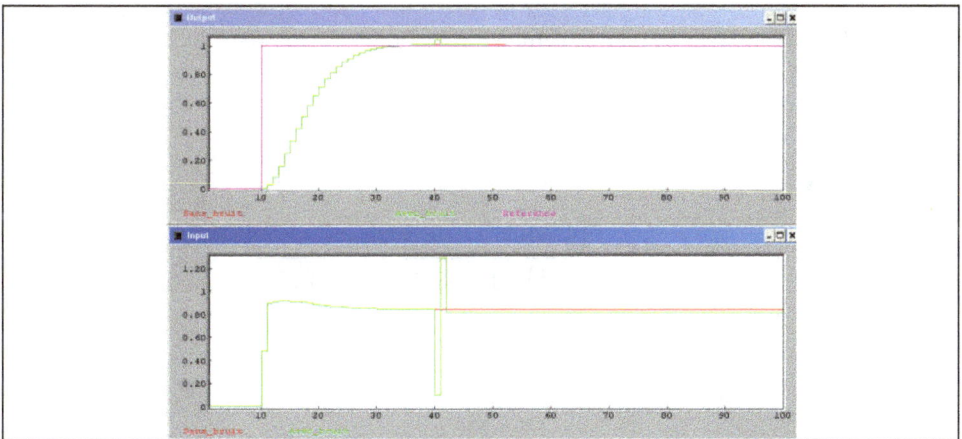

Fig. 11. Step response under and without disturbance.

2.2.3 Reaction temperature control

The temperature is a parameter representative for the chemical and petrochemical processes, with the transfer of heat, in temperature control systems, the mathematical model will be calculated for heat transfer of the product which will be heated or cooled, the structure of an SRA for temperature is given in figure (12) (Ben Abdennour, 2001).

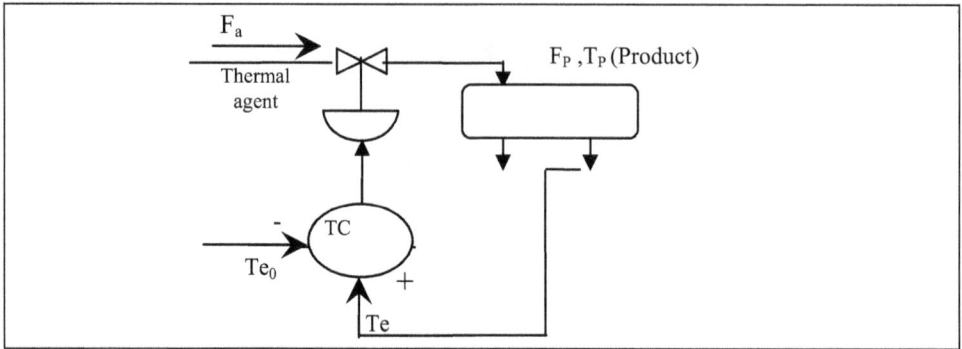

Fig. 12. Temperature control
Where
F: flow of heating agent
Ta: temperature of heating agent
Fp: product flow
Tp: temperature of product
Te: temperature alloy
TE0: required value for the temperature

According to the estimation of polynomial degrees, one could deduct that A is a second degree polynomial and B is of the first degree, the results of the identification and control by WinPim and WinReg respectively are:

Model
Structure of model of identification system: ARX
Identification method: recursive least squares
Adaptation algorithm parametric decreasing gain
Te=5s
Delay: D=0

$$H\left(q^{-1}\right)=\frac{B\left(q^{-1}\right)}{A\left(q^{-1}\right)}=\frac{0.00597q^{-1}}{1-1.683q^{-1}+0.707q^{-2}} \qquad (47)$$

Controller

Control method: Poles placement

$$R\left(q^{-1}\right)=160.81q^{-1}-276.03q^{-2}+118.475q^{-3} \qquad (48)$$

$$S\left(q^{-1}\right) = 1 - q^{-1} \tag{49}$$

$$T\left(q^{-1}\right) = 167.504 - 288.712q^{-1} + 124.463q^{-2} \tag{50}$$

Reference model

$$Am\left(q^{-1}\right) = 1 - 0.446q^{-1} + 0.05q^{-2} \tag{51}$$

$$Bm\left(q^{-1}\right) = 0.442 + 0.161q^{-1} \tag{52}$$

The Nyquist diagram with the location of poles in closed loop of figure (13) are taken in two different cases of treatment, they show the difference between a controller without pre-specification of performance given above, and a robust controller that is given later by adding a few performances (Ogata, 2001).

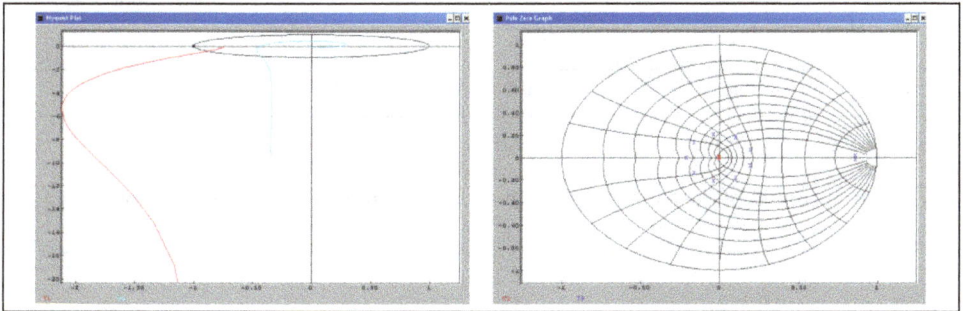

Fig. 13. Nyquist diagram and closed-loop poles.

The performances specification in tracking and further control to clarify the dominant auxiliary poles of the closed loop are successively represented in table (2).

Nr	Traking Double		Regulation Double		HR Simple		HR Double		HS Simple	
	w_0	xi	w_0	xi	T	n	w_0	xi	T	n
1	0.3	1	0.3	0.99						
2	0.3	1	0.3	0.99			0.4	0.7	5	2
3	0.3	1	0.3	0.99						
4	0.3	1	0.3	0.99	6	1	0.5	0.1	5	1
5	0.3	1	0.3	0.99			0.5	0.5		
6							0.5	0.55	5	1
7	0.3	1	0.3	0.99			0.5 / 0.5	0.6 / 0.65	5	2
8	0.3	1	0.3	0.99			0.45 / 0.45 / 0.45	0.9 / 1 / 1	1	1

Table 2. Performances specification.

Robustness margins which correspond to the previously specified performances are reported in table (3), the goal here is to change the performance so that the obtained margins can be maintained in the range of the known robustness margins.

Nr	ΔG (dB)	$\Delta \Phi$ (°)	$\Delta \tau$ (s)	ΔM (dB)
1	6.22	63.0	5.42	-5.83
2	4.37	55.6	5.75	-8.06
3	6.22	63.0	2.33	-7.50
4	6.71	32.0	4.05	-7.76
5	5.5	30.2	2.55	-7.30
6	7.99	61.8	8.37	-4.80
7	6.85	60.7	10.66	-5.45
8	6.04	54.8	12.12	-6.13

Table 3. Successive improvement of robustness margins.

The new polynomial of the controller with the reference model are given as follows:

Controller

$$R\left(q^{-1}\right) = 187.56q^{-1} - 256.13q^{-2} - 30.06q^{-3} + 87.18q^{-4} + 77.22q^{-5} - 78.80q^{-6} + 1629q^{-7} \quad (53)$$

$$S\left(q^{-1}\right) = 1 - 1.16q^{-1} - 0.39q^{-2} + 0.35q^{-3} + 0.34q^{-4} - 0.14q^{-5} \quad (54)$$

$$T\left(q^{-1}\right) = 167.504 - 288.712q^{-1} + 124.463q^{-2} \quad (55)$$

Reference model

$$Am\left(q^{-1}\right) = 1 + 1.06q^{-1} + 0.35q^{-2} \quad (56)$$

$$Bm\left(q^{-1}\right) = 0.32 + 0.71q^{-1} \quad (57)$$

To simulate the behavior of the controller, the given step response of the reference model must be used by considering the existence of a disturbance step type, with amplitude of 2.5 * 10-3%, applied at time 40s, a graphical representation of the step response with and without disturbance is shown in figure (14), in which the assurance of performance in tracking and control can be observed.

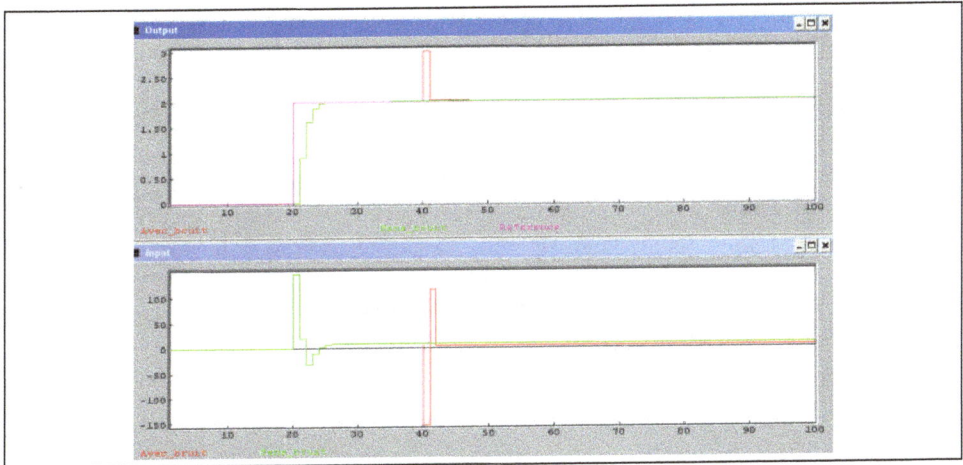

Fig. 14. Step response under and without distanrbance.

3. Conclusion

The realized study case on the pyrolysis reactor validates the research developed in this chapter and provides a guarantee for the successful implementation of the control solutions proposed for such plants (Azzouzi, 2009a). On one hand, there are analyzed the theoretical and practical resources offered by modern Automatic Control in order to achieve the effective solutions for the control of this chemical process, and on the other hand, there are presented mechanisms related to the design and to the implementation of systems for data acquisition, identification, control and robustness (Azzouzi, 2008).

4. References

Azzouzi, M. & Popescu, D. (2008). Optimisation d'un réacteur de pyrolyse par SiSCon, *Conférence Internationale Francophone d'Automatique CIFA'2008*, Bucharest, June 2008

Azzouzi, M. (2008). Systèmes numériques pour la commande avancée des installations pétrolières et pétrochimique, PhD Thesis, Politehnica University of Bucharest, Bucharest

Azzouzi, M. (2009a). Medium optimization approach on petrochemical plant, *The 17th Conference on Control Systems and Computer Science, CSCS-17*, Vol. 2, pp159-163, Bucharest, May 2009

Azzouzi, M. (2009b). Comparative study between SisCon and MatLab in pyrolysis reactor identification, *Journal of Applied Functional Analysis (JAFA)*, pp255-261

Ben Abdennour, R. (2001). *Identification et commande numérique des procédés industriels*, pp. 210-240,Technip, Paris

Borne, P., Tanguy, G. D., Richard, J. P., Rotelle, F. & Zambittakis, C. (1993). *Analyse et régulation des processus industriels; tome 2: Régulation numérique*, pp. 267-275, Technip, Paris

Bozga, G. & Muntean, O. (2000). *Reactoare chimice (reactoare omogene)*, vol. I, pp470-480. Tehnica, Bucharest

Harriott, P. (2002). *Chemical reactor design*, pp. 131-142. CRC First Edition, New York

Landau, Y, D. & Zito, Gianluca. (2006). *Digital Control Systems: Design, Identification and Implementation*, pp. 267-291, First Edition of Springer, Paris

Ogata. K. (2001). *Modern control engineering*, pp. 310-431. Prentice Hall fourth Edition, New York

Oustaloup, A. (1994). *La robustesse: Analyse et synthèse des commandes robustes*, pp. 150-102, Hermes, Paris

Popescu, D., Lupu, C., Petrescu, C. & Matescu, M. (2006b). *Sisteme de Conducere a proceselor industriale*, pp. 24-29, Printech, Bucharest

Popescu, D., Stefanoiu, D., Lupu, C., Petrescu, C., Ciubotaru, B. & Dimon, C. (2006a). *Industrial Automation*, Agir, Bucharest

2

A Possibility to Use a Batch of Ashless Additives for Production of Commercial Transmission and Motor Oil

Milena Dimitrova and Yordanka Tasheva
*University "Prof. Assen Zlatarov"- Burgas,
Bulgaria*

1. Introduction

The aim of the present chapter is to study the possibility to use PSA synthesized by the authors for production of commercial transmission and motor oil.

The practically used compounds are usually surfactants, which must be soluble in a medium of non-polar petroleum hydrocarbons. This required the optimal length of the hydrocarbon radical of the main hydrocarbon chain guarantees property. The polar group of the additive, which determines its main function, has certain electron effect. As a result, significant static and dynamic dipole moments are generated, as it has been discussed in the previous chapters. These effects stipulate the affinity of the additives to different border surfaces, especially in their intermolecular interactions when used within a batch [2].

In our previous investigations, the methods for synthesis of PSA by sulphonation, nitration and oxidation have been reported [3, 4,5].

2. Experimental

Taking into account our earlier studies [6,7] on the optimization of the composition of the additives synthesized by us, a batch of petroleum soluble additives was composed on the basis of an oil component and the most effectve of them was found to be the composition containing PSA-SK-N-H: PSA-SK-S-U: PSA-SK-O-U = 10:50:40 referred to further as batch P-1. The denotations and the most important values for additives preparation have been published earlier [6]. With the selected batch of additives, the possibility to prepare operation-conserving lubricants of transmission and motor type was investigated. For this purpose, basic lubricant mixtures were taken from fractions of the corresponding class of viscosity and the batch P-1 was added to them. The standard physicochemical and performance properties of the alloyed lubricants were then determined to estimate their performance quality. The results obtained from the analysis are presented in Table 1.

Table 2 shows the general technical characteristics of the transmission lubricants after the tests carried out. Tests were performed also by the method PPA-2 [8] for fuller assessment of the protective coatings formed by the transmission lubricant. The tests were carried out in presence of copper disks (alloy M-5) since, as it is well known, this metal catalyzes the oxidation of petroleum hydrocarbons especially in presence of air and at higher temperature. Fig.1 illustrates the depositions of acidic products in lubricant samples depending on the oxidation time at three temperature regimes (100, 200 and 300 °C).

Parameters	Basic fraction TM-5/90	Standard regulations TM-5/90 BSS 9797	Parameters at different concentrations in % of P-1						Method of analyses
			0.001	0.002	0.003	0.004	0.005	0.006	
1	2	3	4	5	6	7	8	9	10
1. Kinematic viscosity at 100 °C, mm²/s	18,5	20.0 ± 2	20,0	20,0	20,2	20,5	20,9	21,4	BSS EN ISO 3104
2. Viscosity index	50	95	97	99	103	109	115	116	BSS ISO 2909
3. Pour point, °C	-6	-15	-8	-10	-14	-16	-18	-20	BSS ISO 3016
4. Cloud point, °C	-9	-18	-11	-13	-16	-19	-23	-24	BSS 1751
5. Marten-Penski temperature in open spot, °C	210	210	210	210	210	210	210	210	BSS EN ISO 2592
6. foam-forming, cm³									BSS ISO
-abbility toward foamforming	80/80/80	90/40/90	90/40/90	95/45/90	97/47/97	98/48/98	100/50/100	110/55/110	6247
-stabillity of foam	10/10/10	0/0/0	0/0/0	0/0/0	0/0/0	0/0/0	0/0/0	2/2/2	
7. Stabillity of oxidation:									DIN 51394
-increasing of viscosity at 100 °C, %	31	20	28	25	23	20	18	17	
-lose of mass, %	7	5	7	7	6	6	5	4	
-insoluble in n-pentan, %	4	1	4	4	3	2	1	1	
8. Acid number, mg KOH/ g	0,01	0,04	0,09	0,08	0,07	0,05	0,03	0,02	BSS 1752

Table 1. Physicochemical and exploration characteristics of transmission oil from viscosity class 90.

Property	Value at the test temperature					
	TM-5/90 – commercial			TM-5/90 – basic fraction + 0,005 % P-1		
	100 °C	200 °C	300 °C	100 °C	200 °C	300 °C
1. Ignition temperature in open pot, °C	204	200	190	204	204	196
2. Kinematic viscosity at 100 °C, mm²/s	21,6	22,4	23,8	22,3	22,7	23,0
3. Freezing temperature, °C	-19	-20	-22	-23	-26	-26
4. Mechanical contaminants, %	0,08	0,12	0,17	0,03	0,05	0,09
5. Corrosion tests						
-by „Salt fog", g/m²	77,13	75,24	72,15	35,24	31,18	30,71
-by „Aggressive media" , mg	2838	2809	2759	1838	1724	1656
-by "Humidity", g/m²	0,2235	0,2178	0,2101	0,1754	0,1621	0,1537
-by "Condensed moisture", g/m²						
- Steel 08 grade, rimming, pcs, g/m²	10	11	12	>12	>12	>12
-copper, degree (Cu, bal)	2	3	4	1	1	1

Table 2. General technical characteristics of transmission lubricants after the tests.

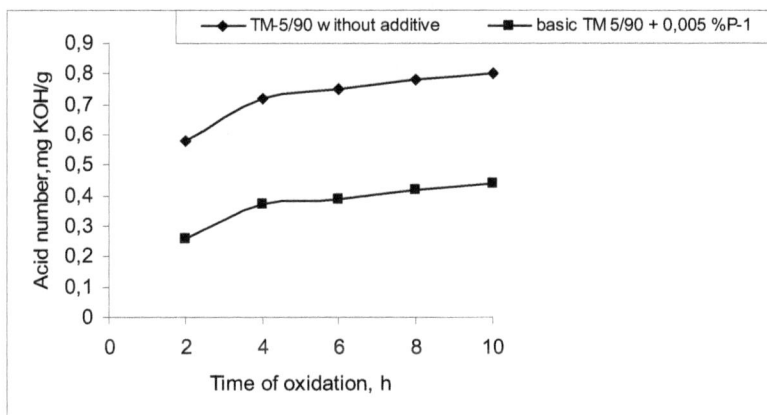

Fig. 1. Depositions of acidic products in lubricant samples depending on the oxidation time at three temperature regimes.

During the proved optimal induction period of oxidation (4 hours), the processes giving acidic products depending on experiment temperature were studied (Fig.2). Simultaneously, the accumulation of carbonyl containing compounds in the oxidized lubricants was established by the intensity of the absorption band at 1720 cm-1 in the IR spectra of their 10% solutions in tetrachloromethane.

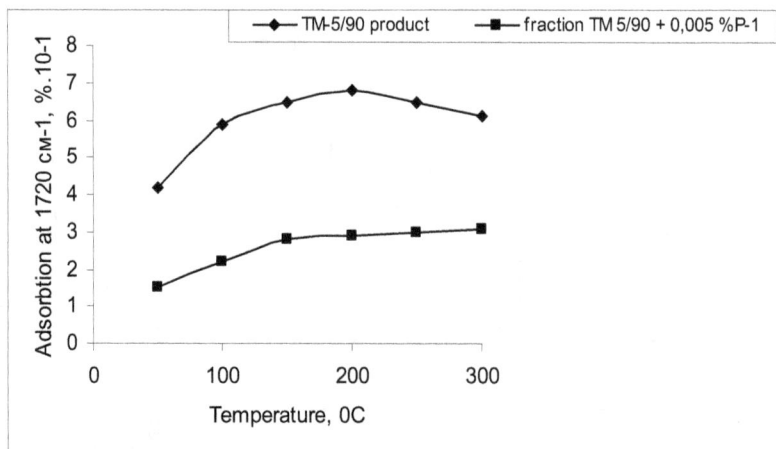

Fig. 2. The intensity of the adsorption band at 1720 cm-1 in the IR spectra of their 10% solutions in tetra chloromethane.

To find the quantity of metal dissolved in the exhaust lubricant as a result of their interaction with the products obtained from their oxidation, analyses of their composition were carried out by atomic absorption spectroscopy (Fig.3).

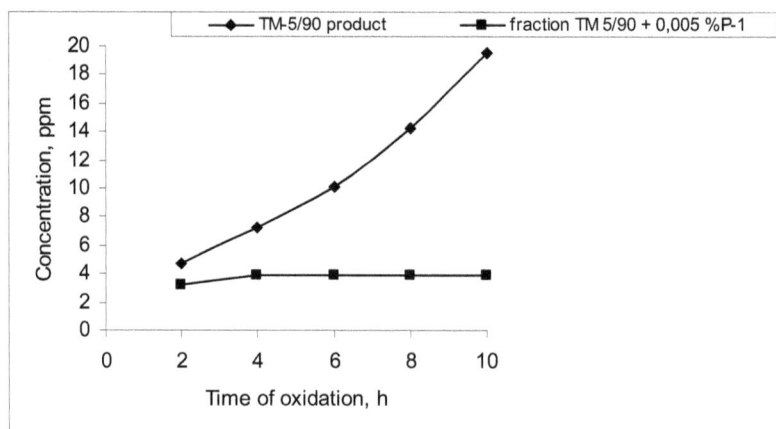

Fig. 3. Quantity of metal dissolved in the exhaust lubricant as a result of their interaction with the products obtained from their oxidation.

As can be seen from Table 1, the use of the batch P-1 in the optimal concentration established to be 0,005 %, the requirements of the Bulgarian national standard can be fulfilled by the transmission lubricant at performance level TM-5. According to BSS 14368-

02, this lubricant can be used to lubricate gears and its properties are like these of the commercial product "Ulita ER". The studies showed that the lubricant obtained has very good anticorrosion and antiwearing properties. As general conclusion it can be stated that the batch of ashless PSA gives the lubricant alloyed polyfunctional properties. The antioxidative stability of the alloyed lubricant is especially important since the transmission oils in the process of operation are subjected to intense stirring under atmospheric medium and in presence of ferrous and non-ferrous metals and alloys. The surfactant properties of he additive affect the amount of foam formed and its stability. The presence of air in the lubricants did not enhance the accumulation of oxidized products in them, All the physicochemical and performance properties measured at the recommended concentration of the additive showed that the transmission lubricant obtained is of the operation-conserving type and possessed enhanced resistance to oxidation. At the same time, it did not change significantly its performance properties during the experiments carried out (table 2).

The studies carried out showed (Fig.1) that the presence of the additive P-1 did not have significant effect on the duration of the induction period of oxidation of the experimental samples. The positive effect observed after the fourth hour of operation was that the intensity of the oxidation processes decreased and the accumulation of acidic products in the exhausted lubricants was almost constant. This resulted from the exhaustion of the easily oxidizable hydrocarbons and, probably, the antioxidative effect of some of the products obtained. Positive effect on the antioxidative protection of the hydrocarbons in the lubricants studied had also the chemiadsorption properties of the ashless PSA participating in the composition of P-1 towards the metal surfaces. When adsorbed on the metal surface, they could impede the catalytic effect they have on the oxidation of the petroleum products.

As can be seen from Fig.2, the most intense oxidation processes were observed at temperatures up to 200 °C. Above this temperature, the acid number remained almost the same. The use of the additive P-1 slowed down the acid number increase rate and limited the accumulation of acidic products (Fig.3).

The studies carried out showed that, without the additive, processes of corrosion-mechanical wearing of all the metals contacting with the lubricant actively occur during the tests according to the PPA-2 method. In the presence of the additive P-1 these processes almost stopped after the fourth hour of operation. Probably, the additive formed chemiadsorption compounds on the metal surfaces which protect them from the interaction with the corrosion aggressive products. As a result, the concentration of metals in the exhausted lubricant sharply decreased and remained almost the same after the induction period.

After 2 months of operation of the lubricant with P-1 and the commercial additive, it was found that probably strong enough chemiadsorption film was formed on the metal surface, which decreased the intensity of their corrosion-mechanical wearing. This effect is much more pronounced with the batch of ashless PSA P-1 at low concentration, compared to the metal containing additives of the previous generation. Due to the good lubricating characteristics of the more viscous lubricant usually used only in summer, the results on the corrosion-mechanical wearing were worse and the protection of the lubricated surfaces was better. (Figs.4 - 7).

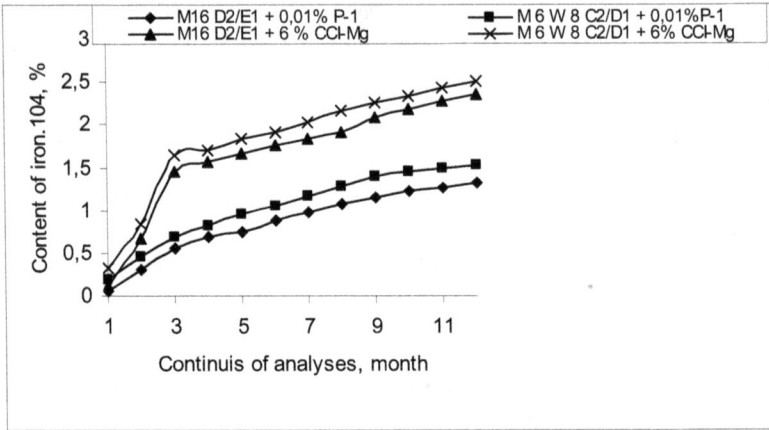

Fig. 4. The wearing of the parts of the piston-cylinder system was monitored by measuring the concentrations of iron.

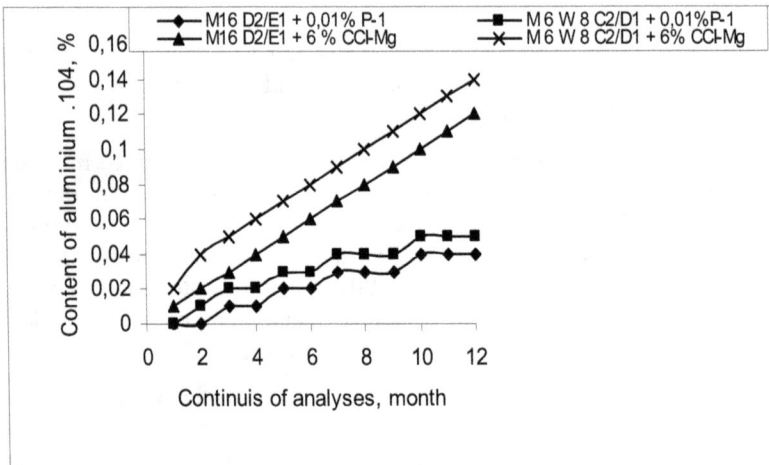

Fig. 5. The wearing of the parts of the piston-cylinder system was monitored by measuring the concentrations of aluminium.

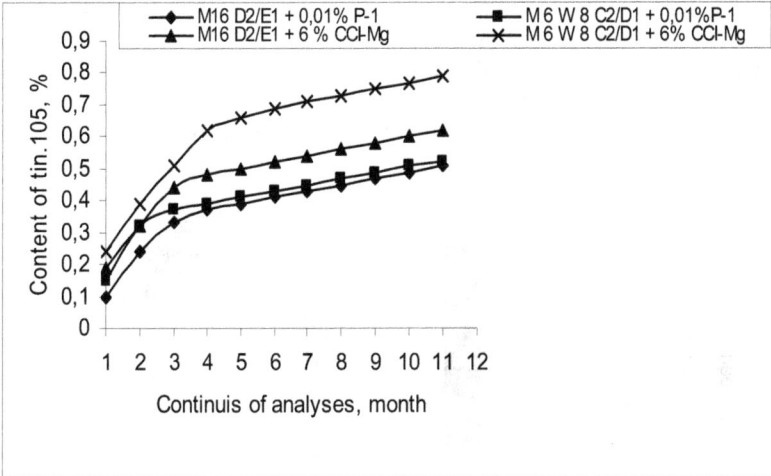

Fig. 6. The wearing of the parts of the piston-cylinder system was monitored by measuring the concentrations of tin.

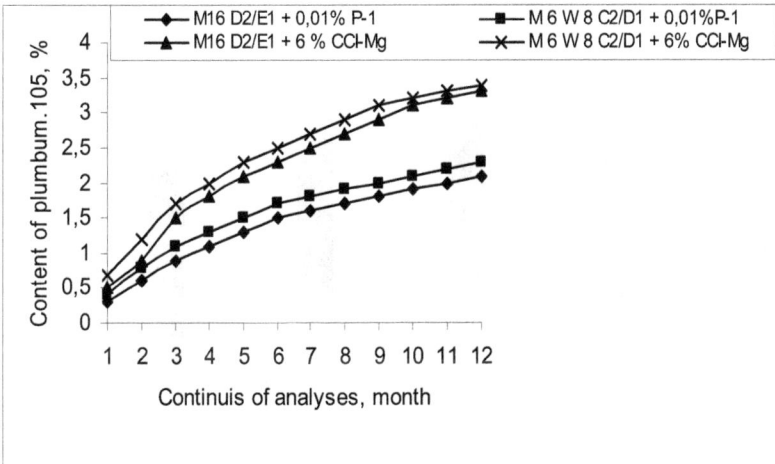

Fig. 7. The wearing of the parts of the piston-cylinder system was monitored by measuring the concentrations of lead.

The results illustrated in figs.8 and 9 confirmed the higher effectiveness with regard to the anti-wearing properties possessed by the ashless additives included in the combined batch P-1. The wearing in the individual cylinders of the motor was found to be different. Probably, in this case, the liquid film formed in the combustion chamber due to the over-fractioning of the fuel during the formation of the fuel-air mixture in the cylinders exerted some influence. In the middle cylinders (II) and (III) the wearing was slightly more intense.

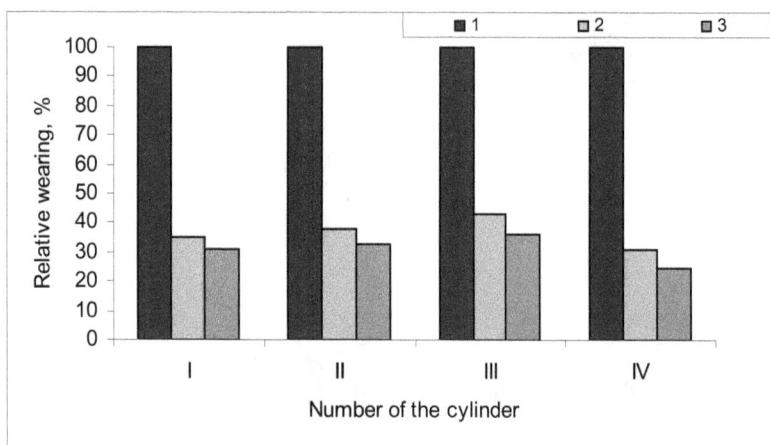

Fig. 8. The results on the relative wearing of the first segment of the pistons of the individual cylinders of the motor „Ch 10,5/13".

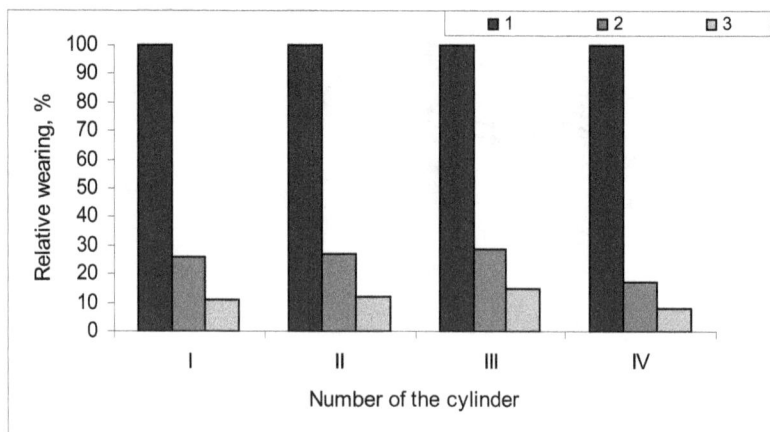

Fig. 9. The results on the relative wearing of the first segment of the pistons of the individual cylinders of the motor „Ch 10,5/13".

It is well known that this type of wearing in presence of a liquid film is of the so called type "corrosion mechanical wearing". The main reason for its occurrence is the washing out of the lubricating film and disturbance of the laws of hydrodynamic friction, as well as decrease of the lubricating film thickness. Nevertheless, this effect of wearing in the presence of the batch P-1 was significantly weaker.

The experiments carried out showed that the nature of the compounds from which the PSA were prepared gives them surface active properties and improve the protective and other properties of the fuels and lubricants alloyed with them. This is especially important for the use of fuels with high sulphur content used more widely for ship engines. In this case, an important problem arises since the sulphur containing compounds and the products of their combustion accelerate the corrosion-mechanical wearing of the aggregates and, of course, have adverse effect on the environment. In this respect, it appeared to be interesting to study the anti-wearing properties of the motor oils M 6W 8 C_2/D_1 and M 16 D_2/E_1 alloyed with the batch additive P-1. Simultaneously, experiments were carried out also with the magnesium-containing additives synthesized by the authors and implements in the industry in the prescribed concentration of 6% [5]. The experiments were carried out with four-cycle four-cylinder motor „Ch 10,5/13" mounted in many lightweight sea vessels. These motors use ship diesel fuel LK produced according to Bulgarian national standard BSS 12833-82 and contains up to 1.4% sulphur. The wearing of the parts of the piston-cylinder system was monitored by measuring the concentrations of iron, aluminium, tin and lead found in the lubricant according to BSS ISO 9778-00 and BSS ISO 15430-02. These are the main metals used to manufacture the parts of the motors. This method is widely used practically both for determination of the moment to replace the lubricants and oils and establishing of emergency cases of accelerated wearing of some units in the motors. The results obtained from the studies of the operation-conserving lubricants are shown in Figs.4-7.

Figs.8 and 9 show the results on the relative wearing of the first segment of the pistons of the individual cylinders of the motor „Ch 10,5/13" determined by their weight after 1 year service using different lubricants.

3. Conclusion

The studies on the use of the batch of ashless petroleum soluble additives P-1 for the production of transmission lubricant carried out showed that the addition of 0.005% P-1 to basic lubricant mixtures gave transmission lubricant complying with BDS 14368-02. The transmission lubricant obtained can be used to lubricate gears and is fully comparable to the commercial product "Ulita ER". The results obtained from the physicochemical analyses showed that the operation-conserving motor lubricants alloyed with the batch mentioned above fully comply with the requirements of BSS 15021 and BSS 9797. It was found that the optimal concentration of the batch P-1 is 0.010%. The ashless PSA included in the batch impart depressing and very good protective properties of the alloyed lubricants.

4. References

[1] Brian L. J., 2001, pp.1204, *Lubricate Engineering*.
[2] Clash T., 2000, pp.375, *Computational chemistry*, v. 7.

[3] Cracknel R., 2003, *Lubricate Engineering*, v. 21, p.17.

[4] Dalton J.C., L.E. Friedrich, 2005, *Journal of Chem. Educ.*, № 15, p.872.

[5] Dewar J.S., J.P. Stewart, 2003, *Journal Amer. Chem. Soc.*, № 97, 2003, p. 127.

[6] Yoshida T., J.Japarash, 2001*Tribology Trans.*, № 2, p. 158.

[7] Werner I., 2001, *Ind. Engineering Chem.*, v. 5, № 57, p. 17.

[8] Dimitrova M., 2008, *PhD Thesis, University „Prof. Dr A. Zlatarov" – Burgas.*

3

Petrochemicals: Cellulosic Wastes as an Alternative Source of Feedstock

Solomon Gabche Anagho[1] and Horace Manga Ngomo[2]
[1]University of Dschang
[2]University of Yaounde I
Cameroon

1. Introduction

Petroleum, the major feedstock into the petrochemical industry is a depleting natural resource. The rapid growth in world population, coupled with the increasing standard of living means that the petroleum reserves, which although they appear vast (Cheschire & Pavitt, 1978), are still limited and fast running out. Secondly, petroleum is the main source of fuel for the increasing number of automobiles, with the effect that the greenhouse gases generated have already exceeded threshold values. The continuous exploitation of petroleum has led to immense degradation of the environment as is observed in oil spills, and erosion in oil producing areas. There is the need to look for alternative sources of fuel and petrochemical feedstock.

Another problem associated with the increasing world population is the need to grow more food to match the population growth. This need gives rise to increased agricultural activities and food processing which result in the generation and accumulation of large amounts of organic wastes in the form of biomass residues, and consequently, municipal refuse which exacerbates environmental degradation. Appropriate disposal of these wastes is a major problem that requires urgent and long lasting solutions.

Over the years, land filling and incineration have been used to dispose of municipal and agricultural wastes. These methods did not raise any major concerns in the past because of the extremely low rates of waste generation at the time. Today, the large amounts of the greenhouse gases, particularly CO and CO_2 from incineration, and ground water contamination from landfills can no longer be ignored.

These problems are best solved by applying modern waste management techniques. Good management practices aim at reducing the rate of exhaustion of the resource base, rather than increase or even sustain the rate of consumption (Mitchell, 1995). These waste management techniques entail that, waste generation is avoided as much as possible by either reusing waste, combustion to heat energy, and then disposing of the resulting residue by the least environmentally damaging methods.

Studies have shown that the biomass component in waste, predominantly of cellulosic materials, is a heavy energy carrier. For example, Ngomo (2004) estimated that in

Cameroon, 1,148.77 PJ of energy, equivalent to 41,325 million kWh of electricity could be harvested annually from the theoretical harvestable biomass wastes arising from agricultural waste, forest residue, and livestock dung.

As early as 1917, the need to obtain chemicals from cellulosic materials had already been eminent. For example, Palmer & Cloukey (1918) in producing alcohol from various species of hardwoods studied the effect of moisture content in the wood on alcohol yield. By destructively distilling beech, yellow birch and maple that had been subjected to seasoning for between 4 and 18 months they observed that beech gave high alcohol yields when the moisture content was high, while excess moisture lowered the yields for yellow birch and maple.

Palmer (1918) used phosphoric acid to catalyse the destructive distillation of hardwoods. Working at both atmospheric pressure, and at elevated pressures of 412 – 712 kN/m^2, they saw that phosphoric acid increased the yields of alcohol at atmospheric pressure. Hawley (1922) continued the work of Palmer by studying the effect of various chemicals as catalyst. Working at atmospheric conditions, he found phosphoric acid to be unstable; and that promising yields of alcohol were obtained by using sodium carbonate, sodium silicate, and calcium carbonate. In an attempt to produce liquid fuels from cellulose, Fierze-David (1925) arrived at the following results and conclusions: hydrogen had no effect on the dry distillation of wood even under high pressure and that it acted just as an inert gas improving on the mechanism of the reaction; copper hydroxide, iron hydroxide and nickel hydroxide catalysed the distillation of cotton cellulose at 450 - 470°C and 150 – 245 atm. to give a distillate consisting of aldehydes, ketones, phenols, cyclic glycols, and fatty acids.

As reported in Henze et al (1942), Bowen et al (1925) re-iterated the importance of catalyst in the liquefaction of cellulose by concluding that cotton yarn did not undergo any appreciable reduction when treated with hydrogen at 440°C and 12,360 - 13,390 kN/m^2. However, when impregnated with nickel salts, almost the whole material was converted to liquids and gases under the same pressure and temperature conditions. The liquid produced was an opaque viscous tar, containing carboxylic acids, phenols and neutral oils. Frolich et al (1928) also observed that high temperature, high pressure and catalyst were necessary for converting sulphite pulp to liquid and gases. The liquid products consisted of phenols, saturated and unsaturated hydrocarbons.

From all these historical works, and the more recent one by Boocock et al (1980), we can deduce that:

- Cellulosic materials can be converted to a mixture of chemicals consisting of gases, liquids and solids;
- High pressure and high temperature in the presence of catalyst are essential for the conversion and the yield. In addition, the actual liquid yield depends on the extent of pressurizing, heating and the type of catalyst;
- The liquid mixture consists of a variety of substances, which can be used for the production of other chemicals.

Other more recent works applied high temperature and high pressure, but not the use of catalyst, to give the same types of products (El-Saied, 1977; Kaufmann and Weiss, 1975).

Cellulosic materials can be converted to oils and other chemicals by the methods of gasification, pyrolysis, hydrogenation, and hydrolysis and fermentation. Gasification converts the cellulosic materials to syngas (synthetic gas), which is a mixture of CO, H_2, CO_2 and CH_4. The reaction takes place at high temperatures of over 700°C without combustion, and under a controlled amount of oxygen and/or steam. Gasification has the advantages that the syngas produced is a more efficient fuel than the original cellulosic material because it can be combusted at a higher temperature; syngas can be combusted directly in fuel cells and gas engines, and it can be used to produce methanol.

In pyrolysis, the cellulosics undergo an irreversible chemical change caused by heat in the absence of oxygen to give char, organic liquids and water. The reaction is endothermic. The heating value of the pyrolysis products is the sum of the heating value of original cellulosic material and net energy added during the pyrolysis process. In hydrolysis/fermentation, the pre-treated cellulosic material is first hydrolysed to glucose. The product is then converted to a mixture containing alcohols by fermentation, followed by distillation to obtain purer alcohols. The process of hydrogenation converts cellulosic materials to gaseous, liquid and solid products. The cellulosic material suspended in a slurry fluid, and in the presence of hydrogen is subjected to high pressure and heated to obtain the hydrogenation products.

The objectives of this work are to:

- convert cellulosic wastes to chemicals by the process of hydrogenation;
- deduce the mechanism for cellulose hydrogenation;
- deduce a rate expression for the process;
- assess the feasibility of the hydrogenation process by doing an energy economy analysis.

Cellulose, a high molecular weight polymer, with the formula $(C_6H_{10}O_5)_n$ is a structural polysaccharide derived from beta-glucose (Updegraff, 1969). It is the primary structural component in the cell wall of plants and trees. In plants, it is highly imbedded in lignin, which prevents it from being easily accessible.

Cellulose is insoluble in water and in neutral solvents such as gasoline, benzene, alcohol, ether and carbon tetrachloride. It is almost insoluble in dilute acids. In the presence of mineral acids of moderate strength, it degrades into glucose (Cheremisinoff, 1980). Oxygen is removed from cellulose as H_2O and CO_2 in the presence of high reducing agents such as H_2 and CO (Boocock et al, 1979). The close association of cellulose and lignin in plant biomass requires the use of drastic reaction conditions such as pressures above 100 atmospheres and temperatures of 250 – 400°C for their processing (Cheremisinoff and Morresi, 1977).

In batch systems, the reducing agents are generated in-situ by adding a little amount of water. Under the reaction conditions, the cellulose rings are cleaved by the reducing agents. Edewor (1980) represented the liquefaction reaction of cellulose broadly as involving the following steps:

Initiation: this is the shift reaction in which water reacts with a little amount of cellulose to give H_2 and CO_2. This is followed by the generation of free radicals.

In the liquefaction step, the remaining cellulose is cleaved by the free radicals obtained from the H_2 and CO_2 to give a variety of products.

A deduction of the detailed reaction mechanism will be given in the kinetics part below (Section 4).

The major difference between hydrogenation and gasification or pyrolysis is that hydrogenation takes place in a liquid medium. The liquid used is called **slurry fluid**. The slurry fluid is used because, it minimises the charring of the cellulosic materials in favour of its conversion to liquid products. That is, the slurry fluid tends to shift the equilibrium towards the formation of liquid products, thereby reducing the amount of char produced. According to Kaufman and Weiss (1975), a slurry fluid should be thermally stable, have low vapour pressure, and good solvent activity. It should also be cheap and readily available. Water is not a good slurry fluid despite its cheapness and ready availability because it generates excessive steam pressures at high temperatures.

To illustrate the production of bituminous oils from cellulosic wastes, we report the experimentation procedures in the studies of Anagho et al (2004, and 2010) and Edewor (1980).

2. Experimental

2.1 Equipment

- Autoclave: half-litre, high pressure, magnetically-driven, and heated from the exterior by a 1,600 watt heater.
- Infrared spectrophotometer: Model Acculab 10, using NaCl discs.
- Gas chromatograph: Hittachi GC model Carle GC 9700 (Shimadzu, Kyoto, Japan).
- Product separation was obtained using a high temperature hydrocarbon column, Chromosorb G, AW, maintained at 180°C.
- The carrier gas was nitrogen flowing at 40 ml/min, and detection was by flame ionization.

2.2 Chemicals and materials

- The cellulosic materials included leaves, grass, cow dung, and municipal refuse. They were dried and pulverized.
- Gasoil was used as the slurry fluid.
- Hydrogen gas, supplied from a cylinder was used along with water to initiate the reaction.

2.3 Experimental procedure

2.3.1 Liquefaction

Each run used 100 g of slurry fluid, 20 g of dry pulverized cellulosic material, 1.2 g of water sealed into the autoclave, and pressurized to 30 kg/cm^2 by hydrogen. Heating was carried out for 3 hours. During heating, temperature and pressure were recorded over some time intervals. After cooling down to room temperature, the reactor was opened and the masses of solids (char) and liquid determined by carrying out a material balance on the reactor.

2.3.2 Distillation and analyses of products

The freshly produced hydrogenation product was fractionated into five main fractions as shown in Table 1.

The freshly prepared hydrogenation product (Sample A) and the distillation fractions were analysed using infrared spectroscopy and gas chromatography to determine the types of chemicals generated from the liquefaction.

The Lassaigne's sodium fusion test was also carried out on all the product samples. It is a qualitative test used to show the presence of oxygen, nitrogen, sulphur and halogens in organic molecules. In this test, the element to be identified is first converted into an inorganic ionisable form which can be readily identified by inorganic tests.

In this study, a little quantity of sodium metal was put into a test tube containing a sample of synthetic fuel and heated to red hot to produce the ionisable compounds.

A representative equation for the test is:

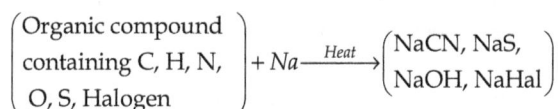

$$\left(\begin{array}{l}\text{Organic compound} \\ \text{containing C, H, N,} \\ \text{O, S, Halogen}\end{array}\right) + Na \xrightarrow{\text{Heat}} \left(\begin{array}{l}\text{NaCN, NaS,} \\ \text{NaOH, NaHal}\end{array}\right)$$

The cyanide was analysed using iron (II) sulphate, while the hydroxide was by anhydrous copper sulphate (Anagho et al, 2010).

Sample Label	Sample name	Conditions of Fractionation
A	Freshly prepared sample, before fractionation	
B	1st Fraction of hydrogenation product	Boiling point: 100°C Pressure: 760 mmHg
C	2nd Fraction of hydrogenation product	Boiling point: 200°C Pressure: 760 mmHg
D	3rd Fraction of hydrogenation product	Boiling point: 250°C Pressure: 760 mmHg
E	4th Fraction of hydrogenation product	Boiling point: 300°C Pressure: 760 mmHg
F	Bituminous residue, dissolved in tetralin	

Table 1. Fractionation products from synthetic fuels obtained from cellulose hydrogenation.

3. Results

3.1 Products yield

The masses of the different streams and for the various cellulosic materials are presented in Table 2 below.

Cellulosic material	Solid product (g)	Liquid product (g)	Gaseous product (g)
Leaf	9.00	7.80	3.20
Cow dung	12.00	5.80	2.20
Groundnut husk	10.00	9.80	0.20
Municipal refuse	8.50	9.30	2.20

Average yield: Solid = 49%; Liquid = 40%; Gaseous = 11% (Anagho et al, 2004)

Table 2. Product yield from the hydrogenation of 20 g of cellulosic material.

The table shows that on the average, 40 per cent by mass of the dry cellulose is converted to synthetic oils by hydrogenation. Also, the amount of solid left for disposal is less than 50 per cent by weight of the dry cellulosic material.

3.2 Temperature and pressure evolution during heating

During heating, the temperature rose rapidly from room temperature to about 250°C, and this took about 75 minutes. As heating was continued, the temperature rose slowly to a value of between 300 and 325°C. It then remained constant with further heating.

The pressure in the reactor also increased with time of heating, but in a slightly different manner. Three sections could be identified (Anagho, 2004). The first section corresponded to the first 75 minutes, where there was very little increase in pressure. The increase in pressure may be attributed to pressure–temperature change at constant volume only. This period is regarded as the heat-up time. The second section was the period of rapid rise in pressure and was regarded as the reaction period, and it lasted 100 minutes on the average. The third was asymptotic. The rapid increase in pressure during the reaction period indicates that reaction is by free radical mechanism.

From the reaction time, the reaction rate for each cellulosic material was calculated. The heat-up time, the reaction time and the reaction rate are presented in Table 3 below. The rate of conversion of cellulose was calculated as the ratio of the amount of cellulose converted in gmol cellulose/m³ solution to the reaction time. (The molar mass of cellulose was taken as 162 g/mol.).

3.3 Analysis of hydrogenation products

IR spectroscopy analyses of the raw cellulose samples showed the presence of the O-H, C-C, C-O and C-N bonds for all the samples. The newly produced substances did not contain O and N as they lacked the characteristic O-H and C-O bands between 3600 and 3100 cm⁻¹ and 1300 and 1000 cm⁻¹ respectively. This indicates that hydrogenation removes the oxygen from cellulose.

Cellulosic material	Heat-up time (min)	Reaction time (min)	Reaction rate (mol/m³s)
Leaf	75.00	100.00	8.57
Cow dung	75.00	100.00	8.57
Groundnut husk	75.00	125.00	6.82
Municipal refuse	100.00	125.00	6.82

Average rate = 7.695 mol/m³min   = 0.128 mol/m³s. (Anagho et al, 2004)

Table 3. Reaction parameters for cellulose hydrogenation.

The presence of the saturated C=C bond and aromatics as depicted by the IR bands at 1600 and 1400 cm⁻¹, coupled with the GC analysis suggest that many types of compounds, such as aromatics, saturated and unsaturated hydrocarbons, both straight chained and cyclic were produced. When stored under light for some time, sediments formed in the originally homogeneous mixture, while oxygen and nitrogen reappeared to show that the hydrogenation products were unstable (Anagho et al, 2004).

The implication of these observations is that the hydrogenation of cellulose leads to the cleavage of the cellulose ring, followed by the removal of the oxygen in it. Hydrogenation also gives rise to unstable hydrocarbons, which when stored for some time regain oxygen to give saturated and unsaturated oxygenated compounds.

4. Kinetics of cellulose hydrogenation

4.1 Mechanism of the hydrogenation process and justification of the products type

As stated above, the cleavage of the cellulose ring during the hydrogenation process leads to the removal of oxygen, followed by the production of a mixture of products consisting of straight chain, branched chain and cyclic hydrocarbon compounds, some of which are saturated and others unsaturated. The process of hydrogenation takes place under high temperature and pressure, and leads to the cleavage of the cellulose ring to give products which are later rearranged to aromatic, naphthenic and unsaturated hydrocarbon compounds. This suggests that cellulose hydrogenation takes place through a free radical mechanism (Anagho, 1987). The cellulose ring is cleaved by the free radicals CO·, O· and H· obtained from CO_2 and H_2 generated in-situ from water and cellulose by the water gas shift reaction (Macrae, 1966a). These processes are shown by Equations 1 to 4 as follows:

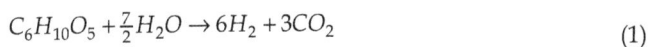

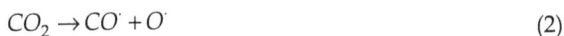

$$C_6H_{10}O_5 + \frac{7}{2}H_2O \rightarrow 6H_2 + 3CO_2 \tag{1}$$

$$CO_2 \rightarrow CO^· + O^· \tag{2}$$

$$H_2 \rightarrow 2H^{\cdot} \tag{3}$$

$$\text{(4)}$$

$$+ \quad H_2O + CO_2 + O_2 + OH^{\cdot}$$

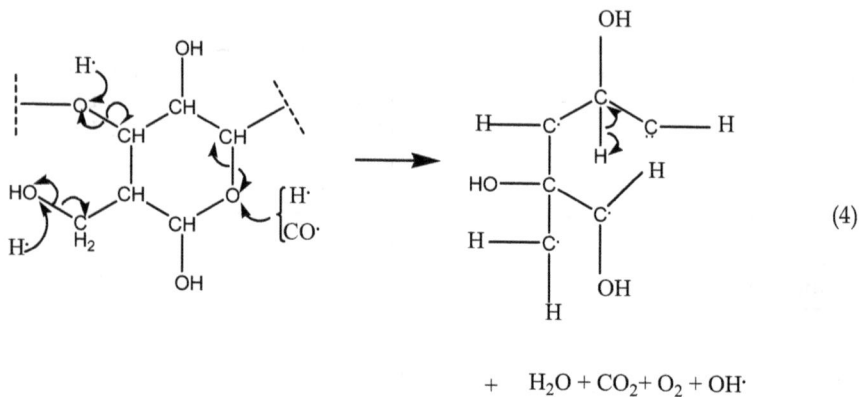

The cleaved ring in Equation (4) further breaks down in various ways into a number of radicals (Equations 5 to 7):

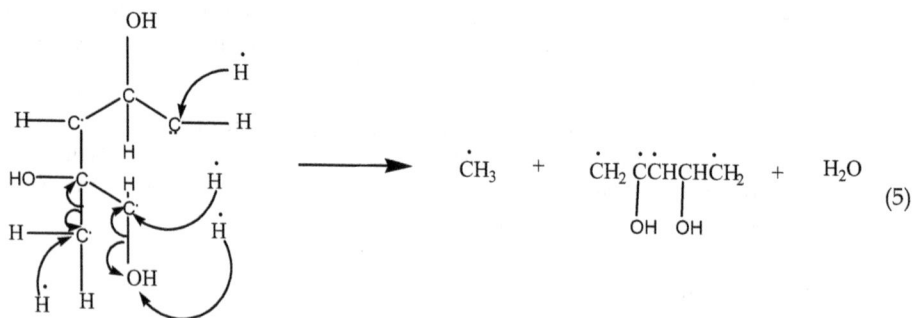

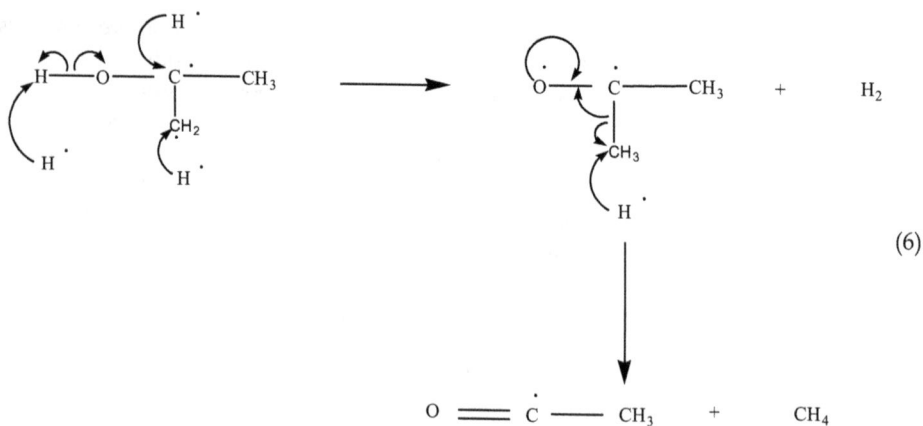

$$\dot{C}H_3 \quad + \quad \dot{C}H_2\,\dot{C}CHCH\dot{C}H_2 \quad + \quad H_2O \tag{5}$$
$$\quad\quad\quad\quad\quad\quad\;\; \underset{OH\;\;OH}{|\;\;\;|}$$

$$\tag{6}$$

$$O = \dot{C} - CH_3 \quad + \quad CH_4$$

And,

$$+ \quad H_2O \qquad (7)$$

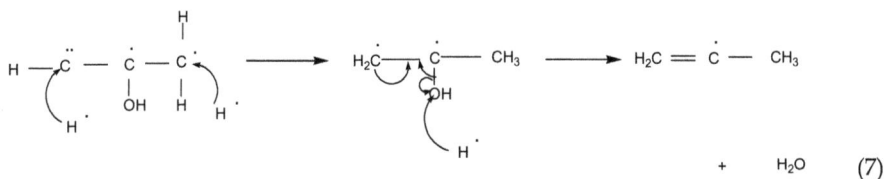

Other radicals likely to have been generated are $\dot{C}_2H_5, \dot{C}_3H_7, \dot{C}_4H_9$ and $\dot{C}_6H_5$.

During the termination reaction, the radicals recombine in diverse ways to give the many products identified by the infrared and GC analyses. For example, the small sized radicals combine to give H_2, H_2O, CH_4, and C_6H_6 as follows:

$$\dot{H} + \dot{H} \rightarrow H_2 \qquad (8a)$$

$$\dot{H} + \dot{O}H \rightarrow H_2O \qquad (8b)$$

$$\dot{C}H_3 + \dot{H} \rightarrow CH_4 \qquad (8c)$$

$$\dot{C}_6H_5 + \dot{H} \rightarrow C_6H_6 \qquad (8d)$$

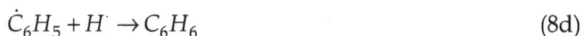

Aldehydes, ketones and alcohols are also formed as shown by Equations 9 to 11.

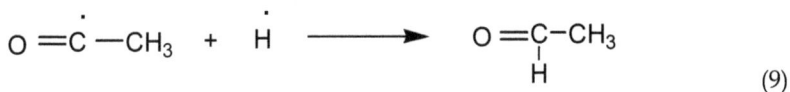

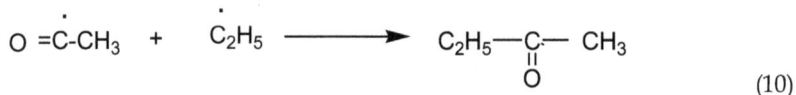

$$(9)$$

$$(10)$$

$$(11)$$

This is followed by the elimination of the OH group to form aliphatic compounds (Equations 12 and 13).

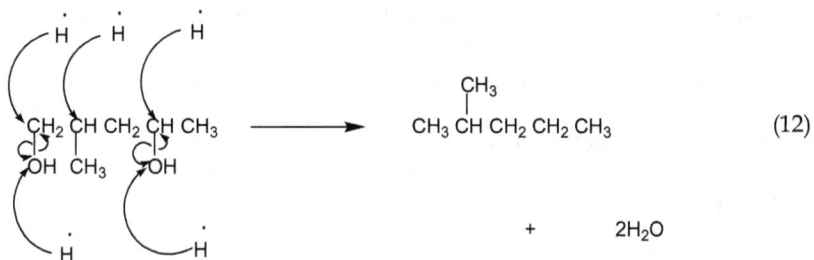

$$(12)$$

$$+ \quad 2H_2O$$

$$H-O-\overset{\overset{\displaystyle H}{|}}{\underset{\underset{\displaystyle CH_3}{|}}{C}}-CH_3 \quad\longrightarrow\quad CH_3\,CH_2\,CH_3 \quad + \quad H_2O \qquad (13)$$

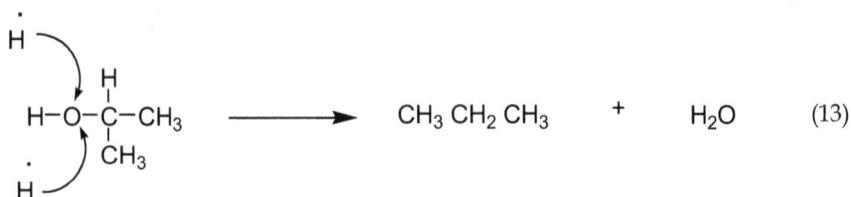

It is the elimination of the OH group from these oxygenated compounds that explains why the IR analyses on the freshly produced oils show the absence of oxygen. The saturated hydrocarbons, cyclo-hydrocarbons and aromatics can be formed when alkyl radicals combine as shown below in Equations 14, 15 and 16 respectively.

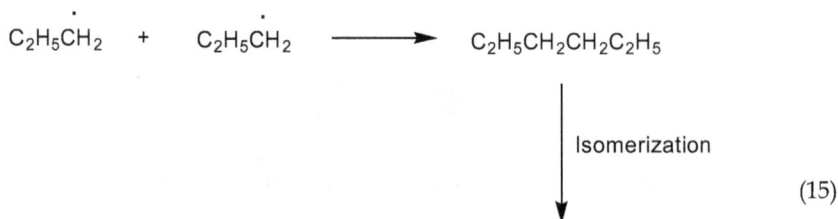

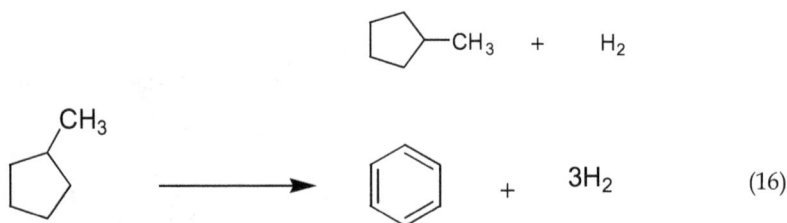

$$H_2C=\overset{\displaystyle \cdot}{C}-CH_3 \quad + \quad \overset{\displaystyle \cdot}{C}H_3 \quad\longrightarrow\quad H_2C=\overset{\overset{}{|}}{\underset{\underset{\displaystyle CH_3}{|}}{C}}-CH_3 \qquad (14)$$

$$C_2H_5\overset{\displaystyle \cdot}{C}H_2 \quad + \quad C_2H_5\overset{\displaystyle \cdot}{C}H_2 \quad\longrightarrow\quad C_2H_5CH_2CH_2C_2H_5$$

Isomerization

(15)

$$\longrightarrow\quad -CH_3 \quad + \quad H_2$$

$$\longrightarrow\quad + \quad 3H_2 \qquad (16)$$

These equations show that the hydrogenation process can be likened to the processes of catalytic cracking and reforming of petroleum in which a variety of saturated, unsaturated, aromatic and naphthenic compounds are produced [Macrae, 1966b].

On storing for some time, the original hydrogenation product (Sample A) and fractions obtained from the atmospheric distillation of Sample A (Samples B to E) were seen to be unstable. They became darker in appearance and sediments were formed at the bottom of their containers. The heavier fractions obtained from the atmospheric distillation of the hydrogenation products exhibited these characteristics the most because it was noticed that the rate of darkening of the products and the formation of sediments appeared faster with the heavier fractions.

IR analyses of these samples led to the following observations:

All samples contained aromatics, with the heavier samples exhibiting more of this characteristic.

The heavier oils and bitumen fractions now contained O-H and C-O bonds to show the presence of oxygenated compounds. These groups were initially absent from the oils when they were just produced, to show that the bituminous oils were unstable and consequently very reactive. They picked up oxygen from the atmosphere during storage. The lighter oil fractions still did not contain oxygen.

The functional groups identified suggest the presence of substances such as benzene, anthracene, phenanthrene, alkenes and other saturated and unsaturated compounds. These compounds are olefins, aromatics and paraffins, which are all essential chemicals, used for the production of many bulk chemicals in the synthesis industry (Macrae, 1966b).

The above observations, that the oils are very reactive and contain substances similar to the essential chemicals used for synthesizing bulk chemicals suggest that the bituminous oils from cellulose or biomass hydrogenation can be used as precursors or feedstock into the petrochemical industries. They can be used as a major substitute or supplement for petroleum whose reserves are continuously being depleted.

The presence of unsaturated and aromatic compounds indicate that the severe reaction conditions for the hydrogenation process [Anagho et al., 2004] led to further breakdown and re-arrangement of the oils produced. This process could be likened to the processes of catalytic cracking and reforming.

4.2 Rate expression for cellulose hydrogenation

As stated earlier, the diverse nature of the products obtained from cellulose hydrogenation coupled with the uncontrollable rise in reactor pressure during the reaction period suggests that cellulose hydrogenation is by free-radical mechanism. In this section we deduce a mechanism for the reaction, from which a rate expression will be obtained. To determine the mechanism, the following assumptions are made:

a. The initiation reaction is the generation of free radicals from CO_2 and H_2 obtained from the in-situ reaction of water and cellulose. The equations for the reactions are equations (2) and (3):

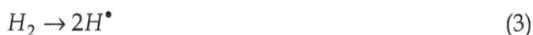

$$CO_2 \rightarrow CO^\bullet + O^\bullet \tag{2}$$

$$H_2 \rightarrow 2H^\bullet \tag{3}$$

Since the bond energy of the C-O bond in CO_2 is 745 kJ/mol, and that for the H-H bond in H_2 is 436 kJ/mol (Arylward and Findlay, 1974), reaction 3 is considered the only initiation reaction.

b. The cleavage of the cellulose ring can take place at any of the C-C or C-O bonds. Although the C-C bond energy of 348 kJ/mol is lower than 358 kJ/mol for the C-O bond, the cleavage takes place at the C-O bond, because the values are quite comparable in magnitude, and the hydrogen has a higher affinity for oxygen than for carbon. So if A is taken as a cellulose ring or molecule, and $B^\bullet$ the radical after the removal of O, then,

$$A + 2H^\bullet \rightarrow B^\bullet + H_2O \tag{17}$$

$B^\bullet$ can be represented as

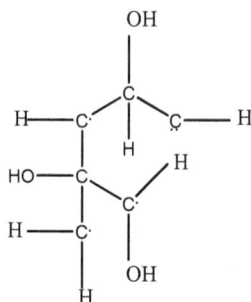

c.   The cleavage of the cellulose ring is the rate determining step, and our interest is in this step. Therefore, the radical $B^\bullet$ does not undergo any further bond breakage into smaller radicals. Instead, the only other propagation reactions are those in which the radical $B^\bullet$ sequentially gains $H^\bullet$ radicals.

From these assumptions, the following mechanism is deduced.

Initiation:

$$H_2 \xrightarrow{k_0} 2H^\bullet \tag{18}$$

Propagation:

$$A + 2H^\bullet \xrightarrow{k_1} B^\bullet + H_2O \tag{19}$$

$$B^\bullet + H^\bullet \xrightarrow{k_2} BH^\bullet \tag{20}$$

$$BH^\bullet + H^\bullet \xrightarrow{k_3} BH_2^\bullet \tag{21}$$

$$BH_2^\bullet + H^\bullet \xrightarrow{k_4} BH_3^\bullet \tag{22}$$

Termination:

$$BH_3^\bullet + H^\bullet \xrightarrow{k_5} BH_4 \tag{23}$$

$$BH_2^\bullet + BH_2^\bullet \xrightarrow{k_6} H_2B - BH_2 \tag{24}$$

$$H^\bullet + H^\bullet \xrightarrow{k_7} H \tag{25}$$

Using Equation (19), and assuming elementary reaction, the rate of cellulose decomposition is

$$-r_A = k_1[A][H^\bullet]^2 \tag{26}$$

The net rates of consumption of the various radicals are:

$$-r_{H^\bullet} = k_0 P_{H_2} - k_1[A][H^\bullet]^2 - k_2[B^\bullet][H^\bullet] - k_3[BH^\bullet][H^\bullet]$$
$$-k_4[BH_2^\bullet][H^\bullet] - k_5[BH_3^\bullet][H^\bullet] - k_7[H^\bullet]^2 \tag{27}$$

$$-r_{B^\bullet} = k_1[A][H^\bullet]^2 - k_2[B^\bullet][H^\bullet] \tag{28}$$

$$-r_{BH^\bullet} = k_2[B^\bullet][H^\bullet] - k_3[BH^\bullet][H^\bullet] \tag{29}$$

$$-r_{BH_2^\bullet} = k_3[BH^\bullet][H^\bullet] - k_4[BH_2^\bullet][H^\bullet] - k_6[BH_2^\bullet]^2 \tag{30}$$

$$-r_{BH_2^\bullet} = k_4[BH_2^\bullet][H^\bullet] - k_5[BH_3^\bullet][H^\bullet] \tag{31}$$

Applying the steady state approximation that the net rate of consumption of the radicals is zero, and then, expressing all other concentrations in terms of [A] and $[H^\bullet]$, we have that

$$-r_{H^\bullet} = -r_{B^\bullet} = -r_{BH^\bullet} = r_{BH_2^\bullet} - r_{BH_3^\bullet} = 0 \tag{32}$$

And

$$k_2[B^\bullet] = k_1[A][H^\bullet] \tag{33}$$

$$k_3[BH^\bullet] = k_2[B^\bullet] = k_1[A][H^\bullet] \tag{34}$$

$$k_4[BH_2^\bullet] = k_5 BH_3^\bullet] \tag{35}$$

$$k_3[BH^\bullet][H^\bullet] = k_4[BH_2^\bullet][H^\bullet] + k_6[BH_2^\bullet]^2 \tag{36}$$

$$k_5[BH_3^\bullet][H^\bullet] = k_4[BH_2^\bullet][H^\bullet] \tag{37}$$

Since $[H^\bullet]$ is much greater than $[BH_2^\bullet]$, the addition of $H^\bullet$ radicals to $BH_2^\bullet$ radicals occur faster than the addition of $BH_2^\bullet$ radical onto itself. This means that k_6 is negligible before k_4. From this, we have that,

$$k_4[BH_2^\bullet] = k_5[BH_3^\bullet] = k_3[BH^\bullet] = k_1[A][H^\bullet]$$

Substituting for the radicals in Equation (27) when $-r_{H^\bullet} = 0$ gives

$$[H^\bullet]^2 \{5k_1[A] + k_7\} = k_0 P_{H_2}$$

Hence,

$$[H^\bullet]^2 = \frac{k_0 P_{H_2}}{5k_1[A] + k_7} \tag{38}$$

Substituting in Equation (26) gives

$$-r_A = \frac{k_0 k_1 P_{H_2}[A]}{k_7 + 5k_1[A]} \tag{39}$$

After the cleavage of the cellulose ring, the termination reaction becomes dominant, and hence, k_1 becomes negligible before k_7; so the denominator of Equation (39) reduces to k_7. Also, hydrogen pressure in the reactor is considered large and hence constant.

Therefore,

$$-r_A = \frac{k_0 k_1}{k_7} P_{H_2}[A]$$

And hence,

$$-r_A = k'[A] \tag{40}$$

Where,

$$k' = \frac{k_0 k_1}{k_7} P_{H_2}$$

Equation (40) shows that cellulose hydrogenation is first order with respect to cellulose concentration. The reactor pressure increases as the cellulose reacts, to signify that the pressure in the reactor, P is proportional to the concentration of cellulose decomposed.

$$-r_A = kP \tag{41}$$

Equation (41) shows that the rate of cellulose hydrogenation is first order with respect to the total reactor pressure.

5. Economics of cellulose hydrogenation

In the former sections, we established that cellulosic wastes can be upgraded to synthetic fuels by hydrogenation. In this section we will be establishing that such a process is economically feasible. This will be done by carrying out an energy economy analysis. In an energy economy analysis, every cost incurred during a manufacturing process is converted to energy equivalent units.

5.1 Unit cost of production

Benn et al (1980) in an energy economy study considered that the unit cost on the production of synthetic oils was dependent on the three components energy, capital and labour.

$$C = X_f P_f + X_c P_c + X_l P_l \tag{42}$$

Where $X_f P_f$ = energy cost; $X_c P_c$ = capital cost and $X_l P_l$ = labour cost.

In this expression,

C = Total cost per unit of product,
X_f = quantity of fuel (i.e. energy) needed per unit of product obtained,
P_f = price per unit of energy,
X_c = quantity of capital per unit of product,
X_l = quantity of labour per unit of product, and
P_l = price per unit of labour.

In the plant for cellulosic waste hydrogenation, labour cost, $X_l P_l$ includes the cost of transporting the wastes to the plant and the labour in the other aspects in the plant.

$X_f P_f$ is actually the on-site cost of the net energy requirement (NER). NER is part of the gross energy requirement (GER) of the process needed to give the required product less the calorific value (CV) of the product. That is,

$$NER = GER - CV \tag{43}$$

The calorific value is the intrinsic energy of the product from the process. The ratio, NER/CV is the fraction of the product oil (in energy equivalents) that is used up during the conversion process. That is,

$$\frac{NER}{CV} = X_f \tag{44}$$

Hence, the net energy fraction recoverable from the process is $1 - X_f$.

As a first approximation, the cost of fuel for the process is equal to the total cost of producing the synthetic fuel. This is because more than 85% of the required fuel for the plant is generated locally. That is, $C \approx P_f$

Hence,

$$P_f = X_f P_f + X_c P_c + X_l P_l$$

$$\therefore C = P_f = \frac{X_c P_c + X_l P_l}{1 - X_f} \tag{45}$$

The equation shows that the cost per unit of product is the ratio of the nonfuel costs to the net energy yield. The equation also shows that no fuel is brought in from out of the plant, and that fuel inputs can be costed at the on-site value.

The values of X_f and C are determined from the energy equivalents of equipment and inputs into the plant. These components include capital equipment, electricity, chemicals, furnace, boiler and hydrogen plants, organic wastes and waste collection.

When the cost of a plant of a given capacity is known, then the cost of another of a different capacity can be found by using the seven-tenths rule given as:

$$\frac{Cost_2}{Cost_1} = \left(\frac{Capacity_2}{Capacity_1} \right)^{0.7}$$

(46)

The subscript, 1 stands for the reference plant, while subscript 2 stands for the new plant.

5.2 Energy requirements of plant and inputs

In the manufacture of the components of the capital equipment and chemicals, many processes are involved, each of which requires an energy input. The cost of the component depends on the total energy cost incurred during its manufacture. Edewor (1979) reported that Casper et al carried out an evaluation of the ratio of the input energy to the pound sterling for a number of products from manufacturing processes. This ratio is called the energy intensity of the product, and it is a measure of the cost of the equipment. They gave the energy intensity of heavy engineering products as 106.28 MJ/£ (1968), and for chemicals as 339.13 MJ/£ (1968). These figures are upgraded to 2011 costs by using cost indices as follows:

Energy intensity (MJ/£ (2011)) = [Energy intensity (MJ/£ (1968)] x

[Cost index (1968)/Cost index (2011)]   (47)

The energy equivalent, also called the energy requirement of any capital equipment is then obtained by multiplying the energy intensity by the cost.

The energy equivalents or energy requirements for the components of a 1000 tonne per day plant for cellulose hydrogenation are presented in the table below.

Plant Component	Energy requirements (x 10^3 MJ/day)
Capital equipment	295
Electricity	1,429
Chemicals	348
Furnace, boiler and H_2 plants	6,443
Organic wastes	22,107
Waste collection	53
Total = GER (Gross energy requirement)	30,675

Table 4. Energy requirements for the components of a 1000 tonne per day cellulosic waste hydrogenation plant.

5.3 Process energy factor and thermal efficiencies

Although the main objective in the hydrogenation of cellulose is to produce oil products, three product streams are obtained: oils, gases and char. The gas yield is only 15% of the cellulose fed, while that of char is significant. The gases produced are consumed completely in the plant. This means that the process produces only oil and char. Hence, the GER into the process is shared between oils and char. If D is the fraction of the GER that goes to produce oils only, then,

D = (Energy output from process through oil)/ (Total process energy output)

That is,

$$D = \frac{M_o(CV_o)}{M_o(CV_o) + M_c(CV_c)} \tag{47}$$

Where, M_o and M_c are the masses of oils and char respectively, while CV_o and CV_c are their calorific values.

For a unit mass of oil produced,

$$NER = \frac{(GER).D}{M_o} - CV_o \tag{48}$$

The process energy factor is

$$X_f = \frac{NER}{CV_o} \tag{49}$$

X_f is a very significant quantity in the energy economy analyses. It is a measure of the ratio of the total energy demand for the process to the total energy yield from it.

For $X_f \geq 0.6$, the process is considered to be energy demanding, and hence, heavily dependent on energy. It is therefore an expensive process, and may not be economically feasible. Conversely, for $X_f \leq 0.4$, the process is energy rewarding, and consequently, inexpensive.

The thermal efficiency of the process with respect to the oils is defined as the ratio of the energy output of the process through the oil product to the gross energy requirement.

$$\eta_{oil} = \frac{M_o(CV_o)}{GER} \tag{50}$$

By considering the char also, the efficiency becomes

$$\eta_{oil+char} = \frac{M_o(CV_o) + M_c(CV_c)}{GER} \tag{51}$$

Table 5 gives the values of GER, NER, X_f, and thermal efficiencies for the 1000 tonne per day cellulose hydrogenation plant. The plant was designed by Anagho (1987) for the treatment of municipal waste in Lagos. The design was based on the Kaufmann and Weiss (1975) 36 tonne per day plant that was used to recycle municipal waste in Manchester, UK.

Property	Magnitude
GER (x 10^3 MJ/tonne)	53.95
NER (x 10^3 MJ/tonne)	16.30
X_f	0.439
$1 - X_f$ = recoverable fraction	0.561
Thermal efficiency with respect to oils, η_{oil}	35.30%
Thermal efficiency with respect to oil and char, $\eta_{(oil + char)}$	69.50%

Table 5. Energy requirement, energy factor and thermal efficiency for 1000 tonne per day cellulose hydrogenation plant.

The energy factor for the plant is X_f = 0.439. Although this value is smaller than 0.60, it is slightly greater than the maximum value for the plant to be economically feasible, it could still be considered as such. The value will get smaller, and the process more competitive when the present petroleum prices get higher.

6. Conclusions

Cellulose hydrogenation is a complex process in that it takes place under very drastic conditions of high temperature and high pressure, producing a mixture of chemical products. This study shows that the complex reaction can be modelled as a first order rate process with respect to the total reactor pressure. That is,

$$-r_A = kP$$

(41)

The hydrogenation generates liquid products at a yield of almost 35% by the mass of the dry pulverised cellulose feed. The products obtained are a mixture of saturated, unsaturated paraffinic and aromatic hydrocarbons, which exhibit instability on storage. An energy economic evaluation of a prototype cellulose hydrogenation process shows that it is an economically feasible process.

The high yield of liquid products, the reactivity of the liquid products and the feasibility of the hydrogenation process suggest that we now have a bioresource substitute for petroleum.

Unfortunately, the technology for obtaining chemicals from cellulose is not well developed. Consequently, it is not widely used to meet the objectives of supplementing petroleum and reducing environmental pollution. The feed into the process comes from so diversified sources of cellulose that the design of plants to process them is difficult.

Future research will look at obtaining alcohols as the main chemicals from bioresource and then using them as the feedstock into petrochemical industries. This will be done by thermally breaking down the lignin surrounding the cellulose, followed by hydrolysing and fermenting the resulting sugars into alcohol. This will call for research into the growing of fast growing crops that have no particular food values to human beings. In addition to being fast growing, these crops should not compete for agricultural lands.

7. References

Anagho, S.G. (1987). Kinetic and Economic Study of the Production of Synthetic Oils from some Agricultural Wastes in Nigeria. Ph.D. Thesis, University of Lagos, Nigeria

Anagho, S.G., Ngomo, H.M. & Edewor, J.O. (2004). Bituminous Oils from the Hydrogenation of Cellulosic Wastes: A Kinetic Study. Energy Sources, Vol. 26, No. 4, pp. 415 – 425, (Feb 2004), ISSN: 0099-8312

Anagho, S.G., Ngomo, H.M. & Ambe, F. (2010). Analysis of the Products of Cellulosic Wastes Hydrogenation as Petrochemical Feedstock. *Energy Sources, Part A: Recovery, Utilization and Environmental Effects*. Vol. 32, No. 11, pp. 1052 – 1066, ISSN: 1556-7036

Boocock, D.G.B., Macay, D., McPherson, M. & Thurier, R. (1979). Direct Hydrogenation of Hybrid Poplar Wood to Liquid and Gaseous Fuels. *Can. J. Chem. Eng.* Vol. 57, (Feb 1979), pp. 98 – 101

Boocock, D.G.B., Macay, D., Franco, H. & Lee, P. (1980). The Production of Synthetic Organic Liquids from Wood using a Modified Nickel Catalyst. *Can. J. Chem. Eng.* Vol. 58, (Aug 1980), pp. 466 - 469

Cherchire, J. & Pavitt, K. (1978). Some Energy Futures, In *World Futures*, Freeman, C. & Johada, M. pp. 113. Universe Books, UK

Cheremisinnoff, P.N. & Morresi, A.C. (1976). Solid Waste as a Potential Fuel, In *Energy from Solid Waste*, pp. 1-32, Dekker, USA

Cheremisinnoff, P.N. (1980). *Wood for Energy Production*, Ann Arbor Science, USA

Crawford, R.L., (1981). *Lignin Biodegradation and Transformation*. John Wiley and Sons, ISBN 0-471-05743-6, USA

Edewor, J.O. (1979). Ph.D. Thesis, University of Manchester Institute of Science and Technology (UMIST), Manchester, UK

Edewor, J.O. (1980). Refuse–Derived Oils. A report prepared for the Greater Manchester County, by UMIST, Manchester, UK

El-Saied, H. (1977). Liquefaction of Lignohemicellulosic Waste by Processing with Carbon Monoxide and Water. *J. Appl. Chem. Biotechnol.* Vol. 27, pp. 443 – 462

Fierz-David, H.E. (1925). The Liquefaction of Wood and Cellulose, and some General Remarks on the Liquefaction of Coal. *Chemistry and Industry*. Vol. 44. (Sept. 1925), pp. 942 – 944

Frolich, P.K., Spalding, H.B. & Bacon, T.S. (1928). Destructive Distillation of Wood and Cellulose under Pressure.*Ind. Eng. Chem.* Vol. 20, No. 1, (Jan 1928), pp. 36 - 40

Hawley, L.F. (1922). Effect of Adding Various Chemicals to Wood Previous to Distillation. *Ind. Eng. Chem.* Vol. 14, No. 1, (Jan 1922), pp. 43 - 44

Henze, H.R., Allen, B.B. & Wyatt, W. (1942). Catalytic Hydrogenation of Cotton Hull Fibre. *Journal of Organic Chemistry*, Vol. 7, pp. 48 – 50

Kaufman, J.A. & Weiss, A.H. (1975). Solid Waste Conversion: Cellulose Liquefaction. Report prepared for the US National Environmental Research Centre by the Worcester Polytechnic Institute, USA

Macrae, J.C. (1966a). Coal—Secondary fuels from coal: 2-By gasification, hydrogenation and Fisher Tropsch synthesis. In *An Introduction to the Study of Fuels*. Elsevier Publishing Company, Netherlands, pp. 108 - 126

Macrae, J.C. (1966b). Petroleum Processing – Conversion Processes. In *An Introduction to the Study of Fuels*. Elsevier Publishing Company, Netherlands, pp. 157 - 169

Mitchell, B. (1995). *Resources and Environmental Management in Canada*, Oxford University Press, UK

Ngomo, H.N. (2004). Opportunity for Investment in Sustainable Bioenergy Production in Countries in the Central African Sub-Region: The Case of Cameroon. *Proceedings of World Bioenergy*, Jonkoping, Sweden. (June, 2-4, 2004), pp. 37-38

Palmer, R.C. & Cloukey, H. (1918).Influence of Moisture on the Yield of Products in the Destructive Distillation of Hardwood. *Ind. Eng. Chem.* Vol. 10, No. 4, (April, 1918), pp. 262 – 264

Palmer, R.C. (1918). The Effect of Catalyzers on the Yields of Products in the Destructive Distillation of Hardwoods. *Ind. Eng. Chem.* Vol. 10, No. 4, (April, 1918), pp. 264 - 268

Updegraff, D.M. (1969). Semimicro Determination of Cellulose in Biological Materials. *Analytical Biochemistry*. Vol. 32, pp. 420 - 424

Part 2

Modeling and Simulation

4

Modeling and Simulation of Water Gas Shift Reactor: An Industrial Case

Douglas Falleiros Barbosa Lima[1], Fernando Ademar Zanella[1],
Marcelo Kaminski Lenzi[2] and Papa Matar Ndiaye[2]
[1]Refinaria Presidente Getúlio Vargas – REPAR / PETROBRAS
[2]Universidade Federal do Paraná – UFPR
Brazil

1. Introduction

Recently, refineries finished products hydro treatment became critical due to changes in fuel regulations. These changes are related to the specification of more clean fuels with special focus in sulfur content reduction. In order to achieve these goals more hydro treatment is needed. Hydrogen is the main raw material to hydro treatment units.

In refining plants, a hydrogen generation unit usually is necessary to supply the hydrogen demand to all demanding processes.

The process known as steam reform unit is the most widely adopted technology. In large scale, it has the highest energetic efficiency and the best cost-benefit ratio (Borges, 2009).

In this process the hydrogen conversion is carried out in two reactors in series. The first one, the steam reform reactor converts steam and a hydrocarbon (naphtha or natural gas) into *syngas*. In the sequence, a reactor known as water gas shift reactor (WGSR) converts the carbon monoxide present in *syngas* into carbon dioxide and more hydrogen is generated.

Consequently, the WGSR, an intermediate step of hydrogen generation process, plays a key role in a petrochemical plant due to hydrogen increasing demand.

1.1 Hydrogen generation unit

The hydrogen generation unit, based on the steam reform technology, is responsible for approximately 95% of generated hydrogen (Borges, 2009).

A simplified unit block diagram is showed in figure 1.1.1.

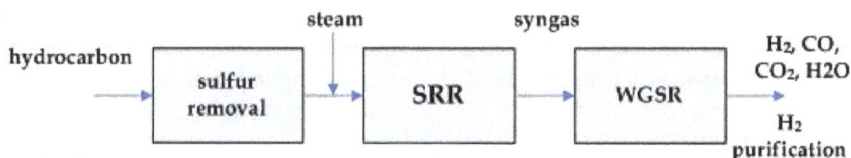

Fig. 1.1.1. Hydrogen generation unit – process block diagram.

First, sulfur is removed from the hydrocarbon stream (usually natural gas), in order to prevent catalyst poisoning and deactivation with the use of a guard bed. Steam is mixed in the main stream in a fixed steam to carbon molar basis. The steam reform reactor (SRR) is a multitubular catalyst filled furnace reactor where the hydrocarbon plus steam are converted into syngas at high temperatures (700°C – 850°C) according to the following reaction:

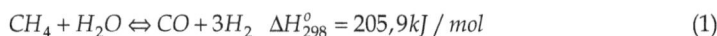

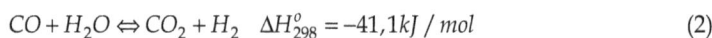

$$CH_4 + H_2O \Leftrightarrow CO + 3H_2 \quad \Delta H^o_{298} = 205,9kJ \: / \: mol \tag{1}$$

$$CO + H_2O \Leftrightarrow CO_2 + H_2 \quad \Delta H^o_{298} = -41,1kJ \: / \: mol \tag{2}$$

The reaction (1) is endothermic and reaction (2) is moderately exothermic. Both are reversible reactions. In the SRR, (1) is the main reaction and generates most of the hydrogen. The reaction (2) due to its endothermic nature occurs in a lower extension in the SRR. The syngas stream composition is CO and H_2, in this process CO_2 and H_2O are also present in gas state. The main purpose of the water gas shift reactor (WGSR) is to carry out the reaction (2) reducing the CO fraction and increasing the hydrogen yield. Finally, the WGSR stream is conducted to a purification section, where hydrogen purity is increased according to the process needs.

2. The water gas shift reaction

2.1 kinetic rate expression

The water gas shift reaction (reaction 2) is a heterogeneous reaction (gas/solid).

According to (Smith et al., 2010) in this kind of application, there are two options in the WGSR step. Using a high temperature shift (HTS) catalyst based reactor or a series of HTS followed by a low temperature shift catalyst based reactor (LTS) with intercooling stage to increase the overall conversion and high purity hydrogen is needed (Newsome, 1980).

The chapter focuses on a HTS ferrochrome catalyst industrial reactor modeling.

The HTS usually is an iron oxide – chromium oxide based catalyst. Also reaction promoters such as Cu may be present in catalyst composition. Operational temperatures vary from 310°C to 450°C. Inlet temperatures are usually kept at 350°C to prevent the catalyst bed temperature from damage. Exit CO concentrations are in the order of 2% to 4%. Industrial reactors can operate from atmospheric pressure to 8375 kPa. Sulfur is a poison for Fe-Cr catalysts.

LTS reactors are copper based catalyst. Typical compositions include Cu, Zn, Cr and Al oxides.

Recent catalysts can be operated at medium temperatures around 300°C. Copper is more sensitive to catalyst thermal sintering and should not be operated at higher temperatures. Sulfur is also a poison to LTS reactors. Typical exit concentration is of 0,1% of CO.

The reaction is operated adiabatically in industrial scale, where the temperature increases along the length of the reactor.

According to Arrhenius law of kinetics, increasing temperature increases the reaction rate. By the other side, the thermodynamic of equilibrium or Le Châtelier principle states that increasing the temperature of an exothermic reaction shifts the reaction to reactants side decreasing its equilibrium conversion. Therefore the water gas shift reaction is a balance between these effects and the reactor optimal operational point takes into account the tradeoff between kinetics and equilibrium driving forces.

In (Chen et al., 2008) experimental data indicates that increasing temperature in HTS will promote the performance of WGSR. For the LTS, the reaction is not excited if the reaction is bellow 200°C. Once the temperature reaches 200 °C the reaction occurs, but the CO conversion decreases with increasing temperature. This fact reveals that that the water gas shift reactions with the HTS and the LTS are governed by chemical kinetics and thermodynamic equilibrium, respectively in industrial conditions.

(Smith et al., 2010) classifies the reaction kinetic models in microkinetic approach and the empirical method.

Basically, the micro kinetic approach explores the detailed chemistry of the reaction. On the other hand, the empirical models are based on the experimental results and are typically expressed in the Arrhenius model and provide an easy and computationally lighter way to predict the rate of reaction. The main disadvantage is the fact that the adjusted model cannot be extrapolated to different composition and types of catalysts.

Many empirical expressions have been reported in literature for HTS. (Newsome, 1980) and (Smith et al., 2010).

An empirical rate expression succesfully used to describe the WGSR in ferrochrome catalysts is a power law type: (Newsome, 1980)

$$r = k_0 \cdot e^{\frac{-Ea}{RT}} \cdot P_{CO}^l \cdot P_{H2O}^m \cdot P_{CO2}^n \cdot P_{H2}^q \cdot \left(1 - \frac{1}{K_{eq}} \cdot \frac{P_{CO2} \cdot P_{H2}}{P_{CO} \cdot P_{H2O}} \right) \tag{3}$$

Where:

r – reaction rate.
Ea – activation energy.
K_0 – pre exponential factor.
K_{eq} – reaction equilibrium constant.
l, m, n, q – estimated parameters by experimental data.
P_y – partial pressure of component y.
R – universal gases constant.
T – absolute temperature.

The reaction equilibrium constant derived from thermodynamics as function of temperature is given by (Smith et al., 2010):

$$\ln\left(K_{eq}\right) = \frac{5693,5}{T} + 1,077 \cdot \ln(T) + 5,44 \cdot 10^{-4} \cdot T - 1,125 \cdot 10^{-7} \cdot T^2 - \frac{49170}{T^2} - 13,148 \tag{4}$$

T – temperature in Kelvin.

Several authors have published estimated parameters for specific catalysts types or catalysts classes. (Newsome, 1980) and (Smith et al., 2010)

Table 2.1.1 summarizes some previous publicated values and authors for HTS catalysts:

Author	Catalyst information	K_o	Ea (kJ/mol)	l	m	n	q
Bohlbro et al. (Newsome, 1980)	Fe_2O_3/ Cr_2O_3 Commercial reduced particle size	-	105,9	0,93	0,24	-0,31	0,00
Bohlbro et al. (Newsome, 1980)	Fe_2O_3/ Cr_2O_3 Commercial Large particle size	-	59,8	0,87	0,26	-0,18	0,00
S.S. Hla et al. (Hla et al., 2009)	Fe_2O_3/ Cr_2O_3/CuO type 1	$10^{2,845}$	111	1	0	-0,36	-0,09
S.S. Hla et al. (Hla et al., 2009)	Fe_2O_3/ Cr_2O_3/CuO type 2	$10^{0,659}$	88	0,9	0,31	-0,156	-0,05
Adams and Barton. (Adams & Barton, 2009)	Fe_2O_3/ Cr_2O_3/CuO Commercial type 1 Same as Hla et al.	725	110	1	0	-0,32	-0,083

Table 2.1.1. Estimated parameters for power law HTS catalysts.

In industrial reactors, catalysts are loaded in pellets. Therefore, intrinsic rate expressions cannot be used in pseudo-homogeneous models without some type of compensation to account for diffusion effects.

According to (Newsome, 1980) two different methods can be applied to model industrial reactors as pseudo-homogeneous reactions. The first was proposed by Bohlbro and Jorgensen consists of estimating the empirical rate expression in laboratory using commercial size catalysts. The diffusion effects remain implicit in the rate expression. First and second rows of Table 2.1.1 illustrate this method. The catalyst in the first row was grounded to avoid diffusion effects. In the second row the same catalyst was used in commercial pellet size. It can be noted slight differences between l, m, n e q parameters. The disadvantage of this method is that the rate expression can model successfully only a reactor with this specific type and size of catalyst.

(Hla et al., 2009) estimates the intrinsic rate parameters for two commercial catalysts as can be seen in rows three and four.

(Adams & Barton, 2009) also model a WGSR reactor using a heterogeneous modeling approach. The intrinsic rate expression is from Hla et al. previous article. The parameters are result for the best fit estimation.

The second approach of pseudo-homogeneous rate modeling consists of using correction factors to compensate for the pore diffusion phenomena in the catalyst, catalyst age, operating pressure and hydrogen sulfide concentration with intrinsic rate expressions. (Singh and Saraf, 1977) is a good example of this method.

2.2 Mathematical modeling

In this section, mathematical expressions for the fixed bed adiabatic catalytic WGSR fundamental principles (conservation equations) are developed.

A basic ideal flow steady state one-dimensional model is presented.

The differential molar balance simplified to a fixed bed reactor can be expressed as equation 5: (Froment and Bischoff, 1990)

$$\frac{dX_a}{dW} = \frac{r_a}{F_{a0}} \tag{5}$$

X_a – component "a" conversion.
W – catalyst weight.
r_a – rate of reaction of component "a".
F_{a0} – molar feed rate of reactant "a".
F_a – molar flow of component "a" leaving the reactor.

The ideal model assumes that concentration and temperature gradients only occur in the axial direction. The only transport mechanism operating in this direction is the overall flow itself, and is considered to be of the plug flow type.

In case of a reactor bed with a fixed cross sectional area (S), the differential molar balance can be rewritten as a function of the reactor differential length as can be seen in equation 6:

$$\frac{dX_a}{dz} = \frac{r_a \cdot \rho_B \cdot S}{F_{a0}} \tag{6}$$

ρ_B – catalyst bulk density.
S – reactor bed constant section area.
z – length of reactor (z axis – axial direction).

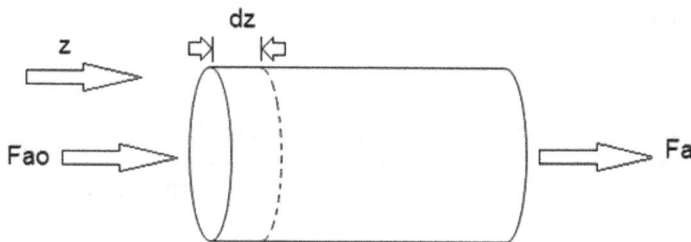

Fig. 2.2.1. Reactor differential volume representation.

The differential equation of energy conservation may be written as equation 7:

$$\sum Fi \cdot Cpi \cdot \frac{dT}{dz} = (-\Delta H_R) \cdot r_a \cdot \rho_B \cdot S - 4\frac{U}{d_t} \cdot (T - T_R) \tag{7}$$

ΔH_R – enthalpy of reaction.
Cpi – specific heat of component "i".
Fi – component "i" molar flow.
T, T_R – reactor temperature and temperature at radius R of internal tube.
d_t – internal tube diameter.
U – overall heat transfer coefficient.

For an adiabatic reactor U equals zero and equation 3 can be simplified to:

$$\sum Fi \cdot Cpi \cdot \frac{dT}{dz} = (-\Delta H_R) \cdot r_a \cdot \rho_B \cdot S \tag{8}$$

The momentum differential equation can be defined as in equation 9:

$$-\frac{dP}{dz} = -f \cdot \frac{\rho_g \cdot u_s^2}{d_p} \tag{9}$$

P – reactor pressure in position z.
f – friction factor.
ρ_g – gas density.
u_s – superficial velocity.
d_p – particle diameter.

The aim of the next topics is to develop this set of equations to model an industrial high temperature WGSR.

2.3 Some design aspects

When designing WGSR (Chen et al, 2008), it is known that a proper selection on certain parameters is of the most importance because the reaction result depends strongly on the combination of these parameters. Typically, the important parameters include the catalyst type, residence time of reactants in a catalyst bed, reaction temperature and feeding reactants ratio or CO/steam ratio.

2.3.1 Types of catalysts

According to the reaction temperature (Chen et al, 2008), the WGSR falls into two categories: high-temperature shift catalyst (HTC) and low-temperature shift catalyst (LTC). The catalyst commonly used in the former is an iron–chromium-based catalyst, whereas a copper–zinc-based catalyst is frequently adopted in the latter.

2.3.2 Residence time

It is known that the catalyst amount in a reactor is highly related to the overall project cost. In other words, if a reaction can be developed with least catalyst, both the operation (or catalyst) cost and the facility (or space) cost can be reduced effectively. The least catalyst amount can be evaluated by determining the residence time of reactants in a catalyst bed.

For both categories of reaction, a residence time of 0.09 s is long enough to establish an appropriate WGSR (Chen et al, 2008), a design factor can be used to compensate deactivation during the catalyst life.

2.3.3 Reaction temperature

The reaction temperature on the design of the WGSR, varies according with the type of catalyst used (Chen et al, 2008), For HTC, increasing reaction temperature, the concentration of CO declines with respect to the temperature, thus the conversion of WGSR increases, but between 400°C and 500 °C the change in this propriety is small, so temperatures between 350°C and 400°C the major reaction occur.

In the case of LTC reaction, for temperatures below 150°C, no reaction take place, increasing to 200°C the conversion rise to above 90%. If temperature is increased further the conversion begins to fall. So, an optimal reaction, for the WGSR with the LTC is obtained in temperatures around 200°C.

The WGSR is an exothermic reaction in nature, from thermodynamics it is known that an increase in reaction temperature will impede the forward reaction for H2 production in the WGSR. Because the behavior, it is realized that the WGSR with the LTC catalyst is governed by chemical equilibrium.

In contrast, for the HTC catalyst, the CO conversion is highly sensitive to the reaction temperature and an increase in temperature is conducive to the hydrogen generation. It follows that the reactions with the HTC are controlled by chemical kinetics (i.e. Arrhenius law).

2.3.4 CO/steam ratio

For both kind of catalysts when the CO/steam ratio is large than 1/4, the performance of the WGSR is sensitive the variation of the ratio. Alternatively, if the ratio is smaller than 1/4, varying the ratio merely has a slight influence on the performance(Chen et al, 2008).

3. Experimental data

Experimental data of an industrial shift reactor of REPAR/PETROBRAS were used for the modeling studies. The reactor performs CO oxidation to CO2 using industrial steam. The reaction occurs in a fixed bed filled with a Fe-Cr-based catalyst, which is shown in Figure 3.1.

The historical data set comprises 4 years of operation. More specifically, samples withdrawn from feed and exist streams were analyzed by chromatography, following the technical standard norm (NBR-14903, 2002). Temperature measurements are performed using K type thermocouples connected to Honneywell STT3000 transmitters positioned according to Figure 3.2. Finally, it is important to mention that feed flow rate measurements were performed using orifice plates connected to a Honeywell STD900 differential pressure transmitter and the reactor pressure drop was measured using manometric pressure coupled to Honeywell STD900 transmitters.

Fig. 3.1. Catalyst.

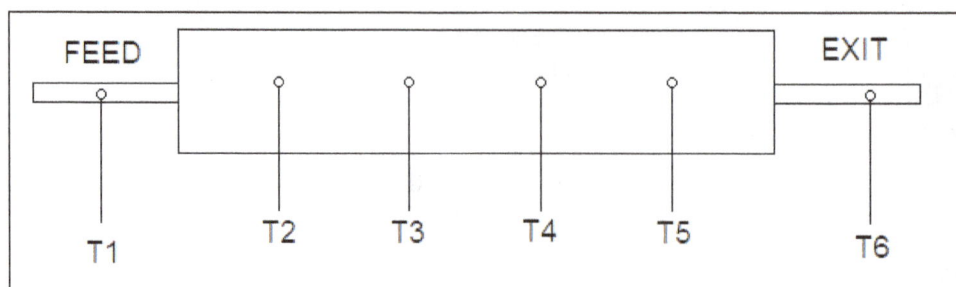

Fig. 3.2. Temperature measurement positioning.

Experimental data of CO conversion are presented in Figure 3.3. These data were normalized by dividing the experimental value by the reactor design conversion value. It is important to state that all experimental data were normalized to the design values in order to avoid numerical convergence issues along the modeling and parameter estimation tasks.

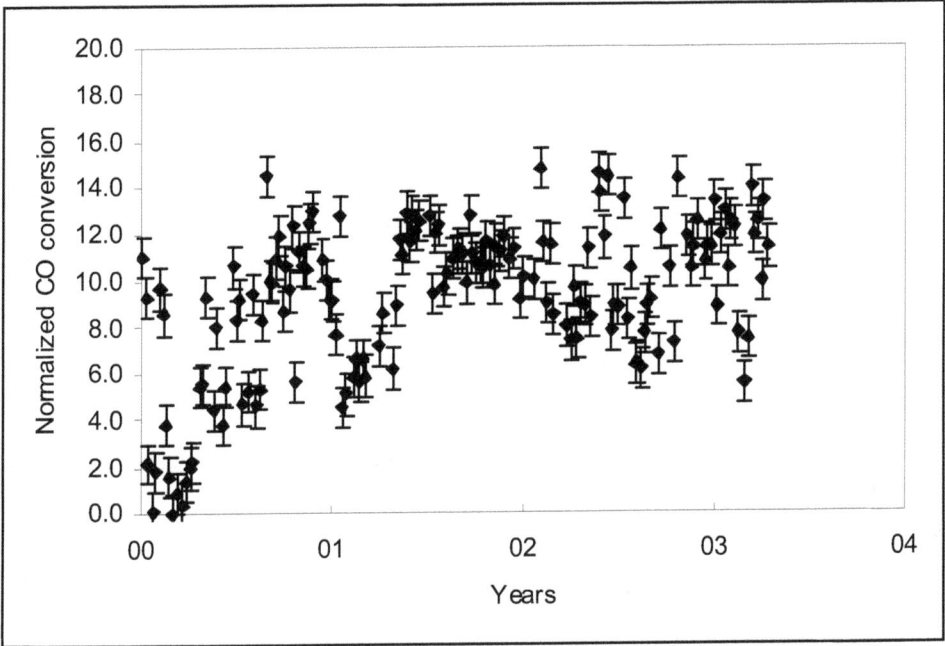

Fig. 3.3. Normalized experimental CO conversion.

The modeling work will focus on CO conversion, however this variable is not directly measured, but actually obtained from chromatographic analysis using Eq. (10), which considers the measurement of CO, CO2 and CH4 in the feed and exit streams.

$$X_{co} = 100 - 100 * \frac{CO_s}{CO_e} * \frac{(CH_{4e} + CO_e + CO_{2e})}{CH_{4s} + CO_s + CO_{2s}} \tag{10}$$

However, it is important to estimate the experimental uncertainty (i) of CO conversion the measurement, i.e., an estimate of the CO conversion variance. This calculation can be done by applying an error propagation approach to Eq. (10), more specifically, an estimate can be obtained by Eq. (11), which considers the variance of CO, CO2 and CH4 concentration measurements, while covariance was considered zero. The variance of each measurement was calculated according to the technical standard norm NBR-14903. This analysis yielded an average standard deviation of 1% of the CO measurement.

$$i = \sqrt{\begin{array}{l}\left(\frac{\partial X}{\partial CH_{4\_e}}\right)^2 * i_{CH_{4\_e}}{}^2 \left(\frac{\partial X}{\partial CO_{\_e}}\right)^2 * i_{CO_{\_e}}{}^2 + \left(\frac{\partial X}{\partial CO_{2\_e}}\right)^2 * i_{CO_{2\_e}}{}^2 + ... \\ ... + \left(\frac{\partial X}{\partial CH_{4\_s}}\right)^2 * i_{CH_{4\_s}}{}^2 \left(\frac{\partial X}{\partial CO_{\_s}}\right)^2 * i_{CO_{\_s}}{}^2 + \left(\frac{\partial X}{\partial CO_{2\_s}}\right)^2 * i_{CO_{2\_s}}{}^2 \end{array}} \tag{11}$$

4. Model parameter estimation and validation

Based on the kinetic expressions, the development of a fundamental model to describe the shift reactor behavior represents an important tool for in-depth studies and performance optimization.

For modeling purposes, the experimental data set was firstly modeled considering a model with a reasonable amount of simplifying hypothesis and literature reported parameters. Afterwards, some of the hypotheses need to be disregarded, allowing an increase in model complexity, leading to further parameter estimation. Different hypotheses were disregarded until the model could successfully describe the experimental data set behavior. Only steady-state analysis was performed.

Parameter estimation was performed using a simplex-based method (Himmelblau et al., 2002) focusing on minimizing the classical least square objective function based on the difference between experimental and predicted CO conversion values. The experimental industrial data set was divided into two groups, the first for parameter estimation and the second for model validation.

Despite the relaxation of some simplifying hypotheses, in all models, both axial and radial diffusion were not considered. This choice occurred due to some features of the studied reactor, for example, it presents a high length to diameter ratio, in order to assure a turbulent flow for different operating conditions. Consequently, effects of a possible external diffusive resistance tend to be reduced. A second common simplifying hypothesis present in all models concerns the absence of pressure drop, therefore an isobaric reactor was considered. This hypothesis was assumed because experimental data revealed a pressure drop less than 3% along the catalyst bed, bellow literature recommendation threshold of 10% to disregard possible pressure drop effects (Iordanidis, 2002).

4.1 Model 1 – Pseudo-homogeneous isothermal reactor

The first modeling attempt represents a very simplified reactor model as given by Eq. (12)

$$\frac{dX_a}{dz} = \frac{r_a \cdot \rho_B \cdot S}{F_{a0}} \tag{12}$$

The reaction rate is based on power law kinetics and is given by Eq. (13)

$$r_a = F_{PRESS} \cdot k_0 \cdot e^{\frac{-Ea}{RT}} \cdot y_{CO}^l \cdot y_{H2O}^m \cdot y_{CO2}^n \cdot y_{H2}^q \cdot \left(1 - \frac{1}{K_{eq}} \cdot \frac{y_{CO2} \cdot y_{H2}}{y_{CO} \cdot y_{H2O}}\right) \tag{13}$$

For modeling purposes, reaction rate considered molar fractions instead of the components partial pressure, according to (Adams & Barton, 2009). According to the authors, a pressure correction factor, Fpress, given by Eq. (14) also needs to be considered as a pressure increase tends to increase the effect of molecular diffusion inside the catalyst pore.

$$F_{PRESS} = P^{\left(0.5 - \frac{P}{250}\right)} \tag{14}$$

where: P is the absolute pressure in bar.

The equilibrium constant, Keq, is given by Eq. (15), obtained by considering temperature dependent heat capacities.

$$\ln\left(K_{eq}\right) = \frac{5693,5}{T} + 1,077 \cdot \ln\left(T\right) + 5,44 \cdot 10^{-4} \cdot T - 1,125 \cdot 10^{-7} \cdot T^2 - \frac{49170}{T^2} - 13,148 \quad (15)$$

Finally, it is worth mentioning that the reaction rate constants were obtained from (Hla et al. ,2009) as presented in Table 4.1.1 and also that the catalyst activity was considered constant.

k_0	Ea	l CO	m H_2O	n CO_2	q H_2
700	111	1.0	0	– 0.36	– 0.09

Table 4.1.1 – Reaction rate exponents.

After solving the reactor model, normalized experimental CO conversion plotted against normalized CO conversion predicted values are shown in Figure 4.1.1. It can be observed that a systematic deviation occurs, probably because of the number of simplifying assumptions. The first hypothesis to be revisited is the reactor isothermal behavior, mainly because of the direct temperature effect on CO conversion and also because of the reaction exothermal behavior.

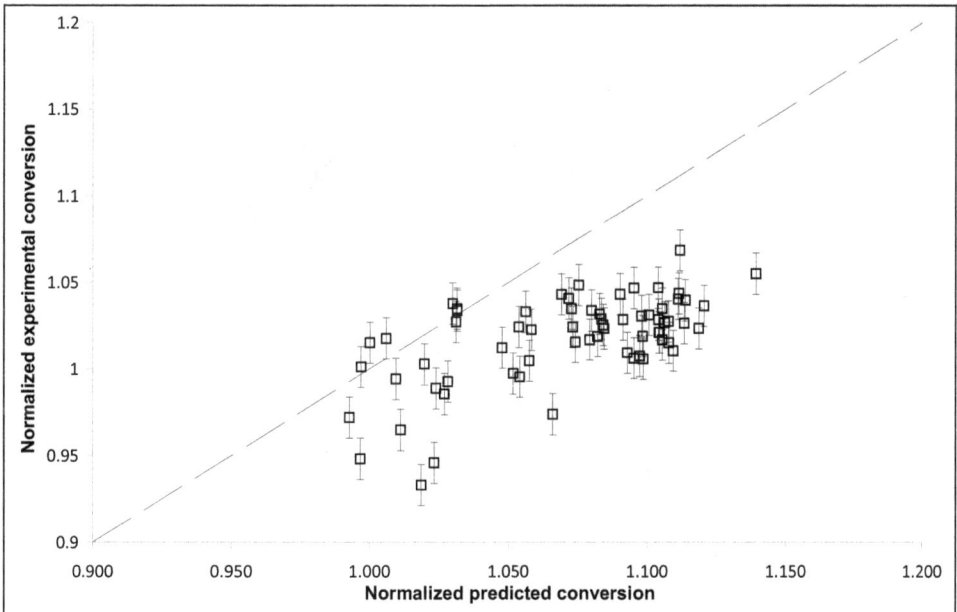

Fig. 4.1.1. Model 1 results.

4.2 Model 2 – Pseudo-homogeneous non-isothermal reactor

This model represents and extension of Model 1. Due to the non-isothermal behavior, the reaction rate is not constant along the reactor length. The exothermal characteristic of the reaction also contributes to performing a heat balance on the reactor, yielding Eq. (16):

$$\sum F_j \cdot c_{pj} \cdot \frac{dT}{dz} = (-\Delta H_R) \cdot r_a \cdot \rho_B \cdot S \tag{16}$$

Consequently, the reactor model is now comprised by a system of ordinary differential equations. Figure 4.2.1 is obtained after plotting normalized experimental CO conversion against normalized CO conversion predicted values. It can be observed, after comparison with Figure 4.2.1, that non-isothermal feature actually resulted in worse model predictions. By analyzing Figure 7, it can be observed that for higher experimental conversion values, systematic higher predictions are obtained. This probably happens because the non-isothermal behavior leads to higher temperature values, consequently not only leading to higher reaction rates overestimating CO conversion, but also affecting diffusion phenomena inside the catalyst particles. Therefore, an alternative modeling approach would be considering an effectiveness factor in order to correct the reaction rate.

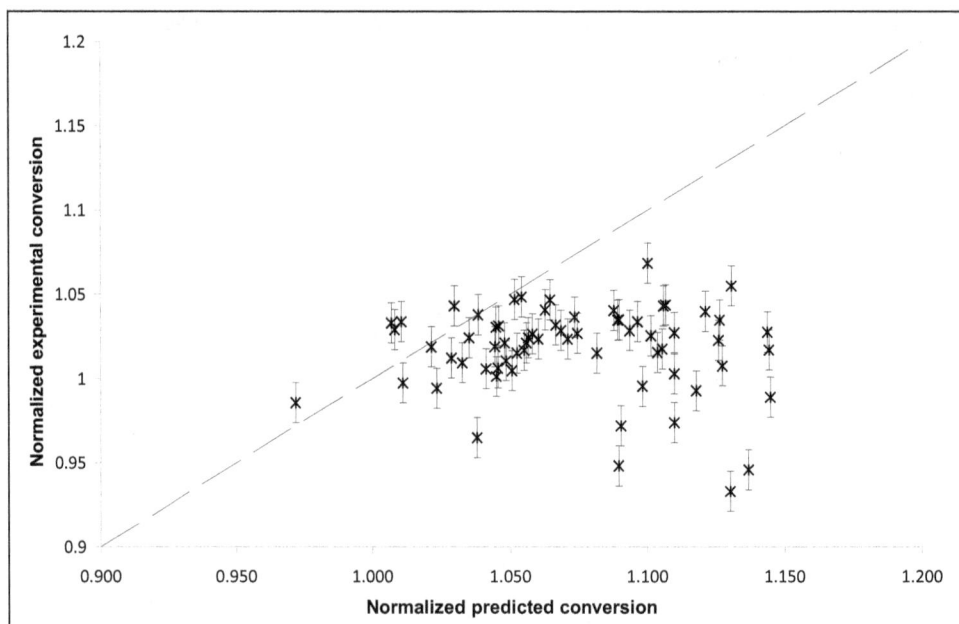

Fig. 4.2.1. Model 2 results.

4.3 Model 3 – Pseudo-homogeneous non-isothermal reactor with reaction rate correction

In this model, the reaction rate (Eq. 13) is corrected by an effectiveness factor as presented by Eq. (17).

$$r_{a-effective} = \eta \cdot r_a \tag{17}$$

It is important to realize that in Model 1 and Model 2 only simulation studies were carried out, no model parameter estimated. Now, parameter η, which remains in the interval (0,1], will be estimated in order to improve data fitting the model predictions. As mentioned before, only part of the experimental data set will be used for parameter estimation using the minimum least square method, while the other part will be used for validation purposes.

After parameter estimation, an optimal value of $\eta = 0.56$ was obtained. Considering this value, the mathematical model was then used to predict CO conversion as presented in Figure 4.3.1. It can be seen a considerable improvement in the model prediction, which was achieved by estimating only one parameter. However, the experimental data set refers to a considerable production horizon; consequently, catalyst activity may have changed, leading to another opportunity for model improvement.

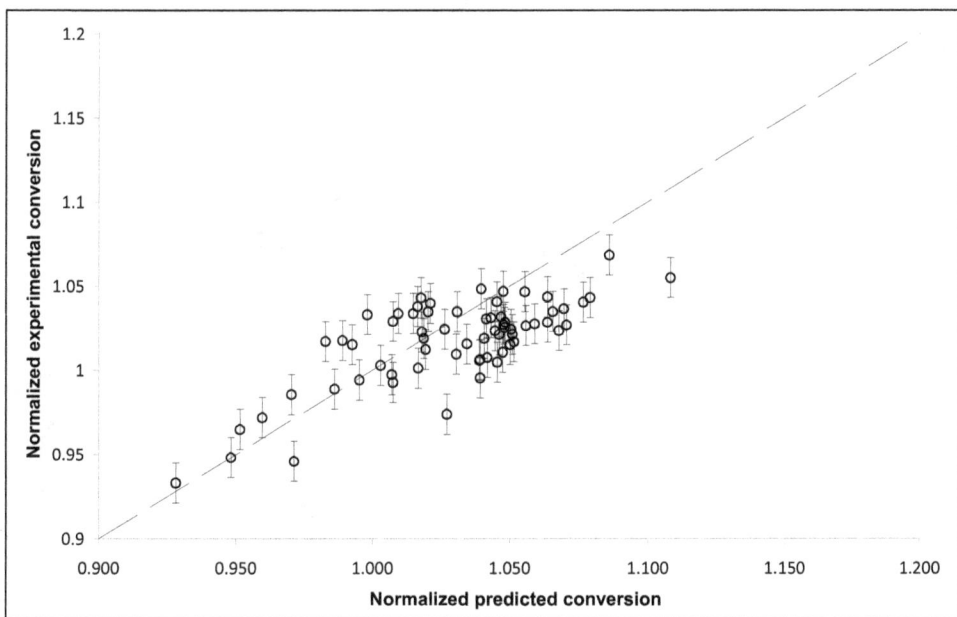

Fig. 4.3.1. Model 3 validation.

4.4 Model 4 – Pseudo-homogeneous non-isothermal reactor with reaction rate correction and catalyst deactivation

This model considers the reaction rate given by Eq. (18)

$$r_{a-effective} = \eta \cdot a(t) \cdot r_a \tag{18}$$

In this effective reaction rate, the catalyst activity, $a(t)$, can be described by a hyperbolic rate expression (Eq. (19)) as previously reported (Keiski et al., 1992):

$$a(t) = \frac{1}{(1 + \alpha \cdot t)^{\frac{1}{3}}} \qquad (19)$$

It must be stressed that α is the only parameter to be estimated in Model 4 as η was kept equal to 0.56. Parameter α was estimated as 10^{-5}. Figure 4.4.1 presents experimental normalized CO conversion plotted against predicted normalized values considering only the validation data set. It can be observed that the use of catalyst deactivation improved model performance, indicating an important role played by catalyst deactivation. However, the intrinsic reaction rate was not yet used for parameter estimation as exponents i, m, n, p values were considered the ones reported in the literature (see Table 4.1.1), consequently, this creates an alternative for model improvement.

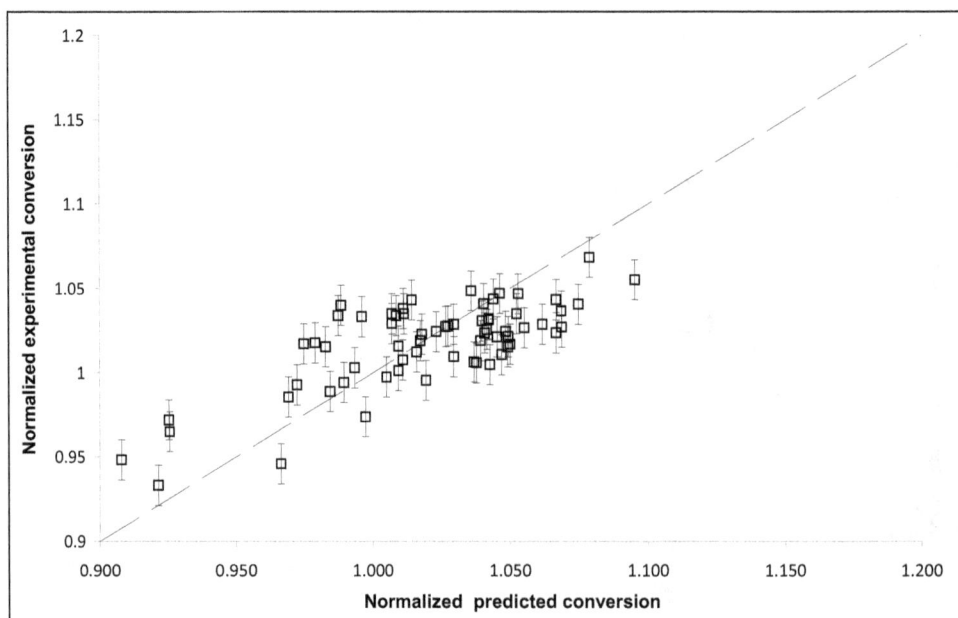

Fig. 4.4.1. Model 4 validation.

4.5 Model 5 – Revisiting model 4 intrinsic reaction rate

It can be observed that the order of H_2O in the reaction rate is equal to 0 as reported in Table 4.1.1. This indicates that H_2O is present in such a large excess that concentration changes does not considerably affect the reaction rate (Hla et al., 2009). However, in the industrial reactor, where the feed composition may change due to the nature of a petrochemical plant, H_2O may be in excess, but not in enough amount to be considered constant throughout the reactor length. Therefore, exponent m needs to be re-estimated, while the others will be kept at the same values. However, changes in m may affect the effective rate, consequently, η value also needs to be revisited and estimated. To sum up, Model 5 basically keeps the same structure of Model 4 only different parameters need to be estimated.

After performing the estimation task, a m value of 0.2 was obtained, simultaneously to a η value equal to 0.575. Figure 4.5.1 presents experimental and predicted values of CO conversion, showing that model predictions improved considering the new m value.

In order to compare the model predictions, the sum of the square difference of experimental and predicted values of CO conversion was performed and Model 2 led to the highest sum being the worst model. The sum of the squares of Model 2 was used as normalization factor and the normalized sum of all models are presented in Figure 4.5.2. It can be concluded that two hypotheses played a key role in the reactor modeling: firstly the isothermal behavior; secondly, the effectiveness factor. Further changes were import to improve, but not considerably, the data fit. It must also be noted that a few number of parameters were estimated, indicating an important likelihood feature of the reactor model. Finally, it is worth stressing that the correlation coefficient between experimental data and model predictions of Model 5 was equal to 0.8.

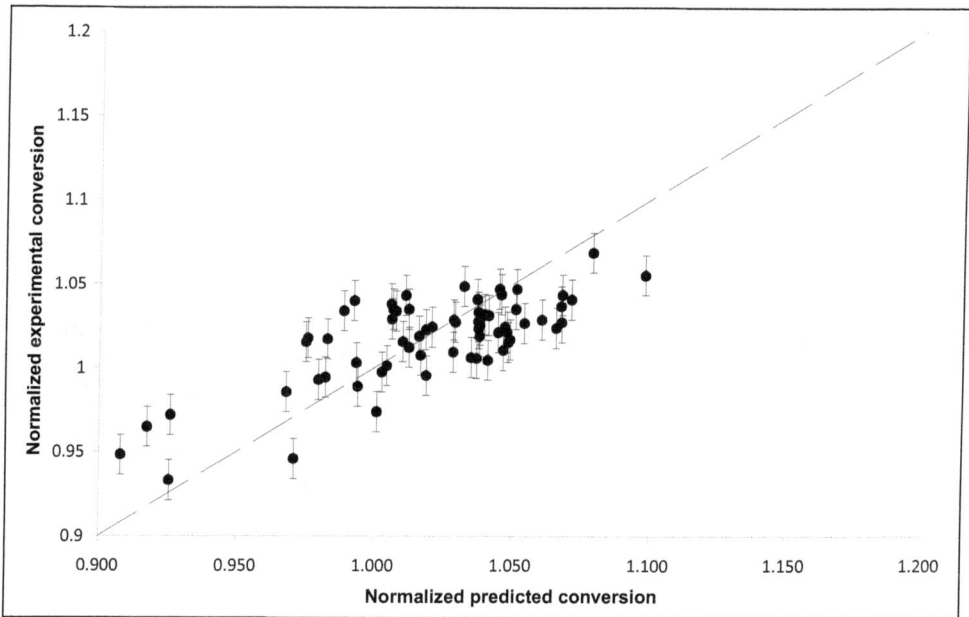

Fig. 4.5.1. Model 5 validation.

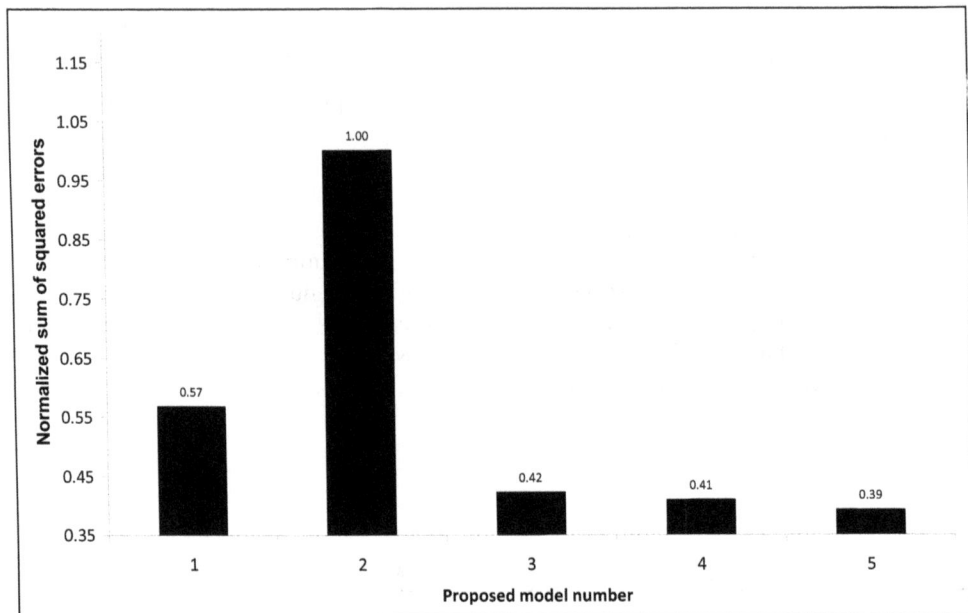

Fig. 4.5.2. Model comparison.

5. A novel approach to optimal process control strategy

Due to the shift reaction features, for a given reactor feed stream composition and flow rate, the temperature of the feed stream can be manipulated in order to maximize CO conversion. Consequently, if the industrial reactor has in-line/on-line composition analyzers as well as flow rate measurement instrumentation, on-line optimization can be successfully implemented. The problem is that this kind of in-line/on-line instrumentation is not only expensive but also may need continuous and excessive calibration and maintenance. Therefore, some question arises: can other real-time measurements be used for CO conversion? How does temperature influence CO conversion correlated? This is an important issue as temperature measurements are usually reliable, accurate, real-time and low cost. Moreover, fixed bed reactors can have temperature instruments installed along the reactor length, providing a temperature profile. Towards this, a novel and alternative approach will be presented in order to overcome this issue, focusing on CO conversion control.

Figure 5.1 presents historical normalized CO conversion data plotted against normalized feed flow-rate values, obtained from PETROBRAS shift reactor unit. As mentioned before, the normalized variables equal to 1 represent the reactor design values. These data refer to the same reactor temperature; however, due to the nature of petrochemical process, raw material composition can fluctuate, explaining the different CO conversion values obtained for the same experimental conditions. As expected, the higher the feed flow rate, the lower the reactor conversion.

In Figure 5.1, one observes, for example, conversions of 1.02 or 1.04, indicating that in these scenarios the experimental values were over the design values. The same observation can be made for the feed flow rate, indicating some operating conditions either bellow or above the design values. Consequently, it is important to emphasize that the reactor feed composition represents an important disturb variable affecting reactor performance, enhancing the importance of well tuned and designed regulatory control loops.

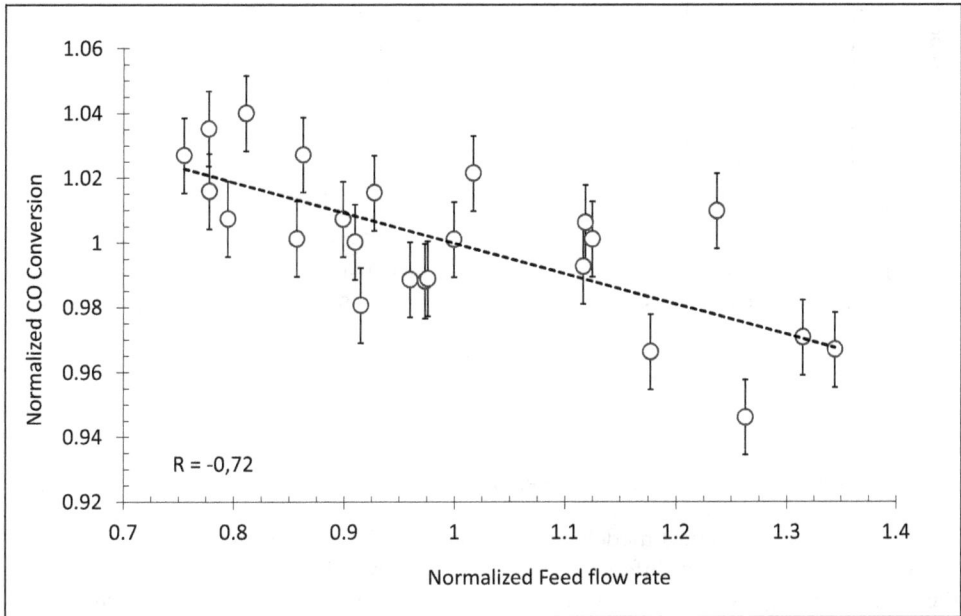

Fig. 5.1. Historical data: CO Conversion versus feed flow rate.

The studied shift reactor has thermocouples installed at the reactor feed and exit streams. It also has four thermocouples equally distributed along the reactor bed. After performing extensive sensitivity analysis, the temperature difference between the thermocouple placed on the reactor exit stream and last thermocouple of the bed provided the highest sensitivity, here denominated Delta_T. After analyzing the historical data set, Figure 5.2 presents experimental data of CO conversion plotted against normalized Delta_T. It can be observed a good correlation between both variables, which corroborate the idea of monitoring CO conversion by using lower cost measurements. Finally, Figure 5.3 presents the relationship between normalized Delta_T and normalized feed flow rate values, as expected indicating strong correlation and also good sensitivity.

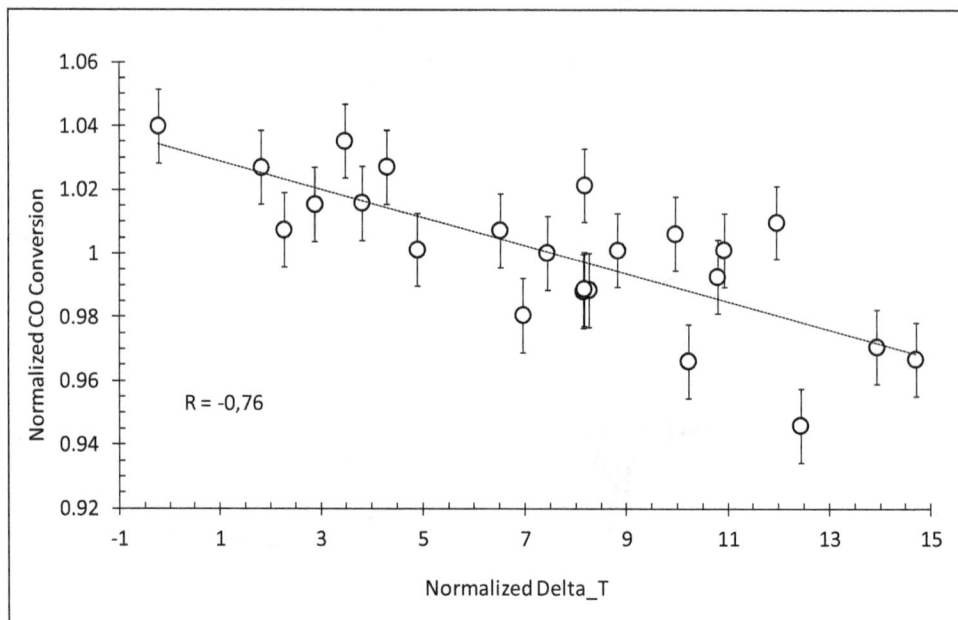

Fig. 5.2. Historical data: CO Conversion versus Delta_T.

The developed mathematical model was used for sensitivity analysis in order to study the CO conversion behavior for different operating conditions. More specifically, this analysis is aimed at providing not only the control feasibility itself, but also the feasibility of leading to operating conditions which may allow optimum conversion values. Figure 5.4 shows the CO conversion behavior for different feed flow rate values, considering in all simulations the feed composition design value. More specifically, the design value of the feed flow rate (FLOW1), the design value increased by 35% (FLOW2) and the design value increased by 55% (FLOW3) were used. Firstly, one can observe that due to the shape of the curve, the feed flow rate temperature can be manipulated in order to reach a maximum CO conversion. Secondly, the simulation results show that keeping the reactor feed temperature constant, for example at 1.03, an increase in the flow rate may lead CO conversion reduction, similar to the behavior of the historical data exhibited by Figure 5.1. It is also important to observe the effect of the temperature of the reactor feed stream on the CO conversion. After careful analysis, it can be also noted that to keep track of CO conversion, for example, at 0.955, the feed flow rate needs to be increased for a feed stream temperature increase. Consequently, feed flow rate and feed temperature can be regarded as important manipulated variables for CO conversion either in servo or regulatory control.

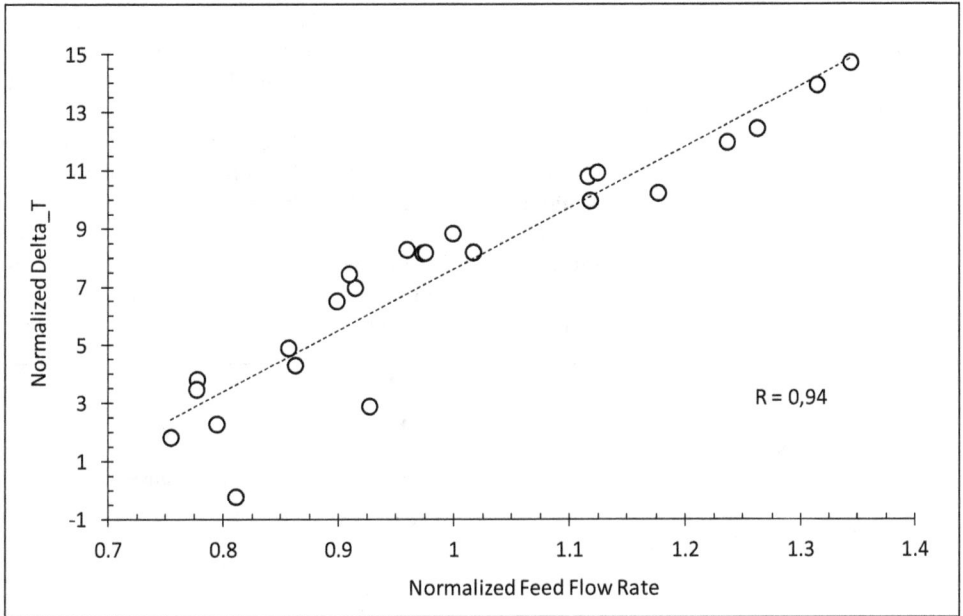

Fig. 5.3. Historical data: Delta_T versus feed flow rate.

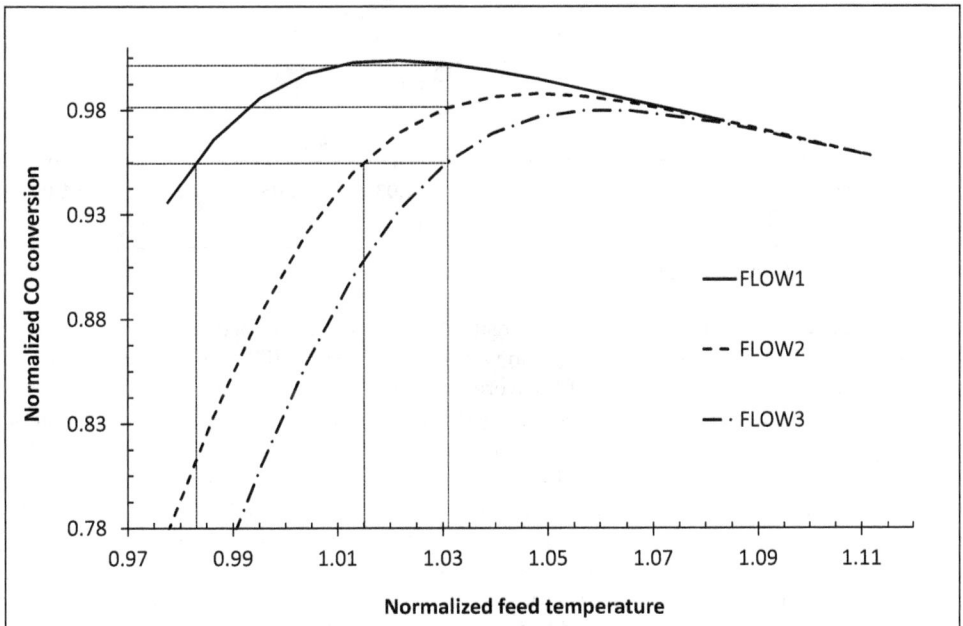

Fig. 5.4. Sensitivity Analysis.

However, it is important not only to provide control feasibility, but also to assure that the control can lead to optimal conversion values. Figure 5.5 presents a sensitivity analysis study focusing on the effect of feed composition. The reactor feed flow rate was kept constant in all simulations, while two different feed compositions were chosen, more specifically, the design feed composition (COMP1) and a feed composition involving an increase in CO amount (COMP2). For each feed composition, the temperature of the feed stream was changed in order to evaluate the correspondent resulting Delta_T values and the maximum conversion was determined and plotted. For COMP1, the maximum conversion is reached at a Delta_T value equal to roughly 3.7. Consequently, for this feed composition and feed flow rate, the feed temperature set-point needs to be roughly 1.01. The same analysis can be performed for composition 2.

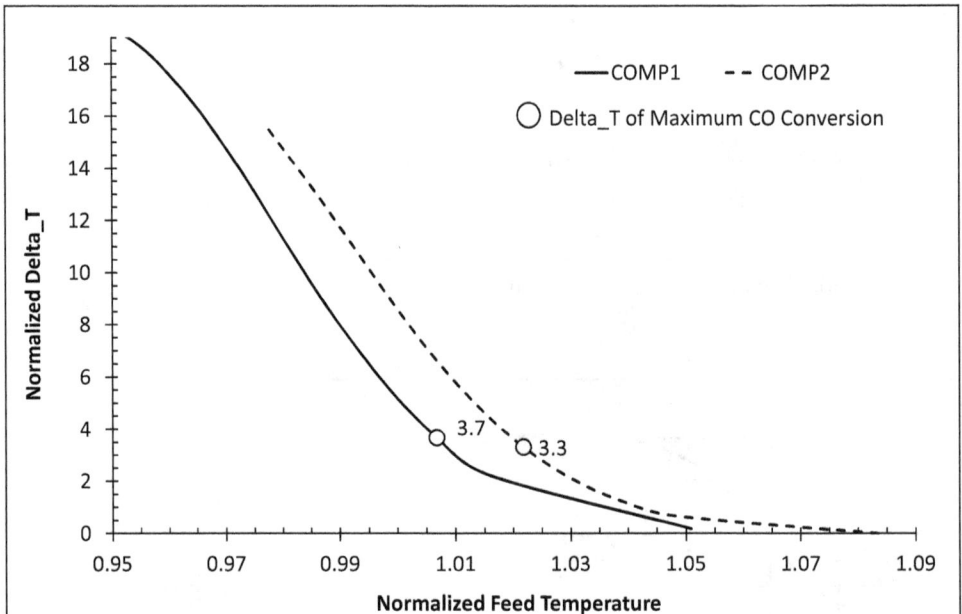

Fig. 5.5. Sensitivity Analysis: Composition.

On the other hand, Figure 5.6 presents a sensitivity analysis study focusing on the effect of feed flow rate. The reactor feed composition was set equal to COMP2 and different values of the flow rate, FLOW1, FLOW2, FLOW3 were chosen. For each feed flow rate value, the temperature of the feed stream was changed in order to evaluate the correspondent resulting Delta_T values. For each feed flow rate, the maximum conversion was also determined and plotted. Considering the feed composition COMP2, for the feed flow rate close to the design value, Delta_T should be equal to 3.3 to lead to the maximum CO conversion, consequently, the feed temperature set-point should be roughly 1.02. Keeping the composition unchanged and increasing the flow rate by 35%, Delta_T should be equal to 3.6 and, therefore, the feed temperature should be increased to approximately 1.05. Increasing the design value of the feed flow rate by 55%, Delta_T should be changed to 4.4, so the feed temperature should be increased to approximately 1.06.

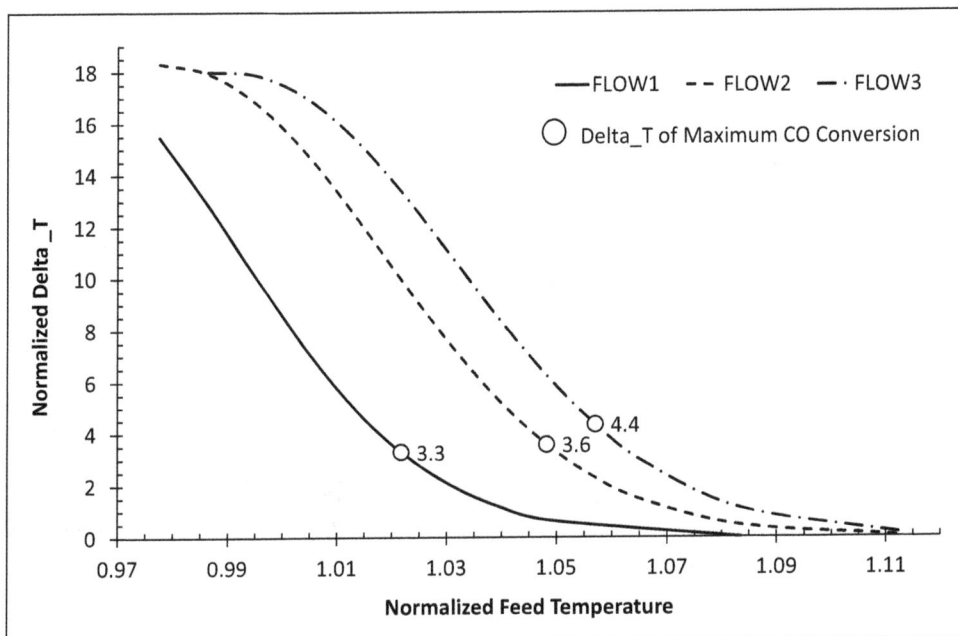

Fig. 5.6. Sensitivity Analysis: Feed Flow.

Based on the historical data and on the sensitivity analysis results, an alternative CO conversion control loop having the following features arises:

- controlling Delta_T represents the same as controlling CO conversion;
- controlling Delta_T can lead to optimal conversion values;
- cascade structure can be used with Delta_T control mastering the loop by providing appropriate set-points to the reactor feed temperature control loop;
- feedforward features may also be considered as if feed flow rate changes the Delta_T value leading to the optimal conversion needs to be updates as observed in Figure 7;
- feedforward features may also be considered if feed analyzers can be installed, as any change in feed composition would lead to a different Delta_T set-point.

6. References

Adams II, Thomas A. and Barton, Paul I. (2009). *A dynamic two-dimensional heterogeneous model for water gas shift reactors*. International Journal of Hydrogen Energy 34, 8877–8891.

Borges, Joana Lopes. (2009). *Diagrama de fontes de hidrogênio*. Dissertação de Mestrado em Tecnologia de Processos Químicos e Bioquímicos. Universidade Federal do Rio de Janeiro. Rio de Janeiro. Brasil.

Chen, Wei-Hsin et al. (2008). *An experimental study on carbon monoxide conversion and hydrogen generation from water gas shift reaction*. Energy Conversion and Management 49, 2801–2808.

Edgar, Himmelblau and Lasdon. (2001). *Optimization of Chemical Processes*. 2nd Edition. New York – McGraw-Hill, pp. 10-11.

Froment, Gilbert F. and Bischoff, Kenneth B. (1990). *Chemical reactor analysis and design*. John Wiley and Sons, ISBN 0-471-51044-0, second edition, pp. 403-404.

Hla, San Shwe et al. *Kinetics of high-temperature water-gas shift reaction over two iron-based commercial catalysts using simulated coal-derived syngases*. Chemical Engineering Journal 146 (2009) 148–154.

Iordanidis, A. A. (2002). *Mathematical Modeling of Catalytic Fixed Bed Reactors*. Ph.D. thesis, University of Twente. ISBN 9036517524, Enschede, The Netherlands.

Keiski, R. L. et al. (1992). *Deactivation of the high-temperature catalyst in nonisothermal conditions*. Applied Catalysis A: General, Vol. 87, pp 185-203.

NBR 14903 (2002) Gás Natural – *Determinação da composição por cromatografia gasosa*. Associação Brasileira de Normas Técnicas.

Newsome, David S. (1980). *The water-gas shift reaction*. Catal. Rev. Sci. Eng., 21(2), 275-318.

Singh, C. C. P. and Saraf, D. N. (1977). *Simulation of high-temperature water-gas shift Reactors*. Ind. Eng. Chem. Process Des. Dev., Vol. 16, No. 3, 313-319.

Smith, Byron R. J. et al. (2010). *A review of the water gas shift reaction kinetics*. International Journal of Chemical Reactor Engineering Vol. 8, Review R4, ISSN 1542-6580.

5

Simultaneous Elimination of Carbon and Nitrogen Compounds of Petrochemical Effluents by Nitrification and Denitrification

Anne-Claire Texier[1], Alejandro Zepeda[2],
Jorge Gómez[1] and Flor Cuervo-López[1]
[1]*Universidad Autónoma Metropolitana-Iztapalapa*
[2]*Universidad Autónoma de Yucatán*
Mexico

1. Introduction

Human activities have resulted in the increase of nitrogen and carbon content in wastewater and groundwater affecting the environment (Bremmen, 2002). The average water consumption in Mexico is close to 0.25 m^3/d, resulting in municipal and industrial wastewater generation between 168 and 232 m^3/s, respectively. Only 12 and 20% of wastewater has received some treatment (Monroy et al., 2000). The increase of nitrogen compounds such as nitrate, nitrite and ammonium in superficial and groundwater has caused several environmental effects such as eutrophication, toxicity to aquatic organisms, loss of biodiversity (Galloway, 1998; Mateju et al., 1992; Schimel et al., 1996); and human health damages as methahemoglobinemia (Mateju et al., 1992; Morgan-Sagastumen et al., 1994); formation of nitrosoamines which are potentially carcinogenic compounds (Cerhan et al., 2001, as cited in González-Blanco et al., 2011) and gastric cancer (Knobeloch et al., 2000; Ward et al., 2005). In the north of Gulf of Mexico, an important hypoxic zone has been detected where oxygen concentration is lower than 2 mg/l due to high nitrate discharges and eutrophication (Alexander et al., 2000; McIsaac et al., 2002). Nitrate concentrations between 7 and 156 mg/l (Antón & Díaz, 2000) and higher than 80 mg/l (Muñoz et al., 2004; Pacheco et al., 2001) have been determined in aquifers of middle and south of Mexico, respectively. These nitrate concentrations are higher than the maximum levels established by Secretary of Environmental and Natural Resources (SEMARNAT), NOM-003-ECOL-1997 (15 and 40 mg total nitrogen/l) (Diario Oficial de la Federación, 1998) and the United States Environmental Protection Agency (USEPA, 2007) (10 mg $N-NO_3^-/l$, 1 mg $N-NO_2^-/l$ and 10 mg $N-NH_4^+/l$). Therefore, it is clear the need of applying effective wastewater treatments for reducing nitrogen contamination.

One of the most important sources of carbon pollution is the petrochemical industry. In 2009, more than 85 million of crude oil barrels were produced in the world in order to satisfy the energy and derivatives demand for industrial uses. At present, Russia is the first crude oil producer followed by Saudi Arabia and USA. Both of them produce almost 30% of the total crude oil. In 2010, Mexico was the seventh petroleum crude oil producer (close to 2.5 million barrel/d) accounting to 0.05 million barrel (eq)/d of derivatives. Monoaromatic hydrocarbons such as benzene, toluene and xylenes (so-called BTX) are also produced from crude oil (Kirkeleit et al., 2006) by catalytic reforming (dehydrogenating process of the aliphatic compounds) and by alkylation and transalkylation processes (Perego & Ingallina, 2004).

A crescent BTX production has been observed all over the world. For instance, in 2000, the USA toluene production was close to 6.5 million of tons. In 2002, xylenes production was around 7 million of tons while in 2005, the USA benzene production was around 7 million of tons with a higher production of toluene and xylenes. BTX are widely used for producing paints, rubber, adhesives, varnishes, agrochemicals, polymers among many other products. In fact, almost 30% of gasoline content corresponds to BTX (Hartley & Englande, 1992; Sander et al., 2010). The average daily gasoline sales in Mexico in 2001 were close to 547,400 barrels (PEMEX Petrochemical, 2001). Accidental spills, evaporation of industrial sources, leakage from storage tanks in gas stations, industrial discharges and car combustion have resulted in environmental pollution with BTX. The great mobility abilities and toxicity of the BTX compounds are of major concern for environment and human health (Coates et al., 2002). These aromatic compounds have a high vapor pressure and are relatively water-soluble. Some of the physical and chemical properties of BTX are illustrated in Table 1. In this sense, BTX presence in soil, air, surface and groundwater has been increasingly reported (Jain et al., 2011; Martínez et al. 2009; Peña-Calva et al., 2004b; Steen-Cristenesen & Elton, 1996).

Compound	Benzene	Toluene	Xylenes		
Chemical structure			o-xylene	m-xylene	p-xylene
Molecular weight (g/mol)	78	92	106		
Density (g/ml)	0.87	0.87	0.87 ± 0.02		
Water solubility (g/l) at 25°C	1.87	0.47	0.17 ± 0.032		
Vapor pressure (mm Hg)	95	28	5.83 ± 0.76		
Partition coefficient octanol/water (log K_{ow})	2.13	2.69	3.04 ± 0.23		
Henry's law constant at 25°C (kPa m³/mol)	0.56	0.67	0.657 ± 0.094		
Polarity	Non polar	Non polar	Non polar		

Table 1. Some physical-chemical properties of benzene, toluene and xylenes (average values for *orto*, *meta*, and *para* isomers).

Simultaneous Elimination of Carbon and Nitrogen Compounds of Petrochemical Effluents
by Nitrification and Denitrification

77

Concentrations (mg/l) ranging from 9-14, 23-81 and 13-171 of benzene, toluene and isomers of xylene have been respectively reported in groundwater (Gersberg et al., 1989). Likewise, considering the non polar characteristic of these compounds, there is an important bioaccumulation of BTX in the lipid fraction of cell membrane. Benzene is the most dangerous of the BTX compounds as it is a known human carcinogen (Huff, 2007). Breathing low levels of benzene can cause drowsiness, dizziness, rapid heart rate, confusion, and unconsciousness whereas higher levels of this compound can result in death (Ashly et al., 1994). It has been reported that benzene may induce damages to the bone marrow and cause anemia (Fishbein, 1985; Vereb et al., 2011), depression in the immune system and leukemia (Huff, 2007). The narcotic and neurotoxic properties of toluene are the major health hazards in humans due to its tendency to accumulate in adipose tissue. Toluene is highly lipophilic affecting the central nervous system (Fornazzari et al., 2003) where it seems to inhibit neuronal transmission by causing changes in protein conformation and hence in the membranal transport (Rosenberg et al, 1988). Acute intoxication from inhalation is characterized by euphoria, hallucinations, dizziness, confusion, headache, ataxia, stupor, and coma. Exposition over long periods of time may produce neuropsychosis, cerebral degeneration with ataxia, peripheral neuropathies, cognitive ability problems, ototoxicity and deafness (Waniusiow et al., 2008). However, there are no epidemiological evidences whether toluene induces cancer in persons exposed to the solvent (Fishbein, 1984; Weelink, et al., 2010). Xylenes seem not to be as toxic as benzene and toluene since their values of LD50 range from 200 to 4000 mg/kg for animals (Fishbein, 1985). According to Mexican legislation (NOM-127-SSA1-1994), (Diario Oficial de la Federación, 1994), the maximum levels established in potable water for BTX are (mg/l): benzene 0.01; toluene 0.3 and isomers of xylene 0.5. In the United States the maximum levels are 0.0005, 1 and 10 mg/l for benzene, toluene and mixed xylenes, respectively (USEPA, 2006, as cited in Farhadian et al., 2008). All of these BTX characteristics clearly illustrate why these compounds represent an environmental and health challenge.

There are different alternatives for nitrogen and carbon compounds removal from contaminated water, such as physical-chemical and biological processes. Gravity separation, volatilization, adsorption, dialysis, inverse osmosis, ultrasonic radiation or chemical reactions as phentom are some of the most used physical-chemical processes (Jacobs & Testa, 2003; Thoma et al., 2006). Nevertheless, most of these processes are not cost-effective, are poorly efficient when highly contaminated streams are treated and generally give undesirable residues (Farhadian et al., 2008). On the other hand, biological treatment processes are economically feasible, result in highly nitrogen and carbon consumption efficiencies and conversion to innocuous products such as N_2, CO_2 and water, respectively. Nitrogen compounds can be biologically removed from water by the coupling of nitrification and denitrification processes (Cuervo-López et al., 2009).

BTX are chemically reduced compounds, thus, their biologic oxidation is thermodynamically favored (Table 7). Many biological attempts have been conducted in order to remove these volatile organic compounds from the environment. Aerobic treatments have been widely proposed for BTX elimination (Duetz et al., 1994; Haigler et al., 1992; Rozkov et al., 1998; Yerushalmi et al., 1999). Removal of BTX mixtures appeared to be favored at aerobic conditions (Deeb & Alvarez-Cohen, 1999; Prenafeta-Boldú et al., 2002); however, negative effects such as inhibition or catabolic repression have also been reported at these conditions

(Chang et al., 1993; Duetz et al., 1997). Nevertheless, aerobic conditions might lead to important BTX stripping to atmosphere due to their volatilization characteristics. In fact, depending on aeration system, losses of BTX have reached up to 30% of their content (Zytner et al., 1994, as cited in Farhadian et al., 2008). In spite of complete removal of these compounds has been achieved (Deeb & Alvarez-Cohen, 2000), high conversion into biomass has also been reported (Chang et al., 1993; Reardon et al., 2000; Reardon et al., 2002). Biomass generation, stripping into atmosphere, besides their relatively low water-solubility seemed to be the most important disadvantages for aerobic treatment utilization. Therefore, anaerobic biological treatments appeared to be an adequate alternative for BTX removal.

Attempts for BTX removal using different electron acceptors such as nitrate (Burland & Edwards, 1999; Major et al., 1988; Martinez et al., 2007; Peña-Calva et al., 2004b; Weelink et al., 2010), sulfate (Craig et al., 1996; Feris et al., 2008) and CO_2 (Corseuil et al., 1998; Lovley, 2000; Morgan et al., 1993; Weelink 2010) have been conducted. There is scarce information describing interaction of BTX mixtures under anaerobic conditions. Benzene appeared to be recalcitrant in most of the cases (Weelink 2010); however, some investigations reported benzene removal under nitrate reducing conditions (Burland & Edwards, 1999). Rapid toluene consumption has been reported under sulfate (Beller & Spormann, 1997), hem (Kane et al., 2002) and methanogenic (Washer & Edwards, 2007) conditions. Most of these studies are focused on the disappearance of the aromatic compounds and the electron acceptors; however, no information about products formation is mentioned as no mass balances are indicated. Emphasize on the dissimilative catabolism has not been made, and further research on this topic is needed. Likewise, scarce consumption rate values are included; however, most of the values reported suggest that the processes are slow due to the microbial growth is slow and scarce. Information about these parameters would be useful for practical applications.

This chapter pretends to succinctly present the current knowledge of biological removal of ammonium, nitrate, and BTX compounds by nitrification and denitrification and describe some environmental factors that affect these microbial processes. Likewise, the goal of this work is to give practical information in order to obtain processes mainly mineralizing. The work is divided into two parts. In the first part, the simultaneous elimination of ammonium and BTX compounds by nitrification is described. In the second part, the simultaneous consumption of nitrate and BTX compounds by denitrification is presented. A general description of microbiology and physiology of nitrification and denitrification is made. Inhibitory effects of BTX compounds on the nitrifying and denitrifying processes are also described. In the case of toluene consumption by denitrification, studies carried out in batch and continuous cultures are also included. The importance of performing a complete evaluation of the biological processes through response variables such as efficiencies, yields, consumption and production rates, mass balances for the development of environmentally acceptable wastewater treatment processes is emphasized.

2. Nitrification

Nitrification has been extensively investigated as a very useful process in the first step of nitrogen removal in biological wastewater treatment. The oxidation of ammonia and nitrite by nitrifying processes generates nitrate for denitrifying processes where nitrate is converted to molecular nitrogen.

Simultaneous Elimination of Carbon and Nitrogen Compounds of Petrochemical Effluents
by Nitrification and Denitrification

79

2.1 Definition of respiratory process

Nitrification is the biological oxidation of ammonium (NH_4^+) to nitrate (NO_3^-) via nitrite (NO_2^-). It is an aerobic respiratory process where nitrifying bacteria use ammonium and nitrite as energy sources, carbon dioxide as a carbon source, and molecular oxygen as the final electron acceptor (Prosser, 1989). The nitrifying process is composed of two consecutive steps: 1) the oxidation of ammonium to nitrite that is mainly carried out by ammonium-oxidizing bacteria (AOB) and 2) the oxidation of nitrite to nitrate that is catalyzed by nitrite-oxidizing bacteria (NOB) (Table 2). As indicated by the free energy change ($\Delta G^{o'}$) values, the oxidation of hydroxylamine (NH_2OH) to nitrite is the main step where the AOB obtain energy. The $\Delta G^{o'}$ is lower for nitrite oxidation than for ammonia oxidation and the consequence is a lower growth yield for NOB than for AOB. Due to the low energy availability for cellular biosynthesis in the respiratory process, growth of nitrifying bacteria is slow and scarce, even in optimal conditions. Growth yields have been found to be 0.08 g cells/g NH_4^+-N for AOB and 0.05 g cells/g NO_2^--N for NOB, respectively (Wiesmann, 1994). Doubling times have been reported to vary between 7 and 24 h for ammonium-oxidizing species and 10 and 140 h for *Nitrobacter* species (Bock et al., 1991).

Reactions	Equations	$\Delta G^{o'}$ (kJ/reaction)
Ammonium oxidation	$NH_4^+ + 0.5O_2 \rightarrow NH_2OH + H^+$	- 8
	$NH_2OH + O_2 \rightarrow NO_2^- + H^+ + H_2O$	- 267
Global reaction	$NH_4^+ + 1.5O_2 \rightarrow NO_2^- + 2H^+ + H_2O$	- 275
Nitrite oxidation	$NO_2^- + 0.5O_2 \rightarrow NO_3^-$	- 74

Table 2. Reactions and $\Delta G^{o'}$ values of ammonium and nitrite oxidizing processes in nitrification.

2.2 Microbiological aspects

Nitrification is carried out by two groups of Gram negative chemolithoautotrophic bacteria that belong to the *Nitrobacteraceae* family: the ammonium- and nitrite-oxidizing bacteria. There are no known autotrophic bacteria that can catalyze the production of nitrate from ammonium (Kowalchuk & Stephen, 2001). The AOB are species of the following genera: *Nitrosomonas*, *Nitrosospira*, *Nitrosolobus*, *Nitrosococcus*, and *Nitrosovibrio*, being *Nitrosomonas* and *Nitrosomonas europaea* the genus and species better studied, respectively. Ammonia can also be transformed by a number of heterotrophic fungi and bacteria but it is generally accepted that chemolithotrophs are the primary nitrifiers in many systems (Kowalchuk & Stephen, 2001). While AOB are considered critical in nitrification, recently it has been reported that two major microbial groups are now believed to be involved in ammonium oxidation: AOB and ammonium-oxidizing archaea (You et al., 2009). *Candidatus* "*Nitrosopumilus maritimus*" and *Candidatus* "*Cenarchaeum symbiosum*" have been shown to be ammonia-oxidizing archaea containing the genes for all three subunits (*amoA*, *amoB*, and *amoC*) of ammonia monooxygenase, the enzyme responsible for ammonia oxidation (Hallam et al., 2006; Könneke et al., 2005, as cited in You et al., 2009). The genera of NOB are: *Nitrobacter*, *Nitrococcus*, *Nitrospina*, and *Nitrospira*. Most of the physiological and biochemical investigations have been carried out with members of the genus *Nitrobacter*. Some strains of

Nitrobacter have been shown to be able to grow heterotrophically but their growth is very slow (Bock et al., 1991).

2.3 Biochemistry aspects

In the ammonium-oxidizing process, the first reaction is catalyzed by an ammonia monooxygenase (AMO) (Eq. 1) and the second one by a hydroxylamine oxidoreductase (HAO) (Eq. 2).

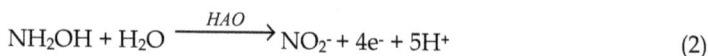

$$NH_3 + O_2 + 2H^+ + 2e^- \xrightarrow{AMO} NH_2OH + H_2O \tag{1}$$

$$NH_2OH + H_2O \xrightarrow{HAO} NO_2^- + 4e^- + 5H^+ \tag{2}$$

In the first reaction of ammonia conversion into hydroxylamine (NH_2OH), one of the oxygen atoms from O_2 is transferred to NH_3, producing hydroxylamine, while the other is involved in H_2O formation (Hollecher et al., 1981). It is generally accepted that ammonia (NH_3) rather than ammonium (NH_4^+) is the real substrate for the enzyme AMO. AMO is located in the cytoplasmic membrane. The broad specificity of the AMO has shown to permit the co-oxidation of numerous organic compounds, including recalcitrant aliphatic, aromatic, and halogenated molecules (Juliette et al., 1993; Keener & Arp, 1993, 1994; McCarty, 1999). The second reaction of the ammonia oxidation, the NH_2OH conversion into NO_2^-, is catalyzed by the complex enzyme system HAO. HAO is located in the periplasmic space (Whittaker et al., 2000). Two of the electrons produced in the second reaction are used to compensate the electron input of the first reaction, whereas the other two are passed via an electron transport chain to the terminal oxidase, thereby generating a proton motive force (Kowalchuk & Stephen, 2001). This proton motive force is used as the energy source for ATP production.

The second step in nitrification is the oxidation of nitrite to nitrate and is catalyzed by the enzyme nitrite oxidoreductase (NOR) (Eq. 3). In this reaction, the oxygen atom is derived from water. NOR is located in the cytoplasmic membrane and is composed of cytochromes *a* and *c*, a quinone and a deshydrogenase dependent on NADH (Aleem & Sewel, 1981; Spieck et al., 1998).

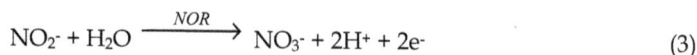

$$NO_2^- + H_2O \xrightarrow{NOR} NO_3^- + 2H^+ + 2e^- \tag{3}$$

2.4 Environmental factors affecting nitrification

Nitrification is affected by environmental factors such as temperature, pH, substrates concentrations (O_2, NH_3, and NO_2^-), and organic matter (Bernet & Spérandio, 2009). Such parameters can cause an effect on both the anabolic (biosynthetic process) and catabolic (respiratory process) processes.

Growth conditions are optimal at 25-30°C and pH 7.5-8.0 for AOB and at 28-30°C and pH 7.6-7.8 for NOB (Bock et al., 1991). In nitrifying processes, ranges of 25-30°C and pH 7.5-8.0 are generally used to obtain a successful nitrification. However, nitrification has been shown to occur in acidic conditions at pH = 5 (De Boer & Kowalchuk, 2001). pH has also an indirect effect on nitrification by acting on the equilibrium ammonium/ammonia (NH_4^+/NH_3, pKa = 9.25 at 25°C) (Hoffman, 2004). At alkaline pH values, the availability of NH_3 is not limited and this would be favorable for nitrification considering that NH_3 is the substrate for AMO (Suzuki et al., 1974). However, it is also known that high concentrations of free ammonia inhibit nitrification (Anthonisen et al., 1976; Vadivelu et al., 2006b). In aerobic nitrifying reactors operated under constant aeration and agitation, NH_3 can be lost from the system by volatilization. On the other hand, the ammonia oxidation to nitrate produces protons and leads to an acidification of the environment. This decrease in the pH value can inhibit nitrification by diminishing the availability of NH_3. pH also establishes the concentrations of nitrite (NO_2^-) and nitrous acid (HNO_2) with a pKa value of 3.34 at 25°C (Whiten et al., 2008). It has been observed that high concentrations of nitrite can inhibit the nitrification process and that could be mainly related to the pH values through the formation of HNO_2. Nitrous acid is known to inhibit nitrification, thus, low pH values must be avoided in nitrifying processes (Anthonisen et al., 1976; Vadivelu et al., 2006a, 2006b). Recently, Silva et al. (2011) reported that the nitrite-oxidizing process was more sensitive to the presence of nitrite than the ammonium-oxidizing process, suggesting that nitrite accumulation in nitrification systems should be controlled to avoid a higher accumulation of nitrite and a decrease in nitrate yield. In consequence, pH value is generally controlled in nitrifying systems and maintained between 7.5 and 8.0 to avoid operational problems.

In the nitrifying respiratory process, oxygen is used as the final electron acceptor. It is generally assumed that oxygen concentrations higher than 2 mg/l are adequate for a successful nitrification (Gerardi, 2002). Nitrification is very sensitive to low dissolved oxygen concentrations. AOB and NOB present different affinities for oxygen (K_{O2}) and NOB are more sensitive to O_2 limitation than AOB (Laanbroek & Gerards, 1993). Therefore, low dissolved oxygen concentration, around 0.2 to 0.5 mg/l, is a possible condition for limiting nitrite-oxidizing activity in nitrification systems (Bernet & Spérandio, 2009). Partial nitrification is used to obtain nitrite instead of nitrate as end product. Partial nitrification is involved in several processes (OLAND, CANON, SHARON, ANAMMOX, denitrification via nitrite) where the shortcut of nitrate could mean considerable savings in demand for oxygen and carbon source (Ahn et al., 2011; Moore et al., 2011; Peng & Zhu, 2006).

The high sensitivity of nitrifying bacteria to the toxic or inhibitory effects of organic compounds is well-documented, and it is known that the stability of nitrifying systems in wastewater treatment can be altered by the presence of organic matter (Schweighofer et al., 1996). Most of the studies on effects of organic compounds on nitrification have used axenic cultures or consortia such as activated sludge as inoculums. It has been shown that the effects mainly depend on the type and chemical structure of the organic pollutant as well as its concentration and hydrophobicity but also the type of culture (axenic or consortium) and the origin of the sludge (Gómez et al., 2000; Zepeda et al., 2006).

2.5 Ammonium and BTX removal

2.5.1 Inhibitory effects of organic matter and BTX on nitrification process

There are numerous studies in the literature on the inhibitory effects of organic compounds on nitrification. Some works with axenic cultures of *Nitrosomonas* sp. or *Nitrobacter* sp. have been focused on the growth inhibition of bacteria (Jensen, 1950; Steinmüller & Bock, 1976), others on the activity of the AMO (McCarty, 1999). There is also information on the effects of different organic substances on the nitrifying respiratory process of microbial consortia (Gómez et al., 2000, Silva et al., 2009; Zepeda et al., 2003, 2006). Various hypotheses have been proposed to explain the negative effects of organic matter on nitrification. In axenic cultures of *Nitrosomonas europaea*, it has been observed that the AMO had a broad substrate range for catalytic oxidation and the inhibitory effects of many organic substances would be due to competition for the active site (Chang et al., 2002; Hyman et al., 1985; Juliette et al., 1993; Keener & Arp, 1993). The enzyme AMO contains copper and iron, and metal chelators such as allylthiourea are inhibitory (Bremmer & Bundy, 1974; Hauck, 1980; Hooper & Terry, 1973, as cited in Bock et al., 1991; Hyman et al., 1990; Zahn et al., 1996). In the literature, it was also mentioned that nitrifying bacteria are susceptible to organic compounds because of a limitation of reducing power (Iizumi et al., 1998; Shiemke et al., 2004), the formation of covalent binding with the enzyme AMO (Hyman & Wood, 1985), or the hydrophobic character of the organic molecules and their effects on biological membranes (Sikkema et al., 1994, as cited by Zepeda et al., 2006; Takahashi et al., 1997). In microbial consortia, the competition between heterotrophs and autotrophs for ammonia and oxygen is another hypothesis commonly mentioned for explaining the nitrification inhibition by organic matter (Hanaki et al., 1990).

Some aromatic compounds are known to inhibit nitrification (Amor et al., 2005; Brandt et al., 2001; Texier & Gómez, 2002). Little attention has been paid to the inhibitory influence of benzene, toluene, and xylenes (BTX) on nitrifying microorganisms and further work is required for understanding how BTX compounds can affect the performance of nitrifying treatment systems. Inhibition of nitrification by toxic chemicals, such as BTX compounds, focuses mainly on the modes of action of inhibitors on the enzyme AMO by using axenic cultures of *N. europaea* (Keener & Arp, 1994; McCarty, 1999). Main results are consequently in relation to the ammonium-oxidizing process and very little information is available on the effects of BTX compounds on the nitrite-oxidizing process. Keener & Arp (1994) in their study on the inhibition of NH_4^+ oxidation by *N. europaea* from various aromatic compounds reported that benzene concentrations of 14.0 and 23.4 mg/l inhibited the NO_2^- production by 60% and 80%, respectively. Dyreborg & Arvin (1995) observed that the level of benzene concentration provoking 100% inhibition of the ammonium oxidation was 10.7 mg/l. In the studies of Zepeda et al. (2003, 2006), the effect of different initial concentrations of benzene, toluene, and *m*-xylene on a nitrifying consortium produced in steady-state nitrification was evaluated in batch reactors. The values for the ammonium consumption efficiency, nitrate yield and nitrification specific rates were determined. These values are needed for characterizing the physiological behavior of the nitrifying sludge in the presence of BTX compounds. Zepeda et al. (2006) observed that at 5 mg C/l of each aromatic compound, there was no significant effect on nitrification efficiency and the ammonium removal efficiency was of 95 ± 7% after 16 h. Benzene and *m*-xylene at 10 mg C/l decreased ammonium consumption efficiency by 57% and 26%, respectively, whereas toluene did not affect the efficiency of the ammonium oxidation process. In all cases, the consumed NH_4^+-N was totally oxidized to NO_3^--N after 16 h. The nitrifying yield was close to 1.0 and nitrite concentration was negligible. This indicated that the process was mainly dissimilative and biomass formation was limited. These results suggested that wastewaters containing up to 5 mg C/l of benzene, toluene, or *m*-xylene would not affect the efficiency of ammonium conversion to nitrate in a treatment system. However, BTX (5-20 mg C/l) induced a significant decrease in the values for specific rates of NH_4^+ consumption (76-99%) and NO_3^- production (45-98%), affecting mainly the ammonium-oxidizing pathway (Table 3). These results indicated that, in the presence of BTX compounds, the nitrification process was inhibited but the nitrifying metabolic pathway was only altered at the specific rate level as nitrate was still the main end product. At 10 mg C/l of BTX compounds, the inhibition order on nitrate production was: benzene > m-xylene > toluene while at 20 mg C/l, the sequence changed to m-xylene > toluene > benzene. The same authors also reported that at 5 mg C/l of BTX compounds, there was no toxic effect on the sludge whereas bacteria did not totally recover their nitrifying activity from 10 to 50 mg C/l. Results indicated that mechanisms involved in the inhibition of nitrification processes by BTX compounds depend on various factors, such as the initial concentration, the chemical structure, and the hydrophobic character of the pollutants. Zepeda et al. (2003, 2006) also observed that the inhibitory effects of BTX compounds seemed to be related to their persistence in the nitrifying cultures. At low concentrations, the chemical structure of the BTX compounds appeared to be the predominant factor whereas at higher concentrations, their hydrophobicity played an important role. The presence of different functional groups as well as the nature of the substituent can influence the metabolism and toxicity of aromatic compounds (O'Connor & Young, 1996). The absence of functional groups may confer to benzene higher stability and persistence to biotransformation while the presence of methyl groups may facilitate the mechanisms of biotransformation for toluene and m-xylene. Moreover, previous studies reported that many cyclic hydrocarbons are toxic and inhibitory to microorganisms because of their hydrophobic character and their devastating effects on biological membranes (Radniecki et al., 2011; Sikkema et al., 1995; Tsitko et al., 1999).

	Initial concn (mg C/l)	Specific rates (g N/g microbial protein-N.h)	
		NH_4^+-N consumption	NO_3^--N production
Control	0	1.389 ± 0.079	0.577 ± 0.030
Benzene	6.5	$0.266 \pm 0.008 \ (-81\%)$[a]	$0.306 \pm 0.024 \ (-47\%)$
	10	$0.226 \pm 0.009 \ (-84\%)$	$0.141 \pm 0.010 \ (-76\%)$
	20	$0.155 \pm 0.004 \ (-89\%)$	$0.090 \pm 0.003 \ (-84\%)$
Toluene	5	$0.317 \pm 0.012 \ (-77\%)$	$0.310 \pm 0.007 \ (-46\%)$
	10	$0.225 \pm 0.006 \ (-84\%)$	$0.221 \pm 0.004 \ (-62\%)$
	20	$0.054 \pm 0.004 \ (-96\%)$	$0.047 \pm 0.007 \ (-92\%)$
m-xylene	5	$0.328 \pm 0.002 \ (-76\%)$	$0.320 \pm 0.002 \ (-45\%)$
	10	$0.219 \pm 0.009 \ (-84\%)$	$0.201 \pm 0.020 \ (-65\%)$
	20	$0.010 \pm 0.002 \ (-99\%)$	$0.009 \pm 0.001 \ (-98\%)$

[a]Percentages of decrease for nitrification specific rates were calculated by using the values obtained in the control culture as references.

Table 3. Nitrification specific rates in the nitrifying cultures in the absence and presence of BTX compounds (from Zepeda et al., 2006).

As it can be seen in Table 4, in nitrifying cultures, binary and ternary mixtures of benzene, toluene, and m-xylene provoked also a significant decrease in the values of specific rates for

NH_4^+ consumption (from 72 to 90%) and NO_3^- production (from 39 to 79%) (Zepeda et al., 2007). The inhibitory effect of BTX compounds on the nitrifying process seemed to be higher when they were present in mixtures than individually. Results showed that benzene, toluene, and m-xylene individually or in mixtures significantly inhibited the nitrifying activity of the sludge by decreasing the nitrification specific rates. However, it was found that the nitrifying sludge can tolerate up to 5 mg C/l of BTX in single solutions or 2.5 mg C/l in mixed solutions, maintaining high the NH_4^+-N oxidation efficiency and the nitrifying yield.

| | Specific rates | | | |
| | NH_4^+-N consumption | | NO_3^--N production | |
	g N/g protein-N.h	Decrease (%)[a]	g N/g protein-N.h	Decrease (%)[a]
Control	1.389 ± 0.079	-	0.577 ± 0.030	-
Single solutions				
Benzene	0.266 ± 0.008	81	0.306 ± 0.024	47
Toluene	0.317 ± 0.012	77	0.310 ± 0.007	46
m-Xylene	0.328 ± 0.002	76	0.320 ± 0.002	45
Mixed solutions[b]				
BT	0.389 ± 0.002	72	0.352 ± 0.006	39
BX	0.194 ± 0.003	86	0.202 ± 0.006	65
TX	0.208 ± 0.007	85	0.173 ± 0.002	70
BTX	0.139 ± 0.012	90	0.121 ± 0.012	79

Table 4. Specific rates of the nitrification cultures in the absence (control) and presence of benzene, toluene, and m-xylene at 5.0 ± 0.5 mg C/l for single solutions and 2.5 ± 0.2 mg C/l for each compound of the mixtures (from Zepeda et al., 2007). [a]Percentages of decrease for nitrification specific rates were calculated by using the values obtained in the control culture without BTX compounds as references. [b]BT, Benzene-Toluene; BX, Benzene-m-Xylene; TX, Toluene-m-Xylene; BTX, Benzene-Toluene-m-Xylene.

As observed by Zepeda et al. (2006, 2007) in the case of BTX compounds, in spite of the inhibitory effects of organic compounds on nitrification, it was demonstrated that in some cases and under controlled experimental conditions, nitrification processes could successfully proceed (Texier and Gómez, 2007). Moreover, microbial consortia under nitrifying conditions have been shown to be able to oxidize simultaneously ammonia and organic compounds, suggesting that nitrifying consortium coupled to a denitrification system may have promising applications for complete removal of nitrogen and organic compounds from wastewaters. This topic is detailed in the following section.

2.5.2 Simultaneous elimination of ammonium, organic matter and BTX removal

Previous studies have shown that the ammonia-oxidizing bacterium *N. europaea* was able to oxidize a broad range of hydrocarbons, including non-aromatic and aromatic compounds, and it is believed that the AMO is participating. The role of the AMO was demonstrated in experiments undertaken with selective inhibitors of the enzyme such as allylthiourea. Table 5 presents some examples of organic substances oxidized by *N. europaea* cultures. In the majority of these studies, kinetic data are missing for ammonium and nitrite oxidation

Simultaneous Elimination of Carbon and Nitrogen Compounds of Petrochemical Effluents
by Nitrification and Denitrification

85

processes as for organic compounds, thus, more investigation is required to understand better the involved mechanisms.

Several investigations have shown that *N. europaea* oxidizes recalcitrant aromatic compounds, and it is also believed that these oxidations are mediated by the AMO (Chang et al., 2002; Hyman et al., 1985; Keener & Arp, 1994; Vannelli & Hooper, 1995). However, in these studies, there was an accumulation of more oxidized aromatic products in the culture medium. For example, in the case of benzene, ring cleavage did not occur and phenol was accumulated in cultures of *N. europaea*. In contrast, Zepeda et al. (2003, 2006) observed that when a nitrifying consortium was used, benzene, toluene, and *m*-xylene were converted to volatile fatty acids. These authors proposed that the aromatic compounds oxidation with ring fission could be the result of the coexistence and participation of both, lithoautotrophic nitrifying bacteria and heterotrophic microorganisms present in the consortium. It can be considered that the high diversity of microorganisms, enzymatic material and possible metabolic pathways which characterizes microbial consortia could contribute to obtain these results. However, further work is required in this direction to elucidate the underlying processes involved in the transformation of BTX by nitrifying consortia.

Organic compound	Main product	Reference
Methane	Methanol	Hyman & Wood, 1983
Ethylene	Ethylene oxide	Hyman & Wood, 1984
Methanol	Formaldehyde	Voysey & Wood, 1987
Carbon monoxide	Carbon dioxide	Tsang & Suzuki, 1982
Methyl Fluoride and Dimethyl Ether	Formaldehyde and methanol-formaldehyde	Hyman et al., 1994
Alcane (up to C_8)	Alcohol	Hyman et al., 1988
Alkene (up to C_5)	Epoxide and alcohol	Hyman et al., 1988
Trans-2-butene	2-butene-1-ol	Vannelli et al., 1990
Methyl sulphur	Methyl sulfoxide	Juliette et al., 1993
Iodoethane	Acetaldehyde	Rasche et al., 1990a
Fluoromethane	Formaldehyde	Rasche et al., 1990a
Bromoethane	Acetaldehyde	Rasche et al., 1990b
Cloromethane	Formaldehyde	Rasche et al., 1990b
Benzene	Phenol	Keener & Arp, 1994
Toluene	Benzyl alcohol and benzaldehyde	Keener & Arp, 1994
Ethylbenzene	Phenethyl alcohol	Keener & Arp, 1994
p-Xylene	4-Methylbenzyl alcohol	Keener & Arp, 1994
Naphtalene	2-Naphtol	Chang et al., 2002
Styrene	Phenylglyoxal	Keener & Arp, 1994
Clorobenzene	4-Clorophenol	Keener & Arp, 1994
Aniline	Nitrobenzene	Keener & Arp, 1994
Nitrobenzene	3-Nitrophenol	Keener & Arp, 1994
Acetophenone	2-Hidroxyacetophenone	Keener & Arp, 1994
Benzyl alcohol	Benzaldehyde	Keener & Arp, 1994
Phenol	Hydroquinone	Hyman et al., 1985
Nitrapirine (aerobic)	6-Acid chloro picolinico	Vannelli & Hooper, 1992
Nitrapirine (anaerobic)	2-Chloro-6-dicloromethyl piridine	Vannelli & Hooper, 1993

Table 5. Oxidation of organic compounds in *Nitrosomonas europaea* cultures.

In their study, Zepeda et al. (2006) reported that the nitrifying sludge was able to completely consume benzene (6.5 mg C/l), toluene (5 mg C/l), and *m*-xylene (5 mg C/l) over 21 h. Specific rates of BTX consumption are presented in Table 6. These results indicate that for initial concentrations below 6.5 mg C/l, toluene and *m*-xylene disappeared faster from the cultures than benzene, which appeared to be most persistent. This is in accordance with the fact that benzene at low concentrations was the highest inhibitory compound on nitrification. Individually as in BTX mixtures, the nitrifying consortium presented the following sequence of biotransformation: *m*-xylene > toluene > benzene. However, in mixtures, significant differences in the values for specific rates of BTX compounds removal were found. The work of Zepeda et al. (2007) emphasizes the importance for considering the possible component interactions in the biotransformation of mixed BTX compounds in nitrification systems.

Up to a concentration of 10 mg C/l, benzene was first oxidized to phenol, which was later totally oxidized to acetate. At a concentration of 5 mg C/l, toluene was first oxidized to benzyl alcohol, which was later oxidized to butyrate while *m*-xylene was oxidized to acetate and butyrate (Zepeda et al., 2006). As it has been previously shown that the AMO oxidized benzene to phenol, then it is likely that benzyl alcohol be an intermediate of toluene oxidation by the AMO (Hyman et al., 1985; Keener and Arp, 1994; Zepeda et al., 2003). It is presumed that *N. europaea* may initiate oxidation of BTX to provide intermediates, such as phenol and benzyl alcohol, which could later be transformed to VFA by heterotrophic microorganisms from the nitrifying consortium. In nitrifying cultures added with benzene, toluene, and *m*-xylene compounds in individual (5 mg C/l) and mixed solutions (2.5 mg C/l for each one), Zepeda et al. (2007) observed that after 24 h, ammonium consumption efficiency and conversion of consumed ammonium into nitrate were close to 100%. These results confirmed that a nitrifying consortium produced in steady state was able to convert benzene, toluene, and *m*-xylene to VFA and simultaneously oxidize ammonium to nitrate through a dissimilative process. This type of nitrifying consortium coupled with a denitrification system may have promising applications for complete removal of both nitrogen and BTX compounds from wastewater streams into N_2 and CO_2.

In the last decades, several nitrifying consortia have been reported to oxidize simultaneously ammonium and various recalcitrant aromatic compounds, such as phenolic compounds (phenol, 2-chlorophenol, *p*-cresol, *p*-hydroxybenzaldehyde) (Amor et al., 2005; Martínez-Hernández et al., 2011; Silva et al., 2009, 2011; Vázquez et al., 2006; Yamaghisi et al., 2001) and BTX compounds (Zepeda et al., 2003, 2006, 2007). In some cases, nitrate and carbon dioxide were the major products from the process. As a consequence, nitrifying processes have been recently proposed as novel alternative technologies for the simultaneous removal of ammonium and aromatic pollutants from industrial wastewaters of chemical complexity such as effluents generated by the petrochemical industry (Beristain-Cardoso et al., 2011, Silva et al., 2011). Nitrification as the initial step in the removal of ammonia from wastewaters might be also used to oxidize simultaneously ammonia and aromatic compounds, allowing their mineralization or the production of intermediates that can be totally oxidized by denitrification (Beristain-Cardoso et al., 2009). However, additional studies in both batch and continuous bioreactors are needed to obtain more information about physiological, kinetic, ecological and engineering aspects in order to control the sludge metabolic capacity in nitrification processes.

	Specific rates (g C/g microbial protein-N.h)
Individual solutions	
Benzene	0.034 ± 0.003
Toluene	0.045 ± 0.003
m-Xylene	0.055 ± 0.001
Binary mixtures[b]	
Benzene	0.005 ± 0.001
Toluene	0.055 ± 0.005
Benzene	0.051 ± 0.005
m-Xylene	0.044 ± 0.005
Toluene	0.012 ± 0.002
m-Xylene	0.053 ± 0.005
Ternary mixtures[b]	
Benzene	0.017 ± 0.002
Toluene	0.039 ± 0.004
m-Xylene	0.050 ± 0.005

[a]The initial concentration of BTX compounds in individual solutions was 5.0 ± 0.5 mg C/l.
[b]The initial concentration for each BTX compound of the mixtures was 2.5 ± 0.2 mg C/l.

Table 6. Specific rates of biotransformation for benzene, toluene, and m-xylene in individual, binary, and ternary solutions in nitrifying cultures (from Zepeda et al., 2007).

3. Denitrification

Denitrification is a biological process that microorganisms use for obtaining energy from the reduction of nitrate to dinitrogen gas (Knowles, 1982; Mateju et al., 1992). Denitrification requires the concomitant oxidation of an electron donor in order to reduce nitrate to N_2 gas. Diverse compounds can be used for this purpose. Organic sources are used in organotrophic denitrification whereas inorganic compounds are used for lithotrophic denitrification. Different organic compounds have been used in organotrophic denitrification such as methanol, glycerol, benzoic acid, acetate, glucose, lactate (Akunna et al., 1993; Cuervo-López et al., 1999; Martínez et al., 2009). Recalcitrant compounds such as p-xylene, benzene, toluene and p-cresol have been successfully removed by organotrophic denitrification (Cervantes et al., 2009; Dou et al., 2008; Hänner et al., 1995; Martínez et al., 2007; Peña-Calva et al., 2004b). Therefore, denitrification seems a good option for treating wastewater and underground water contaminated with BTX as simultaneous removal of nitrogen and carbon compounds can be obtained. Nevertheless, it is essential to remark that for biological wastewater treatment, the denitrification process must be dissimilative and dissipative to prevent biomass generation.

The type of electron source will determine the performance of the denitrifying process in terms of change in Gibbs free energy ($\Delta G^{o\prime}$) and consumption rate (Cuervo-López, et al., 2009). Several respiratory process, including denitrification, with different electron sources and their respective $\Delta G^{o\prime}$ values are illustrated in Table 7. In all cases, the denitrification process is exergonic and appears to be thermodynamically favored when compared with anaerobic processes such as methanogenesis or sulfate reduction.

Electron donor	Equation Denitrification	$\Delta G^{o\prime}$ (kJ/reaction)
Acetic acid	$CH_3COOH + 1.6NO_3^- \rightarrow 2CO_2 + 0.8N_2 + 1.6OH^- + 1.2H_2O$	-843
Ethanol	$C_2H_6O + 2.4NO_3^- + 0.4H^+ \rightarrow 1.2N_2 + 2HCO_3^- + 2.2H_2O$	-1230
Phenol	$C_6H_6O + 5.6NO_3^- + 0.2H_2O \rightarrow 2.8N_2 + 6HCO_3^- + 0.4H^+$	-2818
p-Cresol	$C_7H_8O + 6.8NO_3^- \rightarrow 3.4N_2 + 7HCO_3^- + 0.2H^+ + 0.4H_2O$	-3422
Benzene	$C_6H_6 + 6NO_3^- \rightarrow 3N_2 + 6HCO_3^-$	-2977
Toluene	$C_7H_8 + 7.2NO_3^- + 0.2H^+ \rightarrow 3.6N_2 + 7HCO_3^- + 0.6H_2O$	-3524
Xylene	$C_8H_{10} + 8.4NO_3^- + 0.4H^+ \rightarrow 4.2N_2 + 8HCO_3^- + 1.2H_2O$	-4136
	Aerobic respiration	
Toluene	$C_7H_8 + 9 O_2 \rightarrow 7 CO_2 + 4 H_2O$	-3831
Benzene	$C_6H_6 + 7.5 O_2 + 3 H_2O \rightarrow 6HCO_3^- + 6 H^+$	-3173
	Methanogenesis	
Toluene	$C_7H_8 + 7.5 H_2O \rightarrow 4.5 CH_4 + 2.5 HCO_3^- + 2.5 H^+$	- 131
Benzene	$C_6H_6 + 6.75 H_2O \rightarrow 3.75.5 CH_4 + 2.25HCO_3^- + 2.25 H^+$	-124
	Sulfate reduction	
Toluene	$C_7H_8 + 4.5 SO_4^{2-} + 3H_2O \rightarrow 7 HCO_3^- + 2.5 H^+ + 4.5 HS^-$	- 205
Benzene	$C_6H_6 + 3.75 SO_4^{2-} + 3 H_2O \rightarrow 6 HCO_3^- + 0.325 H^+ + 3.75 HS^-$	-186
	Hem reduction	
Toluene	$C_7H_8 + 94 Fe (OH)_3 \rightarrow 7 FeCO_3 + 29 Fe_3O_4 + 145 H_2O$	- 3398
Benzene	$C_6H_6 + 30 Fe^{3+} + 3H_2O \rightarrow 6 HCO_3^- + 30 Fe^{2+} + 36 H^+$	-3040

Table 7. $\Delta G^{o\prime}$ values of denitrification and several respiratory processes in presence of different electron sources.

3.1 Microbiological aspects

Different taxonomic groups are involved in denitrification including Gram negative and positive bacteria. Their remarkable characteristic is their facultative respiration. Some of them are phototrophic, whereas many others can use organic sources or sulfur compounds. Some of them are illustrated in Table 8.

Denitrifiers able to consume BTX have been isolated from different places such as forest soils, sediments of aquifers, beaches, and contaminated soils with hydrocarbons. Toluene and xylene consumers as *Thauera aromatica* and *Azoarcus evansii* (Anders et al., 1995; Song et al., 1998, as cited in Peña-Calva, 2007), and BTX consumers as *Pseudomonas putida* and *Pseudomonas fluorescences* have been isolated from these sites (Shim et al., 2005). Axenic and enriched cultures of *Rhodococcus pyridinovorans* have been reported to be able to consume *m*-xylene. Enrichment cultures of *Betaproteobacteria* growing with *p*-xylene and nitrate have been reported (Rotaru et al., 2010 as cited in Sander et al., 2010), whereas the elimination of the three xylene isomers can be carried out by microbial consortia (Jung & Park, 2004). Some evidences on benzene consumption have been reported for *Dechloromonas* (Coates et al., 2001), likewise, there are evidences indicating that this strain is able to use BTX mixtures for microbial growth (Chakraborty et al., 2005).

3.2 Biochemistry aspects

The denitrification process could be described as a modular organization in which every biochemical reaction is catalyzed by specific reductase enzymes (Cuervo-López et al., 2009). Four enzymatic reactions take place in the cell as follows: (*i*) nitrate is reduced to nitrite by nitrate reductase *(Nar)*; (*ii*) a subsequent reduction of nitrite to nitric oxide is carried out by nitrite reductase *(Nir)*; (*iii*) afterwards, nitric oxide is reduced to nitrous oxide by the enzyme nitric oxide reductase *(Nor)*; (*iv*) finally, nitrous oxide is reduced to N_2 by the enzyme nitrous oxide reductase *(Nos)* (Lalucat et al., 2006) (Table 9). These reactions take place when environmental conditions become anaerobic (Berks et al., 1995; Hochstein & Tomlinson, 1988). The enzymatic reactions, which are thermodynamically favored, are carried out in the cell membrane and periplasmic space. Small half saturation constant values *(Km)* have been reported for different nitrogen substrates for some denitrifying bacteria, indicating that denitrifying enzymes have a high affinity for their substrate. However, several factors have to be considered, as the presence of small quantities of molybdenum, cooper and hem to ensure the successful enzymatic activity, as they are known cofactors for denitrifying enzymes.

Dominio	Type	Genera	Species	References (as cited in González-Blanco et al., 2011)
Archaea	Organotrophic	*Haloarcula*	*marismortui*	Ichiki et al., 2001
		Halobacterium	*denitrificans*	Tomlinson et al., 1986
Bacteria (Gram +)	Organotrophic	*Bacillus*		Zumft, 1997
Bacteria (Gram -)	Litotrophic	*Beggiatoa*		Kamp et al., 2006
		Thiobacillus	*denitrificans*	Beller et al., 2006
		Paracoccus	*denitrificans*	Ludwig et al., 1993
		Pseudomonas	*eutropha*	Schwartz et al., 2003
		Ralstonia		
	Phototrophic	*Rhodobacter*		Zumft, 1997
	Organotrophic	*Alcaligenes*	*faecalis*	van Niel et al., 1992
		Pseudomonas		Zumft, 1997
		Pseudovibrio	*denitrificans*	Shieh et al., 2004
		Paracoccus	*denitrificans*	Baumann et al., 2006
		Bacillus	*infernus*	Boone et al., 1995
		Aquaspirillum		Thomsen et al., 2007
		Tahuera	*aromatic*	Beristain et al., 2009
		Halomonas	*nitroreducens*	González-Domenech et al., 2008
		Hyphomicrobium	*denitrificans*	Yamaguchi et al., 2003
		Flavobacterium	*denitrificans sp. nov*	Horn et al., 2005

Table 8. Some microbial genera carrying out the denitrifying process (modified from González-Blanco et al., 2011).

Enzyme	Reaction	Km	$\Delta G^{o\prime}$ KJ/reaction	Localization
Nitrate reductase (*Nar*)	$NO_3^- + UQH_2 \rightarrow NO_2^- + UQ + H_2O$	3.8 mM* *Pseudomonas stutzeri*	- 274. 38	Cell membrane and periplasmic space
Nitrite reductase (*Nir*)	$NO_2^- + Cu^{1+} + 2H^+ \rightarrow NO + H_2O + Cu^{2+}$ $NO_2^- + c^{2+} + 2H+ \rightarrow NO + H2O + c^3$	230 µM+ *Alcaligenes xylosoxidans* 53 µM++ *Pseudomonas aeruginosa*	- 76.2	Periplasmic space
Nitric oxide reductase (*Nor*)	$2NO + 2c^{2+} + 2H^+ \rightarrow N_2O + H_2O + 2 c^{3+}$	2.4 nM** *Pseudomonas stutzeri*	- 306.3	Cell membrane
Nitrous oxide reductase (*Nos*)	$N_2O + 2 c^{2+} + 2H^+ \rightarrow N_2 + H_2O + 2c^{3+}$	60 µM*** *Pseudomonas stutzeri*	- 339.5	Periplasmic space

Table 9. Some characteristics of the denitrifying enzymes and its location in the cell. UQH_2: reduced ubiquinone, UQ: ubiquinone, c^{2+}: reduced cytochrome, c^{3+}: oxidized cytochrome. (*Blümle et al., 1991; +Dood et al., 1997; ++Greenwood et al., 1978; **Remde & Conrad, 1991; ***Zumft et al., 1992, as cited in Zumft, 1997).

3.3 Environmental factors affecting denitrification

Low oxygen tension and the presence of nitrogen oxides are required as an inductor of denitrification (Cuervo-López et al., 2009). Denitrifying activity is inhibited in reversible manner under aerobic conditions. Oxygen concentrations ranging from 0.09 to 2.15 mg/l have resulted in a decrease of nitrate consumption and denitrifying rate or accumulation of intermediaries as nitrous oxide (Bonin et al., 1989; Oh & Silverstein, 1999; Wu & Knowles, 1994). Thus, in order to maintain high consumption efficiency of the reducing source and high denitrifying yield values (Y_{N2}, conversion of nitrate to N_2) in denitrifying systems, oxygen concentrations should be maintained lower than 2 mg/l.

Carbon/nitrogen (C/N) ratio is an important factor which determines the denitrifying performance. This variable is related with the accumulation of denitrification intermediaries and could determine the dissimilative reduction of nitrate to ammonium (DNRA) (Cuervo-López, 2009). Several works focused on BTX removal have resulted in nitrite accumulation due to the low C/N ratio established (Burland & Edwards 1999; Dou et al. 2008; Evans et al., 1992). This behavior is in contrast with that reported by Peña-Calva et al. (2004a, 2004b) and Martínez et al. (2007, 2009), where BTX were converted to CO_2 whereas nitrate was reduced to N_2 at C/N ratios close to stoichiometric values. Thus, these results evidenced that complete elimination of both, carbon and nitrogen compounds will outcome with no accumulation of undesirable intermediaries if C/N ratios are close to stoichiometric values. According to Peña-Calva et al. (2004b), the type of electron source has also an important effect in the accumulation of nitrogen intermediaries, as NO_2^- and N_2O were detected when

Simultaneous Elimination of Carbon and Nitrogen Compounds of Petrochemical Effluents
by Nitrification and Denitrification

91

m-xylene was used whereas no intermediaries were observed when toluene was assayed. Another important parameter which affects denitrification, particularly its thermodynamics, is the pH value, as the rate of denitrifying enzymes could be affected. In this sense, batch studies with *Paracoccus denitrificans* indicate important differences in the rate of nitrite and nitrous oxide consumption and production (Thomsen et al., 1994). Another important effect of pH might be observed on the transport mechanisms of substrates as membrane properties could be affected. As a result, changes in consumption efficiency and consumption rate will be observed. Denitrification could be carried out in a temperature range between 5 and 35°C (Lalucat et al., 2006). However, temperature effects can be observed in rates and consumption efficiency of substrates and possibly in the products formation yield. Similarly to pH, effects of low and high temperatures are mainly related with the physicochemical changes in the cell membrane structures, either for lipids and proteins. It has been reported that toluene denitrification was performed with influents at 5°C in a continuous UASB reactor (Martínez et al., 2007). Therefore, for obtaining a constant and acceptable denitrifying rate it is recommendable to conduct biological wastewater treatment at C/N stoichiometric values, neutral pH and temperature values between 20 and 35°C.

Finally, kinetics considerations have to be taken into account for a successful denitrifying process (Cuervo-López et al., 2009). According to these authors, the metabolism of denitrifying cultures can be described by the following equation:

$$q_s = \frac{\mu}{Y_{x/Scx}} + \frac{q_p}{Y_{p/Scp}} \qquad (4)$$

Where q_s is the specific consumption rate of substrate, a constant of first order which its value is dependent on the environmental culture conditions; μ is the specific growth rate; q_p is the specific product formation rate; $Y_{x/Scx}$ and $Y_{p/Scp}$ are substrate conversion or yield for biomass and products respect to the consumed substrate, respectively. The equation summarizes the metabolism of many cell cultures; where $\mu/Y_{x/Scx}$ expresses the biosynthetic process and $q_p/Y_{P/Scp}$ the respiratory process. In denitrification, it is desirable to obtain dissimilative processes where μ and $Y_{X/Scx}$ will be very small for obtaining negligible biomass production but high substrate conversions to molecular nitrogen and bicarbonate. This situation is highly recommended for practical purposes, as operational costs by sludge disposition could be minimized. In this regard, Peña-Calva et al. (2004a, 2004b), Hernández et al. (2008) and Martínez et al. (2007, 2009) have shown that denitrification process with BTX could be clearly dissimilative.

3.4 Nitrate and BTX removal

First evidences indicating toluene consumption by denitrification and its conversion to CO_2 are those obtained by Zeyer et al. (1986). Similar results have been reported by several authors (Alvarez & Vogel, 1995; Chakraborty et al., 2005; Elmén et al., 1997; Kobayashi et al., 2000). However, in these works nitrate consumption and N_2 production have not been mentioned, thus, the occurrence of denitrification process has not been verified. Evans et al. (1991, 1992) have reported toluene consumption and production of CO_2, biomass and intermediaries as benzilsuccinate with a concomitant accumulation of nitrite. Toluene elimination close to 80% was achieved by Schocher et al. (1991). The authors determined that 50% of the aromatic compound was oxidized to CO_2 whereas they assumed that the other fraction was assimilated as biomass. High nitrate concentrations corresponding to low C/N ratio values (close to 0.2) have been used in most of these works, resulting in nitrite accumulation. The three isomers of xylene have been eliminated under denitrifying conditions, where m-xylene was firstly consumed followed by p-xylene and finally by o-xylene (Haner et al., 1995; Kuhn et al., 1988). It has been reported a removal of m-xylene close to 55% but there is no mention about the fate of the aromatic compound or nitrate (Elmén et al., 1997; Evans et al., 1991a). In most of these works enrichment cultures have been extensively used as inocula. Peña-Calva et al. (2004b) evidenced that toluene and m-xylene concentrations ranging from 15 to 100 mg of toluene-C/l and 15 to 70 mg of m-xylene-C/l were used as electron sources for a denitrifying dissimilative respiratory process when sludge without previous acclimation to BTX was used as inoculum. However, the physiologic stability of the sludge was important. Likewise, these authors conducted their assays at C/N ratio close to stoichiometric values (C/N = 1). At these conditions, nitrate consumed was completely converted into N_2, while toluene and m-xylene were completely converted into HCO_3^- within 4 to 7 days of culture. Under these conditions, the efficiency of the denitrification pathway was not influenced by toluene or m-xylene; and the denitrifying yield values were close to 1. However, the authors reported that benzene was recalcitrant. Benzene elimination by denitrification has been reported by using sediments (Major et al., 1988), enriched cultures (Burland & Edwards, 1999) and axenic cultures of *Dechloromonas* RCB (Coates et al., 2001; Kasai et al., 2006, as cited in Weelink et al., 2010). These authors have determined that benzene was oxidized to CO_2 and assimilated into biomass. Dou et al. (2008) have assessed BTX concentrations ranging from 1.3 to 15 mg C/l for each aromatic compound with an enriched mixed bacterial consortium. Similarly to other works, an excess of nitrate (C/N = 0.05) was established for achieving consumption of the aromatic compounds by denitrification. BTX and nitrate were consumed within a period of 20 and 40 days of culture, however, significant nitrite accumulation was observed concomitant to N_2 production. Therefore, benzene is still considered as the most recalcitrant of the BTX under denitrifying conditions (Alvarez & Vogel, 1995; Cunningham et al., 2001; Da Silva & Alvarez, 2002; Peña-Calva et al., 2004b; Phelps & Young, 1999; Weelink et al., 2010).

BTX are usually present in contaminated aquifers as a mixture of them. There are some reports under denitrifying conditions where BTX interactions were evaluated (Alvarez & Vogel, 1995; Ma & Love, 2001). These studies showed that only toluene and xylene were consumed but no benzene. In spite of toluene and xylene were mineralized, reduction of nitrate to molecular nitrogen was not evidenced. The denitrification of different concentrations (50, 70 and 90 mg C/l) of a mixture of BTX using a stabilized sludge without previous contact to BTX was studied by Peña-Calva et al. (2004a). Firstly, toluene was completely consumed and oxidized to HCO_3^- whereas nitrate was reduced to N_2. After this, partial m-xylene consumption (close to 55%) and partial reduction of nitrate to N_2 was observed as NO_2^- and N_2O were accumulated. Benzene was not consumed in any of the cultures. Nevertheless, oxidation of an equal mixture of benzene and toluene coupled to nitrate reduction has been reported in a pure culture of *Dechloromonas* (Chakraborty et al., 2005) and enriched mixed bacteria culture (Dou et al., 2008). However, most studies with consortia have shown that benzene seems to be recalcitrant to biodegradation under denitrifying conditions in both cases: in mixtures and as a sole electron donor. In this sense, it has been suggested that benzene is not consumed due to the lack of oxidizing enzymes (Gülensoy & Alvarez, 1999) and/or to the high structural stability of benzene (Cadwell & Sufita, 2000). Likewise, toxic effects on bacteria cell membranes have been proposed (Sikkema et al., 1994). Limitation in cell transport system for benzene incorporation into the cell has also been proposed (Peña-Calva, 2007). More investigation is required to understand better the involved mechanisms.

Mixtures of BTX are often present in different types of water along with other carbon sources which might influence the rate and extent of its elimination (Lovanh & Alvarez, 2004). Several studies about the elimination of mixtures of readily consumable carbon compounds with BTX have been made. Numerous substrate interactions among BTX and other carbon compounds have been mentioned that affect the elimination of the aromatic hydrocarbons. Under aerobic conditions, high BTX consumption was obtained with mixtures of these compounds (Deeb & Alvarez-Cohen, 1999; Prenafeta-Boldú et al., 2002). However, negative effects such as competitive inhibition with toluene in presence of phenol (Lovanh & Álvarez, 2004) or ethanol (Da Silva et al. 2005) and catabolic repression in presence of succinate or acetate (Lovanh & Álvarez, 2004) have been reported. The effect of alternate carbon sources on BTX consumption associated to microbial growth has been evaluated at denitrifying conditions. Su & Kafkewitz (1994) demonstrated that a nitrate-reducing strain of *Pseudomonas maltophilia* was capable of utilizing toluene and succinate simultaneously. Both substrates were consumed for microbial growth. Data about the effect of organic compounds on oxidation and consumption rate of BTX are seldom found in the literature. In this regard, Gusmao et al. (2006) evaluated toluene, benzene, ethylbenzene and xylenes consumption separately (between 14 and 33 mg/l of each BTEX) or in mixtures (5 mg/l of each hydrocarbon) in the presence of ethanol (566 mg/l) resulting in high toluene removal efficiencies. However, no information about the fate of ethanol or BTEX consumed was made. Evidences of microbial consortia capable of accomplish denitrification with different ratios of acetate-C/toluene-C concentrations at C/N ratio of 1.4 have been reported by Martínez et al. (2007). These authors achieved elimination efficiencies of toluene, acetate and nitrate close to 100%, as well as yields formation of HCO_3^- and N_2 close to 1, indicating complete mineralization of both organic compounds. Likewise, denitrifying process was clearly dissimilative, as no biomass formation was detected.

All these results evidenced the feasibility of denitrification for BTX removal. Nevertheless, in spite of N_2 production was achieved in some of the works; significant nitrite accumulation was also obtained due to the high nitrate concentrations used in the assays. This problem could be overtaken by establishing C/N ratios close to stoichiometric values considering that it was not required an excess of nitrate in order to achieve effective BTX consumption. Moreover, the control and definition of C/N ratio seemed to be an option for negligible biomass production and obtaining dissimilative process during ground and wastewater treatment. Likewise, the use of sludge with no previously contact to BTX but physiologically stable, that is, in denitrifying steady state, could be an economical and less time-costly option when compared to the use of enrichment cultures. This type of denitrifying consortium may have promising applications for complete removal of both nitrogen and BTX compounds from wastewater streams into N_2 and CO_2.

3.5 Kinetic data for BTX consumption

Some kinetic data for BTX consumption obtained in batch denitrifying cultures are illustrated in Table 10, whereas data obtained for mixtures of BTX are mentioned in Table 11. It is difficult to make comparisons among these data considering that operational and environmental conditions, as well as biomass concentration used are different. However, it is possible to see that specific consumption rate values are generally low. Elmen et al. (1997) reported in axenic cultures of *Azoarcus tolulyticus* that specific toluene consumption rate decreased in 25% at toluene concentrations of 102 mg/l, whereas Peña-Calva et al. (2004b) using a denitrifying sludge without BTX acclimation determined a decrease of 21% in specific toluene consumption rate (q_T) at 100 mg toluene-C/l. According to Peña-Calva et al. (2004b), the maximum value of specific *m*-xylene consumption rate (q_X) was reached at 70 mg *m*-xylene-C/l, whereas a significant inhibition of the denitrifying pathway was observed at higher *m*-xylene concentrations, as they observed accumulation of NO_2^- and N_2O. It is important to remark that at the same experimental conditions, the q_X reported by these authors was 6 times lower than toluene consumption rate. Similar results have been obtained by Dou et al. (2008) as they reported that q_T values were higher than q_X values. Likewise, they observed an inhibitory effect of xylenes on biodegradation rate of BTX. These results suggested that wastewaters containing less than 100 mg/l of toluene or *m*-xylene would not affect the removal rate in a treatment system. Nevertheless, generalizations about this situation must be avoided as previous physiological state of sludge, biomass concentration and the rest of culture conditions might affect kinetic data and performance of the respiratory process.

Initial concentration of single BTX in batch assays	Inoculum	BTX removal rate	Reference
70 mg toluene/l	Denitrifying consortium	0.025 mg toluene-C /mg VSS-d	Peña-Calva et al., 2004b
50 mg toluene/l	Enriched mixed culture	1.67 mg/ld	Dou et al., 2008
55 mg toluene/l	*Azoarcus tolulyticus*	0.89 mg toluene/mg biomass-d	Elmen et al., 1997
20 mg toluene/l	Denitrifying consortium	0.007 mg toluene-C /mg VSS-d	Martínez et al., 2009
50 mg benzene/l	Enriched mixed culture	1.46 mg benzene/ld	Dou et al., 2008
70 mg *m*-xylene/l	Denitrifying consortium	0.0037 mg xylene-C /mg VSS-d	Peña-Calva et al., 2004b
50 mg *m*-xylene/l	Enriched mixed culture	1.49 mg *m*-xylene/ld	Dou et al., 2008

Table 10. Some kinetic data obtained for BTX consumption at different denitrifying culture conditions.

Several assays with mixtures of BTX have indicated substrate interactions among these compounds. Differences in consumption rate values have been observed when mixtures of BTX are assayed (Table 11) and compared with the values obtained in assays with the single

BTX compound (Table 10). In this sense, enhancement in xylene and toluene consumption rate between two to three times was determined in mixtures of toluene, benzene and xylene (Peña-Calva et al., 2004b), suggesting a positive interaction in hydrocarbons consumption when they are in mixture (Table 11). In contrast Dou et al. (2008) reported a diminishing in BTX consumption rate when mixtures of BTX concentrations higher than 5 mg/l were assayed and compared with those obtained in single BTX tests (Table 10). On the other hand, reports on mixtures of easy consumable substrate such as acetate and toluene conducted with microbial consortia with no previous contact to BTX, have indicated that at the acetate/toluene ratio of 65/20 the specific consumption rate of toluene value was twice compared to that obtained at 20 mg toluene-C/l alone (Martínez et al., 2009) (Table 11). Therefore, whereas the acetate concentration is clearly higher than toluene the consumption rate for toluene was increased. Authors have suggested that the specific consumption rate of toluene was biochemically enhanced by the presence of acetate. Noteworthy, at these conditions the denitrifying process was clearly dissimilative, as no biomass formation was detected. All these results indicate that substrate interaction among BTX and organic compounds mixtures are complicated as beneficial and deleterious effects on consumption rate values have been observed. More investigation is required on this topic. The wastewater treatment should be focused on obtaining dissimilative process with high consumption rates values. The use of some BTX mixtures or addition of easy consumable substrates might be helpful.

Initial BTX mixture concentration in batch assays (mg/l)	Corresponding single compound concentration in the BTX mixture in batch assays	BTX removal rate	Reference
65 acetate/20 toluene	20 mg toluene/l	0.014 mg toluene-C /mg SSV-d	Martínez et al., 2009
90	39 mg toluene/l	0.021 mg toluene-C/mg SSV-d	Peña-Calva et al., 2004a
300	50 mg toluene/l	1.29 mg toluene/l	Dou et al., 2008
90	41 mg m- xylene /l	0.0032 mg m-xylene-C/mg SSV-d	Peña-Calva et al., 2004a
300	50 mg m- xylene /l	0.92 mg m-xylene /l	Dou et al., 2008

Table 11. Kinetic data of BTX consumption at different concentrations of BTX mixtures at different denitrifying culture conditions.

3.6 Examples of simultaneous elimination of BTX and nitrate in reactors

Different configurations of reactors have been proposed for conducting biological BTX remediation (Farhadian et al., 2008). Some studies on BTX removal by denitrifying bioreactors have been carried out in recent years using different configurations such as horizontal anaerobic immobilized bioreactors (HAIB) (Gusmao et al., 2006, 2007), up flow anaerobic sludge blanket (UASB) reactors (Martínez et al., 2007, 2009) and sequencing batch reactors (SBR) (Hernández et al., 2008).

A horizontal-flow anaerobic immobilized biomass reactor (HAIB) inoculated with a denitrifying consortium was used for treating both a mixture solution of approximately 5.0 mg/l of each BTEX and the hydrocarbons separately dissolved in a solution containing ethanol (Gusmao et al., 2006). Organic matter removal efficiencies were of 95% with benzene and toluene amendments and about 76% with ethylbenzene, m-xylene and the BTEX-mix amendments. Ethanol was removed with an average efficiency of 86% whereas a concomitant nitrate removal of 97% was determined. Another HAIB reactor inoculated with a denitrifying pure culture resulted in nitrate and BTEX removal close to 97% (Gusmao et al., 2007). These systems showed to be an alternative for treating wastewater contaminated with nitrate, BTX and ethanol, however, no data about N_2 formation and BTEX lost due to volatilization or adsorption in the polyurethane used as packing material in the reactors was mentioned. Martínez et al. (2007), reported a successfull denitrification of different toluene-C loading rates (25, 50, 75, 100 and 125) mixed with acetate (accounting for a total carbon loading rate of 250 mg C/ld) in a continuously fed UASB reactor inoculated with denitrifying sludge physiologically stabilized. At these operational conditions, the C/N ratio was adjusted to 1.4. Authors indicated that even at 125 mg toluene-C/ld the biological process was not affected as carbon consumption efficiency values were close to 100%, whereas Y_{HCO3} and Y_{N2} values were higher than 0.71 and 0.88, respectively. This work evidenced that acetate and toluene can be effectively consumed by denitrification at C/N ratios close to stoichiometric values in continuous culture. These results suggest that the simple UASB denitrifying reactor system have promising applications for complete conversion of nitrate, toluene and acetate into N_2 and CO_2 with a minimal sludge production. Likewise, Martínez et al. (2009) reported in a continuous UASB reactor that no sludge acclimation was require for removing toluene from synthetic wastewater. They evidenced that the addition of acetate was useful for improving the specific removal rate of toluene. In order to discriminate whether the acetate effect was on the microbial community or on their metabolic denitrifying activity, these authors conducted molecular biology and ecological studies. Results showed no close effect of acetate on the microbial community composition suggesting that acetate could act as a biochemical enhancer. Thus, practical applications could be derived as no enrichment or previously adapted inocula have to be used in order to achieve successfull BTX elimination by denitrifying respiratory process. Finally, using a sequencing batch reactor (SBR), Hernández et al. (2008) determined that at 70 mg toluene-C/l, toluene and nitrate consumption efficiencies were 84 and 100%, respectively whereas Y_{N2}, Y_{HCO3} values were close to 1, thus, the process was clearly dissimilative. Authors observed that specific toluene (q_T) and nitrate (q_N) consumption rate values were increased in twofold after 17 cycles of operation. They also evaluated the settleabilities properties of the denitrifying sludge by means of sludge volume index (SVI). Under these operational conditions, toluene (r_T) and nitrate (r_N) volumetric consumption rates were 46 mg toluene $L^{-1}d^{-1}$ and 50 mg nitrate $L^{-1}d^{-1}$, respectively, whereas the SVI values remained close to 40 ml/g. Considering that operational troubles due to bad sludge settleability properties are often observed at SVI values higher than 200 ml/g, the results obtained by these authors suggest that SBR could be useful for effectively treating wastewaters contaminated with toluene. However, additional studies in bioreactors are needed to obtain more information about physiological, kinetic, ecological and engineering aspects in order to control the sludge metabolic capacity in denitrification processes for treating BTX removal.

Simultaneous Elimination of Carbon and Nitrogen Compounds of Petrochemical Effluents
by Nitrification and Denitrification

97

4. Conclusion

As described in this chapter, the progress made over the last decades in the understanding of the metabolic capabilities of nitrifying and denitrifying microorganisms for the biotransformation of carbon and nitrogen compounds has allowed the simultaneous elimination of BTX compounds, ammonium, and nitrate from wastewaters by nitrification and denitrification processes. The coupling of nitrifying and denitrifying processes could constitute a technological alternative for the biological treatment of petrochemical effluents. However, more investigation is required on various aspects such as physiology, engineering, and ecology information among others, to control better nitrifying and denitrifying bioreactors for the simultaneous consumption of BTX and nitrogen compounds from industrial effluents of chemical complexity.

5. Acknowledgment

Universidad Autónoma Metropolitana, Consejo Divisional, DCBS. México.

6. References

Ahn, J.H.; Kwan, T. & Chandran, K. (2011). Comparison of partial and full nitrification process applied for treating high-strength nitrogen wastewaters: Microbial ecology through nitrous oxide production. *Environmental Science and Technology*, Vol.45, No.7, pp. 2734-2740

Akunna, J.C.; Bizeau, C. & Moletta, R. (1993). Nitrate and nitrite reductions with anaerobic sludge using various carbon sources: glucose, glycerol, acetic acid, lactic acid and methanol. *Water Research*, Vol.27, pp. 1303-1312

Aleem, M.I.H. & Sewel, D.L. (1981). Mechanism of nitrite oxidation and oxidoreductase systems in *Nitrobacter agilis*. *Current Microbiology*, Vol.5, No.5, pp. 267-272

Alexander, R. B.; Smith, R.A. & Schwars, G. E. (2000). Effect of the stream channel size on the delivery of nitrogen to the Gulf of Mexico. *Nature*, Vol.403, No.17, pp. 758-761

Álvarez, P.J. & Vogel, T.M. (1995). Degradation of BTEX and their aerobic metabolites by indigenous microorganisms under nitrate reducing conditions. *Water Science and Technology*, Vol.31, pp. 15–28

Amor, L.; Eiroa, M.; Kennes, C. & Veiga, M.C. (2005). Phenol biodegradation and its effect on the nitrification process. *Water Research*, Vol.39, No.13, pp. 2915-2920

Anthonisen, A.C.; Loehr, R.C.; Prakasam, T.B.S. & Srinath, E.G. (1976). Inhibition of nitrification by ammonia and nitrous acid. *Journal Water Pollution Control Federation*, Vol.48, No.5, pp. 835–852

Antón, D.J. & Díaz, D.C. (2000). *Sequía en un mundo de agua*. Piriguazú, Ediciones, CIRA-UAEM, San José, Toluca, México

Ashley, D.L.; Bonin, M.A.; Cardinali, F.L.; McCraw, J.M. & Wooten, V.J. (1994). Blood concentrations of volatile organic compounds in a nonoccupationally exposed US population and in groups with suspected exposure. *Clinical Chemistry*, Vol.40, pp. 1401-1404

Beller, H.R. & Spormann, A.M. (1997). Anaerobic activation of toluene and o-xylene by addition to fumarate in denitrifying strain T. *Journal of Bacteriology*, Vol.179, pp. 670–676

Beristain-Cardoso, R.; Pérez-González, D.N.; González-Blanco, G. & Gómez, J. (2011). Simultaneous oxidation of ammonium, p-cresol and sulfide using a nitrifying sludge in a multipurpose bioreactor: A novel alternative. *Bioresource Technology*, Vol.102, pp. 3623-3625

Beristain-Cardoso, R.; Texier, A.C., Razo-Flores, E.; Méndez-Pampín, R. & Gómez, J. (2009). Biotransformation of aromatic compounds from wastewaters containing N and/or S, by nitrification/denitrification: a review. *Reviews in Environmental Science and Biotechnology*, Vol.8, pp. 325-342

Berks, B.C.; Ferguson, S.J.; Moir, J.W.B. & Richardson, D.J. (1995). Enzymes and associated electron transport systems that catalyse the respiratory reduction of nitrogen oxides and oxyanions. *Biochimica et Biophysica Acta*, Vol.1232, pp. 97-173

Bernet, N. & Spérandio, M. (2009). Principles of nitrifying processes, In: *Environmental technologies to treat nitrogen pollution*, F.J. Cervantes, (First Ed.), 23-37, International Water Association publishing, ISBN 1843392224, London, United Kingdom

Bock, E.; Koops, H.P.; Harms, H. & Ahlers, B. (1991). The biochemistry of nitrifying organisms, In: *Variations in autotrophic life*, J.H. Shively & L.L. Burton, (Eds.), 171-200, Academic Press, London, United Kingdom

Bonin, P.; Gilewicz, M. & Bertrand, J.C. (1989). Effects of oxygen on each step of denitrification on *Pseudomonas nautica*. *Canadian Journal of Microbiology*, Vol.35, pp. 1061-1064

Brandt, K.K.; Hesselsoe, M.; Roslev, P.; Henriksen, K. & Sorensen, J. (2001). Toxic effects of linear alkylbenzene sulfonate on metabolic activity, growth rate, and microcolony formation of *Nitrosomonas* and *Nitrospira* strains. *Applied and Environmental Microbiology*, Vol.67, No.6, pp. 2489-2498

Breemen, N.V. (2002). Nitrogen cycle: Natural organic tendency. *Nature*, Vol. 415, pp. 381-382

Bremmer, J.M. & Bundy, L.G. (1974). Inhibition of nitrification in soils by volatile sulphur compounds. *Soil Biology Biochemistry*, Vol.6, No.3, pp. 161-165

Burland, S.M. & Edwards, A. E. (1999). Anaerobic benzene biodegradation linked to nitrate reduction. *Applied and Environmental Microbiology*, Vol.65, pp. 529-533

Cadwell, M.E. & Sufita, J.M. (2000). Detection of phenol and benzoate as intermediates of anaerobic benzene biodegradation under different terminal electron-accepting condition. *Environmental Science and Technology*, Vol.34, pp. 1216-1220

Cervantes, F.J.; Meza-Escalante, E.R.; Texier, A.C. & Gómez, J. (2009). Kinetic limitations during the simultaneous removal of p-cresol and sulfide in a denitrifying process. *Journal of Industrial Microbiology and Biotechnology*, Vol.36, pp. 1417-1424

Chakraborty, R.; O'Connor, S.; Chan, E. & Coates, J. (2005). Anaerobic degradation of benzene, toluene, ethylbenzene and xylene compounds by *Dechloromonas* strain RCB. *Applied and Environmental Microbiology*, Vol.71, No.12, pp. 8649-8655

Chang, S.W.; Hyman, M.R. & Williamson, K.J. (2002). Cooxidation of naphthalene and other polycyclic aromatic hydrocarbons by the nitrifying bacterium *Nitrosomonas europaea*. *Biodegradation*, Vol.13, No.6, pp. 373-381

Chang, M.K.; Voice, T.C. & Criddle, C.S. (1993). Kinetics of competitive inhibition and cometabolism in the biodegradation of benzene, toluene, and p-xylene by two *Pseudomonas* isolates. *Biotechnology and Bioengineering*, Vol. 41, pp. 1057-1065

Coates, J.D.; Chakraborty, R.; Lack, J.G.; O'Connor, S.M.; Cole, K.A.; Bender, K.S. & Achenbach, L.A. (2001). Anaerobic benzene oxidation coupled to nitrate reduction in pure culture by two strains of *Dechloromonas*. *Nature*, Vol.411, pp. 1039–1043

Coates, J.D.; Chakraborty, R. & McInerney, J. (2002). Anaerobic benzene biodegradation-a new era. *Research in Microbiology*, Vol.153, pp. 621-628

Corseuil, H.X.; Hunt, C.S.; Ferreira, R.C.S. & Alvarez, P.J. (1998). The influence of the gasoline oxygenate ethanol on aerobic and anaerobic BTX biodegradation. *Water Research*, Vol.32, pp. 2065-2072

Craig, D.; Phelps, C.D., Kazumi, J. & Young, L.Y. (1996). Anaerobic degradation of benzene in BTX mixtures dependent on sulfate reduction. *FEMS Microbiology Letters*, Vol.145, pp. 433-437

Cuervo-López, F.M.; Martinez, F.; Gutiérrez-Rojas, M.; Noyola, R.A. & Gomez, J. (1999). Effect of nitrogen loading rate and carbon source on denitrification and sludge settleability in UASB reactors. *Water Science and Technology*, Vol.40, No.8, pp. 123-130

Cuervo-López, F.; Martínez-Hernández, S.; Texier, A.C. & Gómez, J. (2009). Principles of denitrifying processes, In: *Environmental technologies to treat nitrogen pollution*, F.J. Cervantes, (First Ed.), 41-65, International Water Association publishing, ISBN 1843392224, London, United Kingdom

Cunningham, J.A.; Rahme, H.; Hopkins, G.D.; Lebron, C. & Reinhard, M. (2001). Enhanced in situ bioremediation of BTEX contaminated groundwater by combined injection of nitrate and sulfate. *Environmental Science and Technology*, Vol.35, pp. 1663–1670

Da Silva, M. & Alvarez, P.J. (2002). Effects of ethanol versus MTBE on benzene, toluene, ethylbenzene, and xylene natural attenuation in aquifer columns. *Journal of Environmental Engineering*, Vol.128, pp. 862–867

Da Silva, L.B.M.; Ruiz-Aguilar, G.M.L. & Álvarez, P.J. (2005). Enhanced anaerobic biodegradation of BTEX-ethanol mixtures in aquifer columns amended with sulfate, chelated ferric iron or nitrate. *Biodegradation*, Vol.16, pp. 105-114

De Boer, W. & Kowalchuk, G.A. (2001). Nitrification in acid soils: micro-organisms and mechanisms. *Soil Biology Biochemistry*, Vol.33, No.7-8, pp. 853–866

Deeb, R.A. & Alvarez-Cohen, L. (1999). Temperature effects and substrate interaction during the aerobic biotransformation of BTEX mixtures by toluene-enriched consortia and *Rhodococcus rhodochrous*. *Biotechnology and Bioengineering*, Vol.62, pp. 526-536

Deeb, R.A. & Alvarez-Cohen, L. (2000). Aerobic biotransformation of gasoline aromatic in multicomponent mixtures. *Bioremediation Journal*, Vol.4, pp. 1-9

Diario Oficial de la Federación. 3 junio de 1998. México

Dou, J.; Liu, X. & Hu, Z. (2008). Substrate interactions during anaerobic biodegradation of BTEX by the mixed cultures under nitrate reducing conditions. *Journal of Hazardous Materials*, Vol. 15, No.2-3, pp. 262-272

Duetz, W.A.; de Yong, C.; Williams, P.A. & Van Andel, J. G. (1994). Competition in chemostat culture between *Pseudomonas* Strains the use different pathways for the degradation of toluene. *Applied and Environmental Microbiology*, Vol.60, No.8, pp. 2858-2863

Duetz, W.A.; Wind, B.; Kamp, M. & van Andel, J. G. (1997). Effect of growth rate, nutrient limitation and succinate on expression of TOL pathway enzymes in response to *m*-

xylene in chemostat cultures of *Pseudomonas putida* (pWWO). *Microbiology*, Vol.143, pp. 2331-2338

Dyreborg, S. & Arvin, E. (1995). Inhibition of nitrification by creosote-contaminated water. *Water Research*, Vol.29, No.6, pp. 1603-1606

Elmen, J.; Pan, W.; Leung, S.Y.; Magyarosy, A. & Keasling, J. D. (1997). Kinetics of toluene degradation by a nitrate-reducing bacterium isolated from a groundwater aquifer. *Biotechnology and Bioengineering*, Vol.55, pp. 82–90

Evans, P.J.; Ling, W.; Goldschmidt, B.; Ritter, E.R. & Young, L.Y. (1992). Metabolites formed during anaerobic transformations of toluene and *o*-xylene and their proposed relationship to the initial steps of toluene mineralisation. *Applied and Environmental Microbiology*, Vol.58, pp. 496–501

Evans, P.J.; Mang, D.T. & Young, Y.L. (1991). Degradation of toluene and *m*-xylene and transformation of *o*-xylene by denitrifying enrichment cultures. *Applied and Environmental Microbiology*, Vol.57, pp. 450-454

Farhadian, M.; Duchez, D.; Vachelard, C. & Larroche, C. (2008). Monoaromatics removal from polluted water through bioreactors- A review. *Water Research*, Vol.42, pp. 1325-1341

Feris, K.; Mackay, D.; Nick de Sieyes, N.; Chakraborty, I.; Einarson, M.; Hristova, K. & Scow, K. (2008). Effect of ethanol on microbial community structure and function during natural attenuation of benzene, toluene, and *o*-xylene in a sulfate-reducing aquifer. *Environmental Science and Technology*, Vol.42, pp. 2289–2294

Fishbein, L. (1984). An overview of environmental and toxicological aspects of aromatic hydrocarbons. I. Benzene. *Science of the Total Environment*, Vol.40, pp. 189-218

Fishbein, L. (1985). An overview of environmental and toxicological aspects of aromatic hydrocarbons. II. Toluene. *Science of the Total Environment*, Vol.42, pp. 267-288

Fornazzari, L.; Pollanen, M.S.; Myers, V. & Wolf, A. (2003). Solvent abuse-related toluene leukoencephalopathy. *Journal of Clinical Forense Medicine*, Vol.10, pp. 93-95

Gerardi, M.H. (2002). Nitrification and Denitrification in the Activated Sludge Process. (Ed.) ISBN 0-471-06508-0, Wiley-Interscience, John Wiley & Sons, Inc., New York

Gersberg, R.; Dawsey, W.J. & Ritgeway, H.F. (1989). Biodegradation of dissolved aromatic hydrocarbons in gasoline contaminated groundwater using denitrification. In: *Petroleum contaminated soils*. Vol. 2 Lewis Publishers. Chelsea Michigan. pp. 211-217

Gomez, J.; Mendez, R. & Lema, J.M. (2000). Kinetic study of addition of volatile organic compounds to a nitrifying sludge. *Applied Biochemistry Biotechnology*, Vol.87, No.3, pp. 189-202

González-Blanco, G.; Beristain-Cardoso, R.; Cuervo-López, F.; Cervantes, F. J. & Gómez, J. (2011). Denitrification applied to wastewater treatment: processes, regulation and ecological aspects. In: *Denitrification*. Editors: Nicolo Savaglio and Raul Puopolo. Nova Science Publishers, Inc. ISBN 978-1-61470-879-7

Gülensoy, N. & Alvarez, P.J. (1999). Diversity and correlation of specific aromatic hydrocarbon biodegradation capabilities. *Biodegradation*, Vol.10, pp. 331-340

Gusmão, V.R.; Chinalia, F.A.; Sakamoto, I.K. & Varesche, M.B. (2007). Performance of a reactor containing denitrifying immobilized biomass in removing ethanol and aromatic hydrocarbons (BTEX) in a short operating period. *Journal of Hazardous Materials*, Vol.139, No.2, pp. 301-309

Simultaneous Elimination of Carbon and Nitrogen Compounds of Petrochemical Effluents
by Nitrification and Denitrification
101

Gusmão, V.R.; Martins, T. H.; Chinalia, F.A.; Sakamoto, I.K.; Thiemann, O.H. & Varesche, M.B. (2006). BTEX and ethanol removal in horizontal-flow anaerobic immobilized biomass reactor, under denitrifying condition. *Process Biochemistry*, Vol.41, No.6, pp. 1391-1400

Haigler, B.E.; Pettigrew, C.A. & Spain, J.C. (1992). Biodegradation of mixtures of substituted benzenes by *Pseudomonas* sp. Strain JS150. *Applied and Environmental Microbiology*, Vol.58, No.7, pp. 2237-2244

Hanaki, K.; Wantawin, C. & Ohgaki, S. (1990). Effects of the activity of heterotrophs on nitrification in a suspended-growth reactor. *Water Research*, Vol.24, No.3, pp. 289-296

Häner, A.; Höhener, P. & Zeyer, J. (1995). Degradation of *p*-xilene by a denitrifying enrichment culture. *Applied and Environmental Microbiology*. Vol.61, pp. 3185-3188

Hartley, W.R. & Englande, A.J. (1992). Health risk assessment of the migration of unleaded gasoline: a model for petroleum products. *Water Science and Technology*, Vol.25, pp. 65-72

Hauck, R.D. (1980). Mode of action of nitrification inhibitors, In: *Nitrification inhibitors-potential and limitations*, J.J. Meisinger, (Ed.), 19-32, American Society of Agronomy, Wisconsin, USA

Hernández, L.; Buitrón, G.; Gómez, J. & Cuervo-López, F.M. (2008). Denitrification of toluene and sludge settleability. *Proceedings of the 4th Sequencing Batch Reactor Conference*, Rome, Italy, April 7-10, 2008

Hochstein, L.I. & Tomlinson, G.A. (1988). The enzymes associated with denitrification. *Annual Review of Microbiology*, Vol.42, pp. 231-261

Hoffman, R.V. (2004). Acidity and Basicity, In: *Organic Chemistry: An intermediate text*, (Second Ed.), 55, ISBN 0-471-45024-3, John Wiley and Sons, Inc, publication, New Jersey, USA

Hollocher, T.C.; Tate, M.E. & Nicholas, D.J.D. (1981). Oxidation of ammonia by *Nitrosomonas europaea*: definitive 18O-tracer evidence that hydroxylamine formation involves a monooxygenase. *Journal of Biological chemistry*, Vol.256, No.21, pp. 10834-10836

Huff, J. (2007). Benzene-induced cancers: abridged history and occupational health impact. *International Journal of Occupational and Environmental Health*, Vol.13, pp. 213-21

Hyman, M.R.; Kim, C. & Arp, D.J. (1990). Inhibition of ammonia monooxygenase in *Nitrosomonas europaea* by carbon disulfide. *Journal of Bacteriology*, Vol.172, No.9, pp. 4775-4782,

Hyman, M.R.; Murton, I.B. & Arp, D.J. (1988). Interaction of ammonia monooxygenase from *Nitrosomonas europaea* with alkanes, alkenes, and alkynes. *Applied and Environmental Microbiology*, Vol.54, pp. 3187-3190

Hyman, M.R.; Page, C.L. & Arp, D.J. (1994). Oxidation of methyl fluoride and dimethyl ether by ammonia monooxygenase in *Nitrosomonas europaea*. *Applied and Environmental Microbiology*, Vol.60, pp. 3033-3035

Hyman, M.R.; Samsone-Smith, A.W.; Shears, J.H. & Wood, P.M. (1985). A kinetic study of benzene oxidation to phenol by whole cells of *Nitrosomonas europaea* and evidence for the further oxidation of phenol to hydroquinone. *Archives of Microbiology*, Vol.143, No.3, pp. 302-306

Hyman, M.R. & Wood, P.M. (1983). Methane oxidation by *Nitrosomonas europaea*. *Biochemical Journal*, Vol.212, pp. 31-37

Hyman, M.R. & Wood, P.M. (1984). Ethylene oxidation by *Nitrosomonas europaea*. *Archives of Microbiology*, Vol.137, pp. 155-158

Hyman, M.R. & Wood, P.M. (1985). Suicidal inactivation and labelling of ammonia mono-oxygenase by acetylene. *Biochemical Journal*, Vol.227, No.3, pp. 719-725

Iizumi, T.; Mixumoto, M. & Nakamura, K. (1998). A bioluminescence assay using *Nitrosomonas europaea* for rapid and sensitive detection of nitrification inhibitors. *Applied and Environmental Microbiology*, Vol.64, No.10, pp. 3656-3662

Jacobs, J.A. & Testa, S.M. (2003). Design considerations for in situ chemical oxidation using high pressure jetting technology. *International Journal of Soil, Sediment and Water*, March/April, pp. 51-60

Jain, P.K.; Gupta, V.K.; Gaur, R.K.; Lowry, M.; Jaroli, D.P. & Chauhan, U.K. (2011). Bioremediation of Petroleum oil Contaminated Soil and Water. *Research Journal of Environmental Toxicology*, Vol.5, pp. 1-26

Jensen, H.L. (1950). Effect of organic compounds on *Nitrosomonas*. *Nature*, Vol.165, No.4207, pp. 974

Juliette, L.Y.; Hyman, M.R. & Arp, D.J. (1993). Inhibition of ammonia oxidation in *Nitrosomonas europaea* by sulfur compounds: thioethers are oxidized to sulfoxides by ammonia monooxygenase. *Applied and Environmental Microbiology*, Vol.59, No.11, pp. 3718-3727

Jung, I.G., & Park, C.H. (2004). Characteristics of *Rhodococcus pyridinovorans* PYJ-1 for the biodegradation of benzene, toluene, *m*-xylene (BTX) and their mixtures. *Journal of Bioscience and Bioengineering*, Vol.97, No.6, pp. 429-431

Kane, S.R.; Beller, H.R.; Legler, T.C., & Anderson, R.T. (2002). Biochemical and genetic evidence of benzylsuccinate synthase in toluene degrading, ferric iron-reducing *Geobacter metallireducens*. *Biodegradation*, Vol.13, pp. 149–154

Keener, W.K. & Arp, D.J. (1993). Kinetic studies of ammonia monooxygenase inhibition in *Nitrosomonas europaea* by hydrocarbons and halogenated hydrocarbons in an optimized whole-cell assay. *Applied and Environmental Microbiology*, Vol.59, No.8, pp. 2501-2510

Keener, W.K. & Arp, D.J. (1994). Transformations of aromatic compounds by *Nitrosomonas europaea*. *Applied and Environmental Microbiology*, Vol.60, No.6, pp. 1914-1920

Kirkeleit, J.; Riise, T.; Bratveit M. & Moen, B.E. (2006). Benzene exposure on a crude oil production vessel. *Annals of Occupational Hygiene*, Vol.50, pp. 123–129

Knobeloch, L.; Salna, B.; Hogan, A.; Postle, J. & Anderson, H. (2000). Blue babies and nitrate-contaminated well water. *Environmental Health Perspectives*, Vol.108, pp. 675-678

Knowles, R. (1982). Denitrification. *Microbiological Reviews*, Vol.46, No.1, pp. 43-70

Kobayashi, H.; Uematsu, K.; Hirayama, H. & Horikoshi, K. (2000). Novel toluene elimination system in a toluene-tolerance microorganism. *Journal of Bacteriology*, Vol.182, No.22, pp. 6451-6455

Kowalchuk, G.A. & Stephen, J.R. (2001). Ammonia-oxidizing bacteria: A model for molecular microbial ecology. *Annual Review of Microbiology*, Vol.55, pp. 485–529

Simultaneous Elimination of Carbon and Nitrogen Compounds of Petrochemical Effluents
by Nitrification and Denitrification

103

Kuhn, E.P.; Zeyer, J.; Eiche, P. & Schwarzenbach, R.P. (1988). Anaerobic degradation of alkylated benzenes in denitrifying laboratory aquifer columns. *Applied and Environmental Microbiology.* Vol.54, pp. 490-496

Laanbroek, H.J. & Gerards, S. (1993). Competition for limiting amounts of oxygen between *Nitrosomonas europaea* and *Nitrobacter winogradskyi* grown in mixed continuous cultures. *Archives Microbiology,* Vol.159, No.5, pp. 453–459

Lalucat, J.; Bennasar, A.; Bosch, R.; García-Valdés, E. & Palleroni, N. J. (2006). Biology of *Pseudomonas stutzeri.* *Microbiology and Molecular Biology Reviews,* Vol.70, No.2, pp. 510-547

Lovanh, N. & Alvarez, P.J. (2004). Effect of ethanol, acetate, and phenol on toluene degradation   activity and tod–lux expression in *Pseudomonas putida* TOD 102: evaluation of the metabolic flux dilution model. *Biotechnology and Bioengineering,* Vol.86, pp. 801-808

Lovley, D.R. (2000). Anaerobic benzene degradation. *Biodegradation,* Vol.11, pp. 107-116

Ma, G. & Love, N.G. (2001). BTX biodegradation in activated sludge under multiple redox conditions. *Journal of Environmental Engineering,* Vol.118, pp. 509-515

Major, D. W.; Mayfiled, C. I. & Barker, J. F. (1988). Biotransfomation by denitrification in aquifier sand. *Ground Water,* Vol.26, pp. 8-14

Martínez, S.; Cuervo-López, F.M. & Gómez, J. (2007). Toluene mineralization by denitrification in an up flow anaerobic sludge (UASB) reactor. *Bioresource Technology,* Vol.98, pp. 1717-1723

Martínez, H.S.; Olguín, E. J.; Gómez J. & Cuervo-López, F.M. (2009). Acetate enhances the specific consumption rate of toluene under denitrifying conditions. *Archives of Environmental Contamination and Toxicology,* Vol.57, pp. 679-687

Martínez-Hernández, S.; Texier, A.C.; Cuervo-López, F.M. & Gómez J. (2011). 2-Chlorophenol consumption and its effect on the nitrifying sludge. *Journal of Hazardous Materials,* Vol.185, 1592–1595

Mateju, V.; Cizinska, S.; Krejei, J. & Janoch, T. (1992). Biological water denitrification. A review. *Enzyme Microbial Technology,* Vol.14, pp. 170-183

McCarty, G.W. (1999). Modes of action of nitrification inhibitors. *Biology and Fertility of Soils,* Vol.29, No.1, pp. 1–9

McIsaac, G. F.; David, M. B.; Gertner, G. Z. & Goolsby, D. A. (2002). Nitrate flux in the Mississippi river. *Nature,* Vol.414, No.8, pp. 166-167

Monroy, O.; Fama, G.; Meraz, M.; Montoya, L. & Macarie, H. (2000). Anaerobic digestion for wastewater treatment in México: State of the technology. *Water Research,* Vol.36, No.4, pp. 1803-1816

Moore, T.A.; Xing, Y.; Lazaenby, B.; Lynch, M.D.J.; Schiff, S.; Robertson, W.D.; Timlin, R.; Lanza, S.; Ryan, M.C.; Aravena, R.; Fortin, D.; Clark, I.D. & Neufeld, J.D. (2011). Prevalence of anaerobic ammonium-oxidation bacteria in contaminated groundwater. *Environmental Science and Technology,* Vol.45, No.17, pp. 7217-7225

Morgan P.; Stephen T.; Lewis T., & Watkinson, R. J. (1993). Biodegradation of benzene, toluene, ethylbenzene and xylenes in gas-condensate-contaminated ground-water. *Environ. Poll.,* Vol.182, pp. 181-190

Morgan-Sagastumen, J.; Jimenez, B. & Noyola, A. (1994). Anaerobic- anoxic -aerobic process with recycling and separated biomass for organic carbon and nitrogen removal from wastewater. *Environmental Technology*, Vol.15, pp. 233-243

Muñoz, H.; Armienta, M.A.; Vera, A. & Ceniceros, N. (2004). Nitrato en el agua subterránea del valle de Huamantla, Tlaxaca, México. *Revista Internacional de contaminación ambiental*, Vol.20, No.3, pp. 91-97

O'Connor, O.W. & Young, L.Y. (1996). Effects of six different functional groups and their position on the bacterial metabolism of monosubstituted phenols under anaerobic conditions. *Environmental Science and Technology*, Vol.30, No.5, pp. 1419-1428

Oh, J. & Silverstein, J. (1999). Acetate limitation and nitrite accumulation during denitrification. *Journal of Environmental Engineering*, Vol.125, No.3, pp. 234-242

Pacheco, J.; Marin, M.; Cabrera, A.; Steinich, B. & Escolero, O. (2001). Nitrate temporal and spatial patterns in 12 water supply wells, Yucatán, Mexico. *Environmental Geology*, Vol.40, pp. 708-715

Pemex, Petroquímica, Informes: www.pemex.com/evolumenventas. html. (2001)

Peng, Y. & Zhu, G. (2006). Biological nitrogen removal with nitrification and denitrification via nitrite pathway. *Applied Microbiology and Biotechnology*, Vol.73, No.1, pp. 15-26

Peña-Calva, A. (2007). Estudio comparativo del efecto de diferentes fuentes de carbono sobre el proceso respiratorio y sedimentabilidad de lodos en un proceso desnitrificante. Tesis de Doctorado en Biotecnología, Universidad Autónoma Metropolitana-Iztapalapa, México, D. F, México

Peña-Calva, A.; Olmos, A.; Cuervo-López, F.M. & Gomez, J. (2004a). Evaluation of the denitrification of a mixture of benzene, toluene and *m*-xylene. *Proceedings of the 10th World Congress Anaerobic Digestion*. pp 1957-1961, Montreal, Canada, August 29 - September 2, 2004

Peña-Calva, A.; Olmos, D.A.; Viniegra, G.G.; Cuervo, L.F. & Gómez, J. (2004b) Denitrification in presence of benzene, toluene, and *m*-xylene. *Applied Biochemistry and Biotechnology*, Vol.119, pp. 195-208

Perego, C. & Ingallina, P. (2004). Combining alkylation and transalkylation for alkylaromatic production. *Green Chemistry*, Vol.6, pp. 274-279

Prenafeta-Boldú, X.F.; Vervoort, J.; Grotenhuis, J.T.C. & van Groenestijn, J.W. (2002). Substrate Interactions during the Biodegradation of Benzene, Toluene, Ethylbenzene, and Xylene (BTEX) Hydrocarbons by the Fungus *Cladophialophora* sp. Strain T1. *Applied and Environmental Microbiology*. Vol.68, No.6, pp. 2660-2665

Prosser, J.I. (1989). Autotrophic nitrification in bacteria. *Advances in Microbial Physiology*. Vol.30, pp. 125-181

Radniecki, T.S.; Gilroy, C.A. & Semprini, L. (2011). Linking NE1545 gene expression with cell volume changes in *Nitrosomonas europaea* cells exposed to aromatic hydrocarbons. *Chemosphere*, Vol.82, pp. 514-520

Rasche, M.E.; Hicks, R.E.; Hyman, M.R. & Arp, D.J. (1990a). Oxidation of monohalogenated ethanes and *n*-chlorinated alkanes by whole cells of *Nitrosomoas europaea*. *Journal of Bacteriology*, Vol.172, pp. 5368-5373

Rasche, M.E.; Hyman, M.R. & Arp, D.J. (1990b). Biodegradation of halogenated hydrocarbon fumigants by nitrifying bacteria. *Applied and Environmental Microbiology*, Vol.56, No.8, pp. 2568-2571

Reardon, K.F.; Mosteller, D.C. & Rogers, J.D. (2000). Biodegradation kinetics of benzene, toluene and phenol as single and mixed substrates for *Pseudomona putida* F1. *Biotechnology and Bioengineering*, Vol.69, pp. 385-400

Reardon, K.F.; Mosteller, D.C.; Rogers, J.B.; Du Teau, N.M. & Kim, K.H. (2002). Biodegradation kinetics of aromatic hydrocarbon mixtures by pure and mixed bacterial cultures. *Environmental Health Perspectives*, Vol.110, No.6, pp. 1005-1011

Rosenberg, N.L.; Kleinschmidt-DeMasters, B. K.; Davis, K.A.; Dreisbach, J.N.; Hormes, J.T. & Filley, C.M. (1988). Toluene abuse causes diffuse central nervous system white matter changes. *Annals of Neurology*, Vol.23, pp. 611-614

Rozkov, A.; Kaard, A. & Vilu, R. (1998). Biodegradation of dissolved jet fuel in chemostat by a mixed bacterial culture isolated from a heavily polluted site. *Biodegradation*, Vol.8, pp. 363-369

Schimel, D.; Alves, D.; Enting, I.; Heimann, M. & Joos, F. (1996). Radiative forcing of climate change. In climate change 1995. In: *The science of climate change*, ed. Hoghton, J.T., Meira-Filho, L.G., Callander, B.A., Harris, N., Kattenberg, A., Maskell, K., 65-131, Cambridge University Press, Cambridge, UK

Schocher, R. J.; Seyfried, B.; Vazquez, F. & Zeyer, J. (1991). Anaerobic degradation of toluene by pure cultures of denitrifying bacteria. *Archives of Microbiology*, Vol.157, pp. 7-12

Schweighofer, P.; Nowak, O.; Svardal, K. & Kroiss, H. (1996). Steps towards the upgrading of a municipal WWTP affected by nitrification inhibiting compounds – a case study. *Water Science and Technology*, Vol.33, No.12, pp. 39-46

Shiemke, A.; Arp, D.J. & Sayavedra-Soto, L.A. (2004). Inhibition of membrane-bound methane monooxygenase and ammonia monooxygenase by diphenyliodonium: implications for electron transfer. *Journal of Bacteriology*, Vol.186, No.4, pp. 928-937

Shim, H.; Hwang, B.; Lee, S.; & Kong, S. (2005). Kinetics of BTX degradation by a coculture of *Pseudomonas putida* and *Pseudomonas fluorescens* under hypoxic conditions. *Biodegradation*, Vol.16, pp. 319-327

Sikkema, J.; de Bont, J.A.M. & Poolman, B. (1994). Interaction of cyclic hydrocarbons with biological membranes. *Journal of Biological Chemistry*, Vol.269, pp. 8022–8028

Sikkema, J.; de Bont, J.A.M. & Poolman, B. (1995). Mechanisms of membrane toxicity of hydrocarbons. *Microbiological Reviews*, Vol.59, No.2, pp. 201-222

Silva, C.D.; Cuervo-López, F.M.; Gómez, J. & Texier, A.C. (2011). Nitrite effect on ammonium and nitrite oxidizing processes in a nitrifying sludge. *World Journal of Microbiology and Biotechnology*, Vol.27, No.5, pp. 1241-1245

Silva, C.D.; Gómez, J.; Houbron, E.; Cuervo-López, F.M. & Texier, A.C. (2009). *p*-Cresol biotransformation by a nitrifying consortium. *Chemosphere*, Vol.75, No.10, pp. 1387-1391

Spieck, E.; Ehrich, S.; Aamand, J. & Bock, E. (1998). Isolation and immunocytochemical location of the nitrite-oxidizing system in *Nitrospira moscoviensis*. *Archives Microbiology*, Vol.169, No.3, pp. 225-230

Steen-Christensen, J. & Elton, J. (1996). Soil and Groundwater pollution from BTEX. Groundwater Pollution Primer. CE 4594: Soil and Groundwater Pollution Civil Engineering, Virginia, pp. 1-10

Steinmüller, W. & Bock, E. (1976). Growth of *Nitrobacter* in the presence of organic matter. *Archives of Microbiology*, Vol.108, pp. 299-304

Su, J.J. & Kafkewitz, D. (1994). Utilization of toluene and xylenes by a nitrate reducing strain of *Pseudomonas maltophilia* under low oxygen and anoxic conditions. *FEMS Microbiol Ecol.* Vol.15, pp. 249-258

Suzuki, I.; Dular, V. & Kwok, S.C. (1974). Ammonia or ammonium ion as substrate for oxidation by *Nitrosomonas europaea* cells and extracts. *Journal of Bacteriology*, Vol.120, No.1, pp. 556-558

Takahashi, I.; Ohki, S.; Murakami, M.; Takagi, S.; Stato, Y.; Vonk, J. W. & Wakabayashi, K. (1997). Mode of action and QSAR studies of nitrification inhibitors: effect of trichloromethyl-1,3,5-triazines on ammonia-oxidizing bacteria. *Journal of Pesticide Science*, Vol.22, No.1, pp. 27-32

Texier, A.-C. & Gomez, J. (2002). Tolerance of nitrifying sludge to *p*-cresol. *Biotechnology Letters*, Vol.24, No.4, pp. 321-324

Texier, A.-C. & Gomez, J. (2007). Simultaneous nitrification and *p*-cresol oxidation in a nitrifying sequencing batch reactor. *Water Research*, Vol.41, No.2, pp. 315-322

Thoma, G.; Gleason, M. & Popov, V. (2006). Sonochemical treatment of benzene/toluene contaminated wastewater. *Environmental Progress*, Vol.17, No.3, pp. 154-160

Thomsen, J. K.; Geest, T. & Cox, R.P. (1994). Mass spectrometric studies of effect of pH on the accumulation of intermediates in denitrification by *Paracoccus denitrificans*. *Applied and Environmental Microbiology*, Vol.60, pp. 536-541

Tsang, D.C.Y. & Suzuki, I. (1982). Cytochrome c554 as a possible electron donor in the hydroxilation of ammonia and carbon monooxide in *Nitrosomonas europaea*. *Canadian Journal of Biochemistry*, Vol.60, pp. 1018-1024

Tsitko, I.V.; Zaitsev, G.M.; Lobanok, A. G.; & Salkinoja-Solonen, M. S. (1999). Effect of aromatic compounds on cellular fatty acid composition of *Rhodococcus opacus*. *Applied and Environmental Microbiology*, Vol.65, No.2, pp. 853-855

USEPA. (May 2007), Nitrates and Nitrites TEACH Chemical Summary, Available from http://www.epa.gov/teach

Vadivelu, V.M.; Yuan, Z.; Fux, C. & Keller, J. (2006a). The inhibitory effects of free nitrous acid on the energy generation and growth processes of an enriched Nitrobacter culture. *Environmental Science and Technology*, Vol.40, No.4, pp. 4442-4448

Vadivelu, V.M.; Keller, J. & Yuan, Z. (2006b). Effect of free ammonia and free nitrous acid concentration on the anabolic and catabolic processes of an enriched *Nitrosomonas* culture. *Biotechnology and Bioengineering*, Vol.95, No.5, pp. 830-839

Vannelli, T. & Hooper, A.B. (1992). Oxidation of nitrapyrin to 6-chloropicolinic acid by ammonia-oxidizing bacterium *Nitrosomonas europaea*. *Applied and Environmental Microbiology*, Vol.58, pp. 2321-2325

Vannelli, T. & Hooper, A.B. (1993). Reductive dehalogenation of the trichloromethyl group of nitrapyrin by the ammonia-oxidizing bacterium *Nitrosomonas europaea*. *Applied and Environmental Microbiology*, Vol.59, pp. 3597-3601

Vannelli, T. & Hooper, A.B. (1995). NIH shift in the hydroxylation of aromatic compounds by the ammonia-oxidizing bacterium *Nitrosomonas europaea*. Evidence against an arene oxide intermediate. *Biochemistry*, Vol.34, No.37, pp. 11743-11749

Vannelli, T.; Logan, M.; Arciero, D.M. & Hooper, A.B. (1990). Degradation of halogenated aliphatic compounds by ammonia-oxidizing bacterium *Nitrosomonas europaea*. *Applied and Environmental Microbiology*, Vol.56, pp. 1169-1171

Simultaneous Elimination of Carbon and Nitrogen Compounds of Petrochemical Effluents
by Nitrification and Denitrification

107

Vázquez, I.; Rodríguez, J.; Marañón, E.; Castrillón, L. & Fernández, Y. (2006). Simultaneous removal of phenol, ammonium and thiocyanate from coke wastewater by aerobic biodegradation. *Journal of Hazardous Materials*, Vol.173, pp. 1773-1780

Vereb, H.; Andrea, M.; Dietrich, A.M.; Alfeeli, B. & Agah, M. (2011). The possibilities will take your breath away: breath analysis for assessing environmental exposure. *Environmental Science and Technology*, Vol.45, pp. 8167-8175

Voysey, P.A. & Wood, P.M. (1987). Methanol and formaldehyde oxidation by an autotrophic nitrifying bacterium. *Journal General of Microbiology*, Vol.33, pp.283-290

Waniusiow, D.; Campo, P.; Cossec, B.; Cosnier, F.; Grossman, S. & Ferrari, L. (2008). Toluene-induced hearing loss in acivicin-treated rats. *Neurotoxicology and Teratology*, Vol.30, pp. 154-160

Ward, M.H.; deKok, T.M.; Levallois, P.; Brender, J.; Gulis, G.; Nolan, B.T. & VanDerslice, J. (2005). Workgroup report: drinking-water nitrate and health-recent findings and research needs. *Environmental Health Perspectives*, Vol.113, No.11, pp. 1607-1614

Washer, C.E. & Edwards, E.A. (2007). Identification and expression of benzylsuccinate synthase genes in a toluene-degrading methanogenic consortium. *Applied and Environmental Microbiology*, Vol.74, No.4, pp. 1367-1369

Weelink, A.A.B., van Eekert, H.A. & Stams, A.J.M. (2010). Degradation of BTEX by anaerobic bacteria: physiology and application. *Reviews in Environmental Science and Biotechnology*, Vol.9, pp. 359–385

Whiten, K.W.; Davis, R.E.; Peck, M.L. & Stanley, G.G. (2008). Química. (Octava, Ed.), Cenage Learning Editores S.A. Anexo A14

Whittaker, M.; Bergmann, D.; Arciero, M. & Hooper, A.B. (2000). Electron transfer during the oxidation of ammonia by the chemolithotrophic bacterium *Nitrosomonas europaea. Biochimica et Biophysica Acta*, Vol.1459, No.2-3, pp. 346-355

Wiesmann, U. (1994). Biological nitrogen removal from wastewater. In: *Advances in biochemical engineering/biotechnology*, A. Fiechter, (Ed.), 113-154, Springer, Verlag, Berlin

Wu, Q. & Knowles, R. (1994). Cellular regulation of nitrate uptake in denitrifying *Flexibacter canadensis. Canadian Journal of Microbiology*, Vol.40, pp. 576-582

Yamagishi, T.; Leite, J.; Ueda, S.; Yamaguchi, F. & Suwa, Y. (2001). Simultaneous removal of phenol and ammonia by an activated sludge process with cross-flow filtration. *Water Research*, Vol.35, No.13, pp. 3089-3096

Yerushalmi, L.; Manuel, M.F. & Guiot, S.R. (1999). Biodegradation of gasoline and BTEX in a microaerophilic biobarrier. *Biodegradation*, Vol.10, pp. 341-352

You, J.; Das A.; Dolan, E.M. & Hu, Z. (2009). Ammonia-oxidizing archaea involved in nitrogen removal. *Water Research*, Vol.43, No.7, pp. 1801-1809

Zahn, J.A.; Arciero, D. M.; Hooper, A.B. & DiSpirito, A.A. (1996). Evidence for an iron center in the ammonia monooxygenase from *Nitrosomonas europaea. FEBS Letters*, Vol.397, No.1, (November 1996), pp. 35-38, ISSN 0014-5793

Zepeda, A.; Texier, A.-C. & Gomez, J. (2003), Benzene transformation in nitrifying batch cultures. *Biotechnology Progress*, Vol.19, No.3, pp. 789-793

Zepeda, A.; Texier, A.-C. & Gomez, J. (2007). Batch nitrifying cultures in presence of mixtures of benzene, toluene and m-xylene. *Environmental Technology*, Vol.28, No.3, pp. 355-360

Zepeda, A.; Texier, A.-C.; Razo-Flores, E. & Gomez, J. (2006). Kinetic and metabolic study of benzene, toluene and *m*-xylene in nitrifying batch cultures. *Water Research*. Vol.40, No.8, pp. 1643-1649

Zeyer, J.; Kuhn, E.P. & Schwarzenbach, R.P. (1986). Rapid microbial mineralization of toluene and 1,3-dimethylbenzene in the absence of molecular oxygen. *Applied and Environmental Microbiology*, Vol.52, pp. 944-947

Zumft, W.G. (1997). Cell biology and molecular basis of denitrification. *Microbiology and Molecular Biology Reviews*, Vol.61, pp. 533-616

6

Simulation of Non-Newtonian Fluids Through Porous Media

José Luis Velázquez Ortega and Suemi Rodríguez Romo
Facultad de Estudios Superiores Cuautitlán/UNAM
Cuautitlán Izcalli, Edo. de Méx.
Mexico

1. Introduction

The past two decades have appeared techniques for simulation of hydrodynamic systems (Mc Namara & Zanneti, 1998; Higuera & Jiménez, 1989; Qian et al., 1992; Doolen, 1990; Benzi et al., 1992), magnetohidrodynamics (Chen, 1991), multiphase and multicomponent fluids (Shan & Chen, 1993), included suspensions (Ladd, 1994) and emulsions (Boghosian et al., 1996), flows chemically reactive (Chen et al., 1995) and multicomponent flows through porous media (Landaeta, 1997), only for mention some. This techniques, don't based on directly in the discreet state of hydrodynamic equations, don't ether tally with the microscopic level of molecular dynamic. These techniques have been named techniques of mesoscopic simulations. In general, these discreet techniques involve collisions of "particles" which retain mass, momentum and some cases energy and consistently, give place a hydrodynamic macroscopic behavior. These methods are known like Cellular Automaton and Lattice Boltzmann.

The Lattice Boltzmann Method (LBM), including the method Cellular Automaton (AC), present a powerful alternative to standard approaches known like "of up toward down" and "of down toward up". The first approximation study a continuous description of macroscopic phenomenon given for a partial differential equation (an example of this, is the Navier-Stokes equation used for flow of incompressible fluids); some numerical techniques like finite difference and the finite element, they are used for the transformation of continuous description to discreet it permits solve numerically equations in the computer.

The second approximations study a microscopic description of the particles, through molecular dynamic's equations. Here the position and speed of each atom or molecule in the system are calculated with the solution of the Newton's movement equations.

Between the two approximations, are the LBM and the AC, which are considered mesoscopic approaches that was mentioned previously (Wolf-Gladrow, 2000).

With regard to the LBM, this has his origin en the AC, the first effort for simulate systems of flow of fluids was made in 1973 by Hardy, Pomeau and de Pazzis (*HPP*) (Hardy et al, 1973), whom showed the model into and this is named with their initials "*HPP*". In this model, all particles have mass unit and speed; the particles are limited to travel in directions

$\vec{e}_i, i = 1,...,4$ of the square lattice. A boolean variable is used to indicate the presence or absence of the fluid's particle in each place of lattice in one direction given. A rule of exclusion is applied, is for it, that a particle only permits travel in each direction lengthwise of link for stage. When diverse particles arrive in some e at same place, these collisions in accordance with collision's rules pre established in the model, the collision is conduct for the conservation of mass and momentum. If this collision isn't presenting, then the particles will continue their travel in straight line. The model HPP is absolutely stable, but present absence of isotropy in some tensorials terms involved in Navier-Stokes equation. In the same way, the results of the model don't recuperate the Navier-Stokes equation (NSE) at microscopic level. This insufficiency is due to inappropriate grade of rotational symmetry of the square lattice (Hardy et al, 1973). In 1986, Frisch, Hasslancher and Pomeau (Frisch et al, 1986) developed that NSE can be recuperated using a hexagonal lattice. Their model is known like FHP model. This model is similar to HPP, only those particles with unitary mass and speed they move in the vertex of discreet hexagonal lattice. A Boolean variable is used to indicate the presence or absence of a particle this direction. The particles are updated in each time step, when happen a collision or are flowing to a neighbouring site, depend of the case. The collisions of the particles are determined for a pre inscription the collision's rules that contain all possible state of collisions. This model presents two problems; the presence of statistic noise and the incapacity for simulate fluids in three dimensions.

With regard to the LBM, it has its basis in concepts of the kinetics of gases theory. Is method has been employed for simulate many physical phenomenon with the use of discrete lattice, the Boltzmann equation and computational tools. For example in dynamic of fluids, the Navier-Stokes equations can't solve directly, but if when recover like consequence. The problem is obtaining the function of particle's distribution, $f_i(\vec{x}, t)$ and starting from these obtains observable variables like viscosity, fallen pressure and Reynolds number. In general, the methods of lattice Boltzmann consist in two operations; the first is denoting the advance of particles at the neighbouring site of the lattice. The second operation, is for simulate particle's collisions. An important advantage is that this method is fully parallel and local; too is simple made the programming. In this can incorporate boundary conditions and complexes geometries (Chen et al, 1994).

2. Basis of kinetics theory and Boltzmann equation

Fort the solution of complex problems of dynamics of fluids, exist traditionally two kinds of points of view: the first is macroscopic, which is considered continuous, with an approach of differential equations in partial derivatives, for example of Navier-Stokes equations used for flow of incompressible fluids and numerical techniques for its solution. The second point of view is microscopic it has its basis in kinetics theory of gases and statistical mechanics.

With regard to the kinetics theory of gases, this mentions that a gas contain many particles in interaction α (the order is 10^{23}), the physical state is describing by their positions $\vec{r}^\alpha = \{r_1^\alpha, r_2^\alpha, r_3^\alpha\}$ their speeds $\vec{v}^\alpha = \{v_1^\alpha, v_2^\alpha, v_3^\alpha\}$ during the time t. The micro state is given by the complete set $\{\vec{r}^\alpha, \vec{v}^\alpha\}$ and each particle can be describing through its trajectory in the space of 6-dimensional by complete extension $\vec{r}^\alpha$ and $\vec{v}^\alpha$, it is named *phase space* (Struchtrup, 2005).

This chapter will shortly explain the basics of the Boltzmann equation, the derivation of the Lattice Boltzmann equation and the connection to the Navier-Stokes equations.

2.1 Boltzmann equation

A fluid can consider like a continuous medium when the fluid is dense, this meaning that the particles are behind the other particles and there isn't space between they, in this case the fundamental equations that conduct the fluid's evolution are of kind Navier-Stokes (NE). One form of determinate if the continuous medium is acceptable, is through of Knudsen's number $(Kn=\lambda/l)$, which is defining like the relation between the free mean trajectory λ (mean distance that a molecule travels through before collisions with other molecule) and the characteristic length l; the model continuous medium is acceptable for a rank of $0.01 \leq Kn \leq 1$. In other words the Knudsen's number should be less that unit, in this form the continuous hypothesis will be valid.

In other way in the real fluid, the particles have strong and continuous interactions. In a gas named rarefied, the particles move during many time like free particles, except when collisions occurs of binary body.

When the Knudsen's number is in the rate of $Kn \geq 0.01$, the fluid could be considered like flow of rarefied gas.

The LBM has its basis in the approximation of a fluid like rarefied gas (or diluted gas) of particles. The rarefied gas can be describing by Boltzmann equation that was derivative the first time in 1872 by Ludwig Boltzmann (Duderstadt & Martin, 1979; Schwabl, 2002; Chapman & Cowling, 1970). The NS equations can derive of Boltzmann equation in a limit.

The Boltzmann equation is

$$\frac{\partial f}{\partial t} \quad + \quad v_\alpha \frac{\partial f}{\partial r_\alpha} \quad + \quad F_\alpha \frac{\partial f}{\partial v_\alpha} \quad = \quad \Omega(f) \tag{1}$$

Rate of change of f   Change due to   Change of f due   Collision of
with respect to time   change in velocity   to the body force F_α molecules
which influences the
velocity of the molecule

Any solution about Boltzmann equation needs an expression for collision term $\Omega(f)$. The complexity of it, carry the search of simple models of collision processes, it will permit to make easy the mathematical analysis. Perhaps collision model more known was suggested simultaneously by Bhatnagar, Gross and Krook (Bhatnagar et al., 1954) and it is known like *BGK*:

$$\Omega_{BGK}(f) \approx \upsilon \left[f^{eq}(\vec{v}) - f(\vec{r},\vec{v},t) \right] \tag{2}$$

Where υ is an adjusted parameter and $f^{eq}(\vec{r},\vec{v},t)$ is the local thermodynamic equilibrium of distribution. This simplification is called the *"single-time-relaxation"* approximation, already that the absence of space dependence, is implicating that $f(t) \rightarrow f^{eq}(\vec{v})$ the exponential manner in time like $\exp(-vt)$. This model keeps important properties of collision term in Boltzmann equation. For example, satisfies *theorem H* and obeys the laws of mass, momentum and energy conservation (for this reason, it keeps the structure of hydrodynamic equations), it keeps the collision invariants

The collision term with the model BGK is writing the following form

$$\Omega_{BGK}(f) = \frac{f^{eq} - f}{\tau} \tag{3}$$

Without knowing the form of $\Omega(f)$ there are however several properties which can be deduced. If the collision is to conserve mass, momentum and energy it is required that

$$\int \begin{bmatrix} 1 \\ \bar{v} \\ v^2 \end{bmatrix} \Omega(f) d\bar{v} = 0. \tag{4}$$

The terms $\psi_i, i = 0,...,4$ where $\psi_0 = 1, \psi_1 = v_1, \psi_2 = v_2, \psi_3 = v_3$ and $\psi_4 = v^2$ are frequently called the elementary collision invariants since $\int \psi_i \Omega(f) d\bar{v} = 0$

Any linear combination of the ψ_i terms is also a collision invariant.

The ordinary kinetics theory of neuter gas, the Boltzmann equation is considered with collision term for binary collisions and is despised the body's force F_α. This simplified Boltzmann equations is an integro - differential non lineal equation, and its solution is very complicated for solve practical problems of fluids. However, Boltzmann equation is used in two important aspects of dynamic fluids. First the fundamental mechanic fluids equation of point of view microscopic can be derivate of Boltzmann equation. By a first approximation could obtain the Navier-Stokes equations starting from Boltzmann equation. The second the Boltzmann equation can bring information about transport coefficient, like viscosity, diffusion and thermal conductivity coefficients (Pai, 1981; Maxwell, 1997).

2.2 Equilibrium distribution function and theorem H

The equation (1) with the collision term described for binary collisions, it isn't lineal, is for that reason that the solution is very difficult. Nevertheless, exists a solution for Boltzmann equation, it isn't trivial and is very important and is known like *distribution function Maxwellian*. For this case the Boltzmann equation presents a non reversible behavior and distribution function lays to distribution Maxwellian, this represent the situation of an uniform gas in stationary state.

For derivate the distribution fuction Maxwellian, supposes the absence of external forces F_α and uniform gas, the distribution function is $f(\bar{r}, \bar{v}, t)$ independent of space coordinates $\bar{r}$, i.e, $f = f(\bar{v}, t)$. Whereas these conditions, Boltzmann equation (1) with binary collision term is

$$\frac{\partial f}{\partial t} = 2\pi \iint (\overline{ff}' - ff') g_0 b \ dbdv'. \tag{5}$$

Where takes on that particle is spherical and symmetrical, for that reason the integral with regard to the angle ε can be evaluated immediately.

The function H is

$$H = \int f \cdot \ln f \cdot d\bar{v}. \tag{6}$$

Equation (6) can differentiate with regard to time $\left[\partial H/\partial t = \int (1 + \ln f)\partial f/\partial t d\bar{v}\right]$, in this form replaces $\partial f/\partial t$ of equation (5) for obtains

$$\frac{\partial H}{\partial t} = 2\pi \iiint g_0 b\left(\overline{ff}' - ff'\right)(\ln f + 1)db d\bar{v}' d\bar{v}. \tag{7}$$

If considers inverse collisions, it means particles with speed $\bar{v}$ and $\bar{v}'$ collided and moved outside with speeds $\bar{\bar{v}}$ and $\bar{\bar{v}}'$. The result of these collisions will be

$$\frac{\partial H}{\partial t} = 2\pi \iiint g_0 b\left(\overline{ff}' - ff'\right)(\ln f' + 1)db d\bar{v}' d\bar{v}. \tag{8}$$

If $d\bar{v}d\bar{v}' = d\bar{\bar{v}}d\bar{\bar{v}}'$; addend the equations (7) y (8), and changing the variables $\bar{v} \leftrightarrow \bar{\bar{v}}$ and $\bar{v}' \leftrightarrow \bar{\bar{v}}'$ obtains and divided by four obtains

$$\frac{\partial H}{\partial t} = \frac{\pi}{2} \iiint g_0 b\left(\overline{ff}' - ff'\right)\left[\ln ff'/\overline{ff}'\right]db d\bar{v}' d\bar{v}. \tag{9}$$

Can notice those terms $\left(\overline{ff}' - ff'\right)\ln\left(ff'/\overline{ff}'\right) \leq 0$ and the others terms in the integral of equation (9) are positives then

$$\frac{\partial H}{\partial t} \leq 0 \tag{10}$$

Previous expression is known like *Boltzmann theorem H* and indicates that H can never increase. How H can't increase but whether lays out a limit, that situation is $\partial H/\partial t = 0$. It is possible if and only if, in equation (9), whether $\left(\overline{ff}' = ff'\right)$. This condition is known like *detailed balance* and can be expressed like $\left(\ln \bar{f} + \ln \bar{f}' = \ln f + \ln f'\right)$

Therefore, whether f^{eq} is an equilibrium distribution, then $\ln f^{eq}$ is an invariant of the collision and it could be $\left\{\ln f^{eq} = \sum_i \alpha_i \psi_i; \quad i = 0,1,...,4\right\}$. Where ψ_i are invariant collisions (equation 4) and α_i is a constant. It writes again like

$$\ln f^{eq} = \ln(\alpha_0') - \alpha_4' \frac{1}{2} m\left\{\left[v_1 - \left(\frac{\alpha_1'}{\alpha_4'}\right)\right]^2 + \left[v_3 - \left(\frac{\alpha_3'}{\alpha_4'}\right)\right]^2\right\}, \tag{11}$$

where $\alpha_0' = \exp(\alpha_0), \alpha_1' = \alpha_1/m, \alpha_2' = \alpha_2/m, \alpha_3' = \alpha_3/m$ *and* $\alpha_4' = 2\alpha_4/m.$

with $\bar{V} = (\bar{v} - \bar{\alpha}')/\alpha_4'$ in which $\bar{\alpha}' = (\alpha_1', \alpha_2', \alpha_3')$ it can write like

$$f^{eq} = \alpha'_0 e^{-\alpha'_4 \frac{1}{2}mV^2}.$$ (12)

The equation (12) is Maxwell distribution function for a gas and describes the *equilibrium state of distribution function* f. The constants can find if replace equation (12) whit conservation laws obtains the common form of Maxwell's distribution function

$$f^{eq} = \frac{\rho}{m}\left(\frac{m}{2\pi k_B T}\right)^{3/2} \exp\left[\frac{-m(\vec{v}-\vec{u})^2}{2k_B T}\right].$$ (13)

The theorem H shows that any distribution function f lays out its equilibrium state f^{eq} for long times.

If a system is not uniform, it is not in thermodynamic equilibrium then can obey law of Maxwell's speeds distribution. However, if "equilibrium absence" is not big, can considered like good approximation, all little volume (microscopic scale) is in equilibrium (considered like subsystem). This is for two reasons. First little portions of gas contain a big number of molecules. Second the necessary time for established the equilibrium in a little volume is brief in comparison with necessary time for that transport processes get equilibrium in little volume with rest of system (it is true when concentration, temperature, etc. gradients are not too much big). In consequence, can suppose that is *local thermodynamic equilibrium* so speed distribution in any volume element (macroscopic) of medium is Maxwellian, although density, temperature and macroscopic velocity change the position (Duderstadt & Martin, 1979; Schwabl, 2002; Bhatnagar et al., 1954; Pai, 1981; Maxwell, 1997; Succi, 2001; Succi, 2002; Cercignani, 1975; Lebowitz & Montroll, 1983).

2.3 From the Boltzmann equation to the lattice Boltzmann equation

Lattice Boltzmann equation can be obtain through two ways, first is through of "cellular automaton" and second starting from Boltzmann equation, it was review previously, for carries out derivation of Boltzmann's lattice equation is necessary the *space time discretization*. Immediately presents brief description of second way, it shows by pace series.

The derivation begins when the Boltzmann equation with the model BGK is writing like

$$\frac{\partial f}{\partial t} + \vec{\xi} \cdot \nabla f = -\frac{1}{\lambda}(f-g).$$ (14)

In the equation (14), $f \equiv f(\vec{x},\vec{\xi},t)$ is distribution function of only particle, $\vec{\xi}$ is microscopic velocity, λ is the relaxation time due to collision, and g is the Boltzmann-Maxwellian distribution function (fM), is important mention that collision term has been transforming in accordance with equation (2).

Hydrodynamic properties of fluid, density ρ, velocity $\vec{u}$ and temperature T, could be calculated starting from momentums function f. For quantify the fluid's temperature, uses the energy density $\rho\varepsilon$ (He & Luo, 1997).

$$\rho = \int f(\vec{x},\vec{\xi},t)d\vec{\xi},$$ (15)

$$\rho\vec{u} = \int \xi\, f(\vec{x},\vec{\xi},t)\,d\vec{\xi}, \tag{16}$$

$$\rho\varepsilon = \int \tfrac{1}{2}(\vec{\xi}-\vec{u}) f(\vec{x},\vec{\xi},t)\,d\vec{\xi}. \tag{17}$$

2.3.1 Time discretization

Equation (14) it formulates in form of ordinary differential equation (ODE)

$$\frac{df}{dt} + \frac{1}{\lambda}f = \frac{1}{\lambda}g, \tag{18}$$

Where $\dfrac{d}{dt} \equiv \dfrac{\partial}{\partial t} + \vec{\xi}\cdot\vec{\nabla}$ is substantial derivative length wise of microscopic velocity $\vec{\xi}$. Equation (18) that mentioned is a lineal ODE of first order, which can be integrated over a time step of δt. Using the integral factor method is obtaining

$$f\left(\vec{x}+\vec{\xi}\delta_t,\vec{\xi},t+\delta_t\right) = e^{-\delta_t/\lambda}\frac{1}{\lambda}\int_0^{\delta t} e^{t'/\lambda}g\left(\vec{x}+\vec{\xi}t',\vec{\xi},t+t'\right)dt' + e^{-\delta_t/\lambda}f\left(\vec{x},\vec{\xi},t\right) \tag{19}$$

Assuming that δ_t is small enough and g is smooth enough locally, and neglecting the terms of order of $O(\delta_t^2)$ or smaller in the Taylor expansion of the right hand side of the equation (19) we obtain

$$f\left(\vec{x}+\vec{\xi}\delta_t,\vec{\xi},t+\delta_t\right) - f\left(\vec{x},\vec{\xi},t\right) = -\frac{1}{\tau}\left[f\left(\vec{x},\vec{\xi},t\right)-g\left(\vec{x},\vec{\xi},t\right)\right] \tag{20}$$

In the equation (20) $\tau \equiv \lambda/\delta_t$ is the dimensionless relaxation time; this equation is very similar to lattice Boltzmann equation. For obtaining the equation is necessary the space velocities discretization, too equilibrium function g it could be consistent with Navier-Stokes equations. The equation (20) is the evolution of distribution function f with discreet time (He & Luo, 1997; Maxwell, 1997).

2.3.2 Approximation of equilibrium distribution

A point of view very import, is the obtaining of equilibrium distribution function, it was mentioned before, of this depends obtaining appropriately the Navier-Stokes equations. Maxwell's distribution that is employing like equilibrium distribution g for unitary particle mass and "D" dimensions is

$$g(\vec{u}) = \frac{\rho}{(2\pi RT)^{D/2}}\exp\left[-\frac{\left(\vec{\xi}-\vec{u}\right)^2}{2RT}\right], \tag{21}$$

It could be expressed the following form

$$g(\vec{u}) = \frac{\rho}{(2\pi RT)^{D/2}}\exp\left(-\frac{\vec{\xi}^2}{2RT}\right)\exp\left(\frac{\vec{\xi}\cdot\vec{u}}{RT} - \frac{\vec{u}^2}{2RT}\right). \tag{22}$$

The function "g" expands through of Taylor series in $\bar{u}$ until third order and is necessary eliminate these

$$g(0,\bar{u}) = \frac{\rho}{(2\pi RT)^{D/2}} \exp\left(-\frac{\vec{\xi}^2}{2RT}\right)\left[1 + \frac{\vec{\xi}\cdot\bar{u}}{RT} + \frac{\left(\vec{\xi}\cdot\bar{u}\right)^2}{2(RT)^2} - \frac{\bar{u}^2}{2RT}\right].$$
(23)

For little velocities (or little Match numbers), this is an exact approximation.

If $g(0,\bar{u}) = f^{(eq)}$, obtains the following equation, it will be using like local equilibrium distribution in nest derivatives (He & Luo, 1997; Maxwell, 1997; Wilke, 2003).

$$f^{(eq)} = \frac{\rho}{(2\pi RT)^{D/2}} \exp\left(-\frac{\vec{\xi}^2}{2RT}\right) \times \left[1 + \frac{\left(\vec{\xi}\cdot\bar{u}\right)}{RT} + \frac{\left(\vec{\xi}\cdot\bar{u}\right)^2}{2(RT)^2} - \frac{\bar{u}^2}{2RT}\right].$$
(24)

2.3.3 Discretization of the velocities

For discretization of the velocities, it will be $-\infty$ a $+\infty$ in both directions "x" and "y" for specific case in two-dimensional model (D2Q9), it will unroll in this part of chapter. The particle momentums of distribution function are very important, because of this depends the consistence of (N-S) equations. in the same way, the isotropy is keeping during the discretization, it is an important property in the symmetry of NE equations, of this form, lattice will be invariant for problem rotations.

Is important mention that an isothermic model is only necessary first momentum. These can be describing by the equation (24) two dimensions like

$$I = \int \psi(\vec{\xi}) f^{(eq)} d\vec{\xi} = \frac{\rho}{(2\pi RT)^{D/2}} \int \psi(\vec{\xi}) \exp\left(-\frac{\vec{\xi}^2}{2RT}\right)$$
$$\times \left[1 + \frac{\vec{\xi}\cdot\bar{u}}{RT} + \frac{\left(\vec{\xi}\cdot\bar{u}\right)^2}{2(RT)^2} - \frac{\bar{u}^2}{2RT}\right] d\vec{\xi}.$$
(25)

The integral of equation (25) has the next form $\int e^{-x^2}\psi(\vec{\xi})dx$, where $\psi(\vec{\xi})$ is the momentum function, it contains powers of velocities components. For recuperate model D2Q9 for square lattice is using the system of Cartesian coordinates and $\psi(\vec{\xi})$ is

$$\psi_{m,n}(\vec{\xi}) = \xi_x^m \xi_y^n,$$
(26)

in previous equation, ξ_x y ξ_y are components of "x" and "y" of velocity $\vec{\xi}$. The integration of equation (25) using these values gives us the following equation for the equilibrium distribution function for two dimensional, 9- velocity LBE model:

$$f_\alpha^{eq} = w_\alpha \rho \left\{ 1 + \frac{3(\vec{e}_\alpha \cdot \vec{u})}{c^2} + \frac{9(\vec{e}_\alpha \cdot \vec{u})^2}{2c^4} - \frac{3\vec{u}^2}{2c^2} \right\}. \tag{27}$$

The w_α are $w_0 = \dfrac{4}{9}$; $w_{1,2,3,4} = \dfrac{1}{9}$; $w_{5,6,7,8} = \dfrac{1}{36}$. and microscopic velocities

$$\vec{e}_0 = (0,0)^T,$$

$$\vec{e}_{1,2,3,4} = \zeta_{1,2}, \zeta_{2,1}, \zeta_{3,2}, \zeta_{2,3} = (\pm 1, 0)^T c, (0, \pm 1)^T c,$$

$$\vec{e}_{5,6,7,8} = (\pm 1, \pm 1)^T c.$$

Here is $c = \sqrt{3RT}$ where c is the sound speed of the model (ThÄurey, 2003).

2.4 From lattice Boltzmann equation to the Navier-stokes equation

The popular problems of kinetics theory is the derivation of hydrodynamic equations, in certain conditions, solution of $f(\vec{r}, \vec{v}, t)$ transport equation is similar the form that can relate directly to continuous or hydrodynamic description. In certain conditions the transport process is like hydrodynamic limit. In 1911 David Hilbert was who proposed the existence Boltzmann equations solutions (named normal solutions), and these are determinate by initial values of hydrodynamic variables it return to collision invariant (mass, momentum and kinetics energy), Sydney Chapman and David Enskog in 1917 were whose unrolled a systematic process for derivate the hydrodynamic equations (and their corrections of superior order) for these variables.

In spite of have been proposed many approximated solutions to Boltzmann equation (including the Grad's method of 13 moments, expansions of generalized polynomial, bimodal distributions functions), however the Chapman-Enskog is the most popular outline for generalize hydrodynamic equations starting from kinetics equations kind Boltzmann (James & William, 1979; Cercignani, 1988).

2.4.1 Chapman-Enskog expansion

For show that ERB can use for describing the fluid's behavior, NS equations are derivate by process are named *Chapman-Enskog's expansion* or *multi-scale analysis*. It depends of *Knudsen's number* it was mentioned at the first part of this chapter; it is the relation between the *free mean trajectory* and the *characteristic length*.

For derivation of NS's equation, the Boltzmann equation divides in different scales for space and time variables. It has his basis in the expansion of parameterε, it is necessary for using the Knudsen's number. In general the expansion is truncating after second order terms. The following representation is for temporal variables

$$t = \varepsilon t_0 + \varepsilon^2 t_1 \tag{28}$$

Time *t* represents local relaxation that is very quickly in fluid for collisions. The diffusion processes are time scale *t₁*. In this only space expansion is considerate, it is represented in the next first order expansion.

$$\vec{x} = \varepsilon \vec{x}_1 \tag{29}$$

The advection and diffusion are considered similar in space scale "x_1". While, the representation of differential operator is similar

$$\frac{\partial}{\partial x_\alpha} = \varepsilon \frac{\partial}{\partial x_\alpha} \quad ; \quad \frac{\partial}{\partial t} = \varepsilon \frac{\partial}{\partial t} + \varepsilon^2 \frac{\partial}{\partial t}. \tag{30}$$

For consistent expansion, is necessary the second order term in space. Momentum equations of f are expanded directly an addition the next form

$$f = \sum_{n=0}^{\infty} \varepsilon^n f^n. \tag{31}$$

The dependence of time f is on account of variables ρ, $\ddot{u}$ and T.

The NS equation can be recovered starting from analysis of lattice Boltzmann equation

$$f_i(\vec{x} + \vec{e}_i, t+1) - f_i(\vec{x}, t) = -\frac{1}{\tau}(f_i(\vec{x}, t) - f_i^{eq}(\vec{x}, t)). \tag{32}$$

Expanding Equation (32) in both space and time up to second order yields

$$\left\{ \left[(\varepsilon \partial_{t0} + \varepsilon^2 \partial_{t1}) + (e_{i\alpha}) \right] + \frac{1}{2} \left[(\varepsilon \partial_{t0} + \varepsilon^2 \partial_{t1}) + (e_{i\alpha}) \right]^2 \right\} \left(f_i^{(0)} + \varepsilon f_i^{(1)} + \varepsilon^2 f_i^{(2)} \right) =$$
$$-\frac{1}{\tau} \left(f_i^{(0)} + \varepsilon f_i^{(1)} + \varepsilon^2 f_i^{(2)} - f_i^{(eq)} \right). \tag{33}$$

The three scales from O (ε^0) to O (ε^2) can be distinguished in Equation (33), and are handled separately. In the following, subsequent expansions of the conservation equations will be performed. Giving as result the continuity equation to ε^0,

$$\partial_{t_0}\rho + \partial_\alpha(\rho u_\alpha) = 0. \tag{34}$$

To ε^1, equation (33) *Eule'r equation*. Where $p = (1/3)c^2\rho$.

$$\partial_t(\rho u_\alpha) + \partial_\beta \rho u_\alpha u_\beta = -\partial_\alpha p. \tag{35}$$

For the hydrodynamics of a liquid with viscosity "ε^2", obtaining

$$\partial_{t_0}\rho u_\beta + \partial_\beta(\rho u_\alpha u_\beta) = -\partial_\alpha p + \boxed{\left(\tau - \frac{1}{2}\right)\frac{1}{3}c^2}\rho(\partial_\alpha u_\beta + \partial_\beta u_\alpha). \tag{36}$$

In the above equation $\upsilon = \frac{1}{3}\left(\tau - \frac{1}{2}\right)c^2$ is the kinematic viscosity

$$\therefore \partial_t \rho u_\alpha + u_\beta \partial_\beta u_\alpha = -\frac{1}{\rho}\partial_\alpha p + \upsilon \nabla^2 u_\alpha. \quad \text{Eq. Navier-Stokes.} \tag{37}$$

(He & Luo, 1997; ThÄurey, 2003).

3. The model

The Lattice Boltzmann Method (LBM), its simple form consist of discreet net (lattice), each place (node) is represented by unique distribution equation, which is defined by particle's velocity and is limited a discrete group of allowed velocities. During each discrete time step of the simulation, particles move, or hop, to the nearest lattice site along their direction of motion, where they "collide" with other particles that arrive at the same site. The outcome of the collision is determined by solving the kinetic (Boltzmann) equation for the new particle-distribution function at that site and the particle distribution function is updated (Chen & Doolen, 1998; Wilke, 2003). Specifically, particle distribution function in each site $f_i(\bar{x},t)$, it is defined like a probability of find a particle with direction velocity $\bar{e}_i$. Each value of the index i specifies one of the allowed directions of motion (Chen et al., 1994; ThÄurey, 2003).

For this work we use D2Q9 and periodic boundary conditions in the inflow and outflow plane and non-slip boundary (bounce-back) conditions on the walls and the porous matrix. Bounce-back conditions were used whenever the fluid hit a node of the porous matrix. Our porous media is represented by blocks that are projections in the plane of actual three dimensional geometries Stability is improved by considering the porous matrix as made out of these blocks and makes the code less noisy as well. To initialize the lattice, a constant body force (F) is used and acts during the simulations, which physically corresponds to a constant pressure gradient. In this work we focus only on externally applied pressure; namely we deal with pressure-driven flows.

Fluid flow in porous media is modeled here by using a modification of the Lattice Bhatnagar-Gross-Krook (LBGK) technique (eq. 32).

Fluid density and velocity are calculated as follows: $\rho = \sum_{i=0}^{b} f_i$; $\rho u = \sum_{i=0}^{b} f_i \cdot e_i$. The method used in this work involves a modification of the LBGK, where instead of constant viscosity, the effective viscosity is used directly as a rate-dependent relaxation time parameter avoiding the calculation of the matrix elements for Ω_{ij} (Boek et al., 2003; Ahronov & Rothman, 1993; Rakotomala et al., 1996). τ Is the relaxation time provided by the kinematic fluid viscosity.

$$\tau = \frac{1}{2} + 3k\left(\frac{dv_x}{dy}\right)^{n-1} \tag{38}$$

To test our algorithm, simulations were run for fluids in a rigid pipe with n = 0.33, 0.56, 1.0, 2.0 and k = 0.001, 0.005, 0.5, 10.0 respectively, in a 200×200 lattice. The steady-state velocity profiles were reached after 30,000 times step. Dimensionless velocity profiles from the LBM and analytical solutions are shown in Figure 1. Our numerical results and analytical solutions are within a 0.9% error (the maximum value being for n>1).

Next step in our model is the simulation of the fluid flow in porous materials (or packed beds). These materials have a portion of space occupied by heterogeneous or multiphase matter.

Velocity profiles

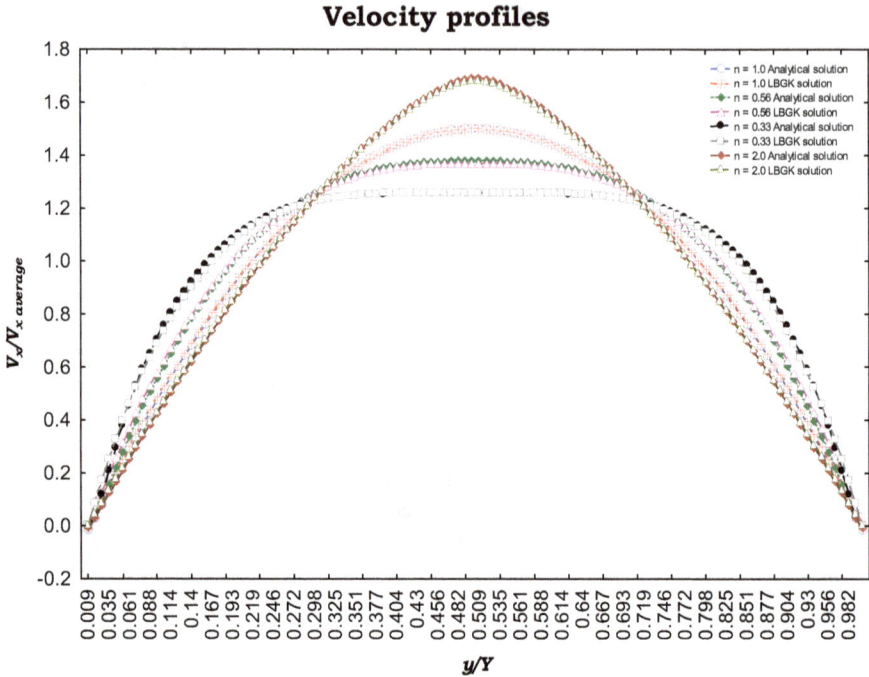

Fig. 1. LBGK velocity profiles and Analytical Navier Stokes velocity profiles in a rigid pipe.

In studying flows through porous media, we extend the LBM so as to study arbitrary and random approaches. Even at a first glance, many porous media, particularly the natural ones, show that the distribution of pores is random (Christakos & Hristopulos, 1997), so randomness is a really important factor to describe natural and man-made porous materials; they usually possess formidably complicated architecture. In modeling, reasonable idealizations have to be assumed (Dullien, 1992; Gibson and Ashby, 1988; Telega and Bielski, 2003).

Simple predictions concerning the flow of non-Newtonian fluids through porous media (Bird et al., 1960) have long been provided. Though Darcy's law for the flow of Newtonian fluids through a porous medium agrees with a model of parallel tubes, analogous behavior is not obvious for non-Newtonian fluids. This model does not take into account the tortuosity of the flow path among others; an additional nonlinearity in this case. As a first step, by assuming that a porous medium can be viewed as a collection of parallel pipes, the relation between flux and force for a non Newtonian fluid can be approached by

$$q = C \left(\frac{dP}{dx} \right)^{1/n} \tag{39}$$

where q is the volumetric rate of flow per unit area. C is a proportionality constant, function of the consistency index k, the porosity ε and the effective permeability K. For $n = 1$, equation (39) reduces to Darcy's law. In practically all cases, eq. (39) is valid as long as the generalized Reynolds number; $Re_{gen} = \left\{ \left[(2\varepsilon/3(1-\varepsilon))D_\rho \right]^n (\bar{v}/\varepsilon)^{2-n} \rho \right\} \Big/ \left[8^{n-1}k(3n+1/n)^n \right]$

does not exceed some value between 1 and 10. Here, D_ρ is a length dimension scale proposed by Collins (Geankoplis, 2000).

Our LBM are such that all experiments performed in this work are in the range $Re_{gen} < 1$ and can provide some output in the relative importance of the different parameters (for the fluid, and the geometrical structure) with statistical validation. We fully comply with the empirical macroscopic scale relation (39) obtained experimentally.

Two different types of porous media are studied here: a) *Arbitrarily generated porous media*. In this case, the authors propose a set of porous media totally constructed by arbitrary choice, fulfilling the following important conditions. At first percolation is guaranteed. Secondly, well defined, interconnected channels are constructed and finally low tortuosity is provided.

b) *Randomly generated porous media*. In this case, the medium is generated by means of a random configuration. The Box-Muller method for generating standard Gaussian pseudo-random numbers is used to obtain the positions of the seeds in the solid matrix, 3 X 3 lattice sites nodes are defined as blocks around each seed; thus providing the porous matrix. This simplified geometry substitutes the projections on a plane of more complex and realistic cases (throats channels, chairs, cylinders, etc) improve stability and avoids noisy results.

The core of the Box-Muller method is a transformation that takes as inputs random variables from one distribution and as outputs random variables in a new distribution function. It allows us to transform uniformly distributed random variables to a new set of random variables with a Gaussian distribution. We start with two independent random numbers, x_1 and x_2, which come from a uniform distribution (in the range from 0 to 1). Then, apply the transformations $y_1 = \sqrt{-2\ln x_1} \cos(2\pi x_2)$ and $y_2 = \sqrt{-2\ln x_1} \sin(2\pi x_2)$ to get two new independent random numbers that have a Gaussian distribution with zero mean and a standard deviation of one. This method of generating random porous media produces similar pore space characterization that several natural media.

Moreover, as it is well known for a Newtonian fluid that standard Darcy law can be used in order to obtain permeability. Namely

$$K = \mu v \Big/ \left(\frac{dP}{dx} \right) \tag{40}$$

Here K is permeability and μ is the viscosity for a Newtonian fluid. In the case of a non-Newtonian fluid with power law viscosity, we introduce the apparent viscosity (instead of the standard viscosity) which is given by the following equation.

$$\mu_{app} = k(dv_x/dy)^{n-1} \tag{41}$$

Thus, effective permeability for non-Newtonian fluids (with power-law viscosity) is calculated in this work as follows

$$K = \mu_{app} v \big/ (dP/dx) = k (dv_x/dy)^{n-1} v \big/ (dP/dx) \qquad (42)$$

4. Local effective permeability

We work with porosities 47.4%, 51.9%, 68.7% and 71.9%, randomly and arbitrarily generated, outputs of our experiments are presented (see figures 2-4). This is done for $n=2$, $n=1$, $n=0.529$ and different pressure gradients. Actually, if $n=1$, the permeability bands do not depend on the pressure gradients as expected in natural systems for porosities 50% or larger. In these Figures, the pore space characterization is shown at the upper left corner of the figure. Secondly, right below, the flow paths in each particular case are shown (from the numerical data, vuggy zones can be detected, among others). Finally, in the right part of each figure, the permeability bands produced by our LBM are introduced. Although qualitative in part, important contributions of our paper are the correlations of the band structures with different pressure gradients (F), fluid parameters (n, the index number), geometrical factors (porosity and tortuosity), and randomness.

In Figure 5, we explore the role of different porosities and pressure gradients in arbitrarily generated porous media, for a non-Newtonian fluid ($n=0.529$). The progressive introduction of high permeability bands as a natural consequence of the geometry, fluid, and flow parameters shows interesting patterns, as can be observed.

Uncertainty is well recognized as an important factor in natural porous media (in fact, it is well known that effective permeability is subject to greater uncertainty than porosity) and numerous studies have employed random methods to model Newtonian flow in subsurface porous media by assuming a given effective permeability probability density function.

It is important to construct models that are able to closely mimic the heterogeneity of actual porous media and sufficiently efficient to allow simulation of flow and transport phenomena. To predict the network flow at core scale (for instance, in hydrocarbon reservoirs, packed beds and aquifers), we propose to construct the permeability probability distribution within our model. Our results are shown in Table 2 in this work. Although it is widely understood that the selection of a particular effective permeability probability density will markedly influence simulation results in applications, only a few studies (this paper, among others) describe the manner in which to construct these effective permeability probability density functions from mesoscale information.

We correlate the effect of randomness, porosity, and fluid parameters with permeability fields and probability distributions predicted by our model, thus improving our understanding of heterogeneous media for applications to natural systems. Normally, at core scales for reservoirs and aquifers; among others, the unknown permeability distribution in the subsurface on all length scales is much needed for practical goals (Sitar et al., 1987; Cooke et al., 1995).

Normally, first step in estimating permeability from thin-sections of natural media is to convert representative digital images into binary images. These binary versions are used for porosity estimation and as conditional data for the stochastic pore-structure realizations.

Flow simulations are then conducted on each pore-structure realization. Our approach then provides the local effective permeability estimate for each bed type (see for instance Figures 2 to 5).

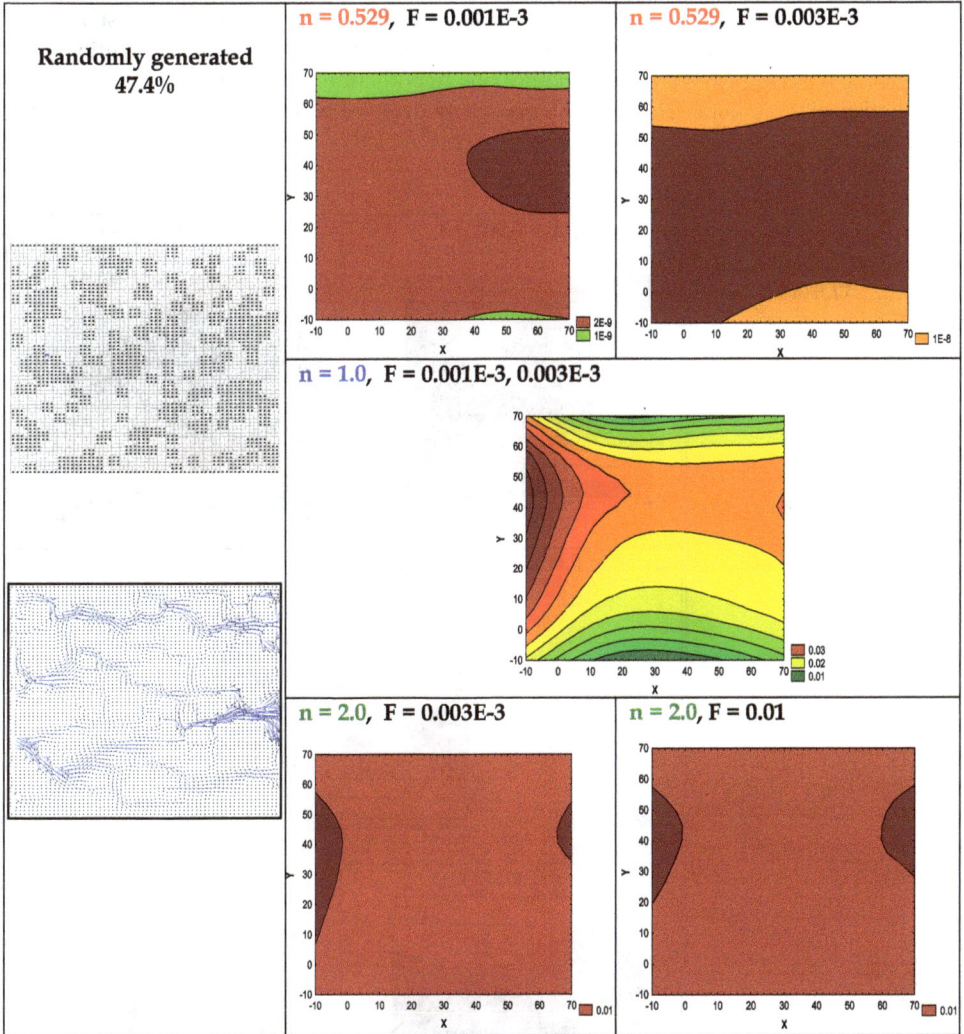

Fig. 2. Randomly generated porous media, porosity 47.4%, flow paths (left) and permeability bands in lattice units (right) for non-Newtonian fluids, n=0.529, n=2 and Newtonian fluids. Here F represents the pressure gradient.

We perform experiments for non-Newtonian fluids and obtain permeability bands (see Figures 2-5) for the following power-law fluids A) n = 0.529, B) n = 1 and C) n = 2. This is done in the following cases: a) arbitrarily generated porous media and b) randomly generated porous media, porosities 47.4%, 51.9%, 68.7% and 71.9%, and a number of pressure gradients, see Figures 2 to 5. Also arbitrarily generated porous media, flow paths, and the obtained permeability fields for a number of porosities, pressure gradients and n=0.529 are introduced in Figure 5. We then perform a null hypothesis analysis over the data by Kolmogorov-Smirnov contrast, obtaining the best fitting effective permeability distribution for each case, as shown in Table 2. Kolmogorov-Smirnov contrast (a non-parametric null hypothesis analysis) is a technique where two data distributions are compared (the one to be tested and another one hypothetically true) and they are accepted to be statistically the same, provided the maximum distance between both is below a certain threshold.

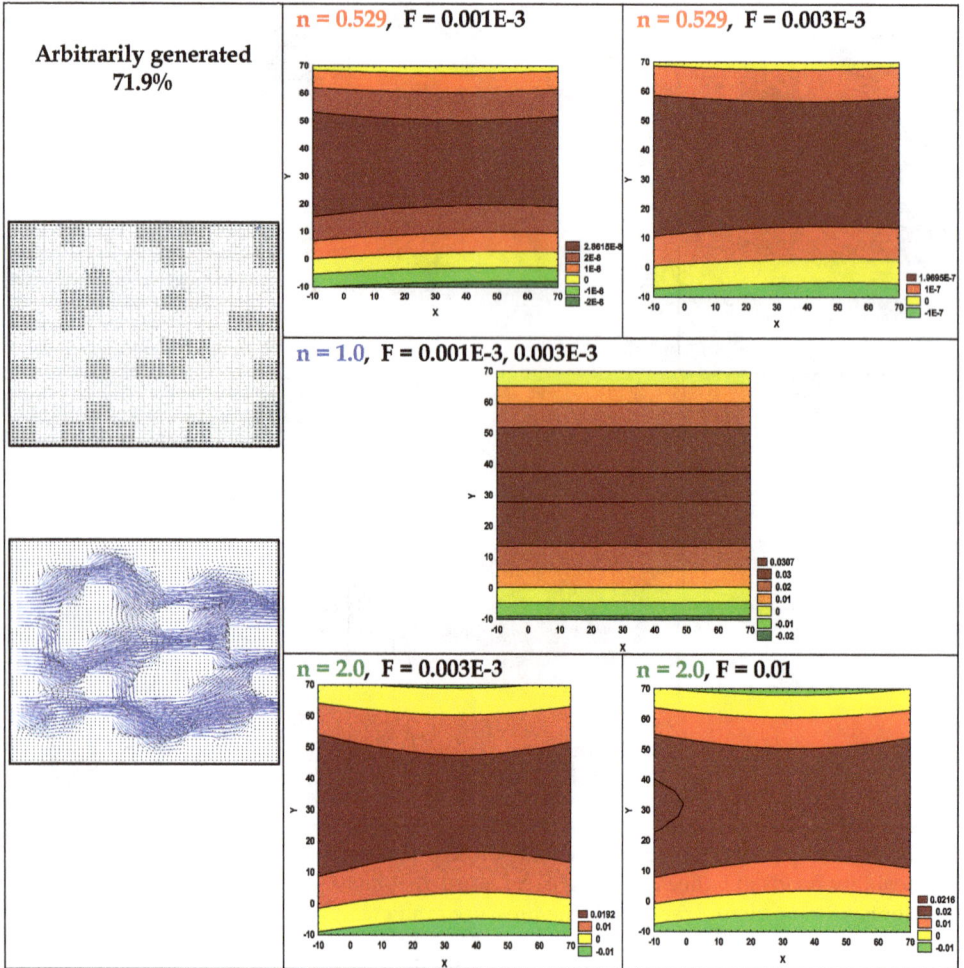

Fig. 3. Arbitrarily generated porous media, porosity 71.9%, flow paths (left) and permeability bands in lattice units (right) for non-Newtonian fluids, n=0.529, n=2 and Newtonian fluids. Here F represents the pressure gradient.

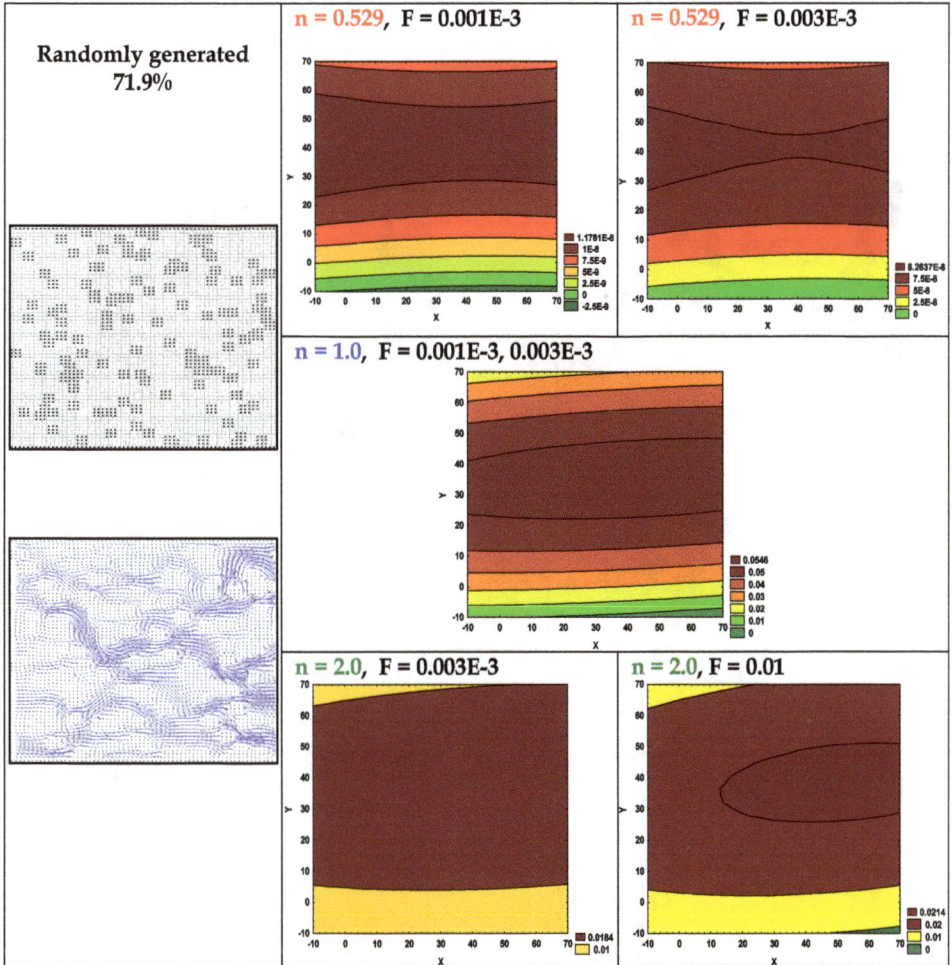

Fig. 4. Randomly generated porous media, porosity 71.9%, flow paths (left) and permeability bands in lattice units (right) for non-Newtonian fluids, n=0.529, n=2 and Newtonian fluids. Here F represents the pressure gradient.

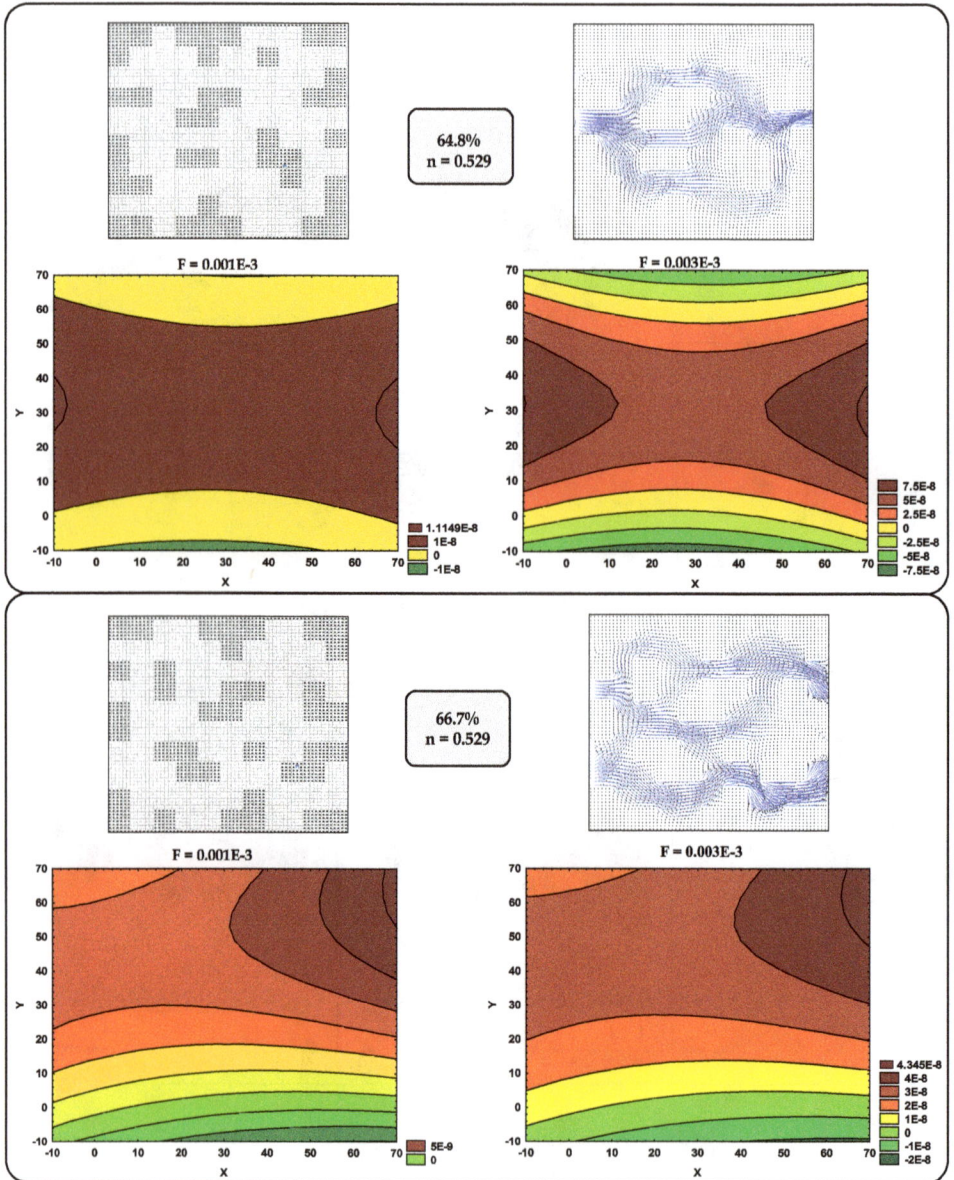

Fig. 5. Arbitrarily generated porous media, flow paths and permeability bands in lattice units for non- Newtonian fluid (n=0.529) and different pressure gradients (F) are shown. We study porosities 64.8, 66.6%, 68.7%, 70.1%, 71.9%.

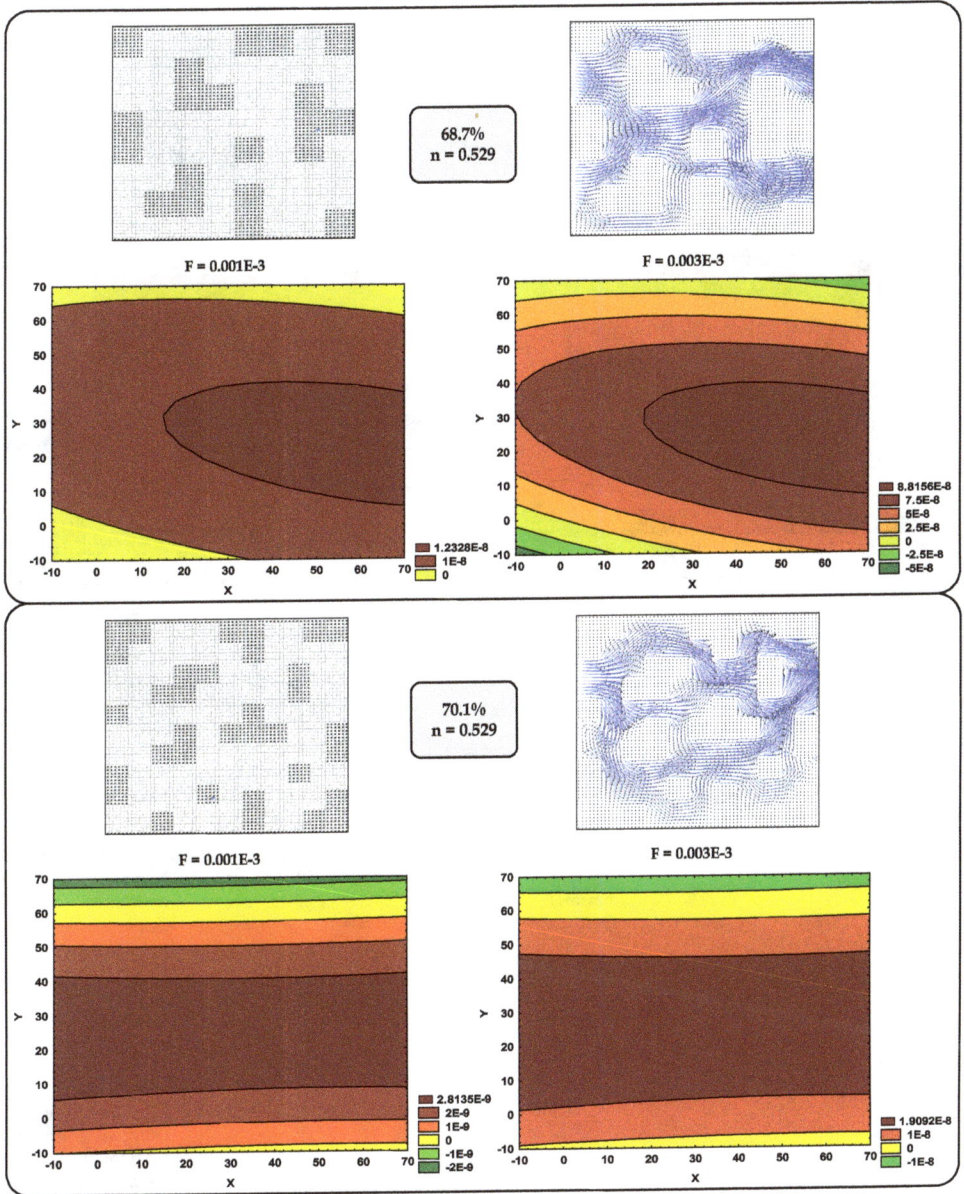

Fig. 5. (Continued)

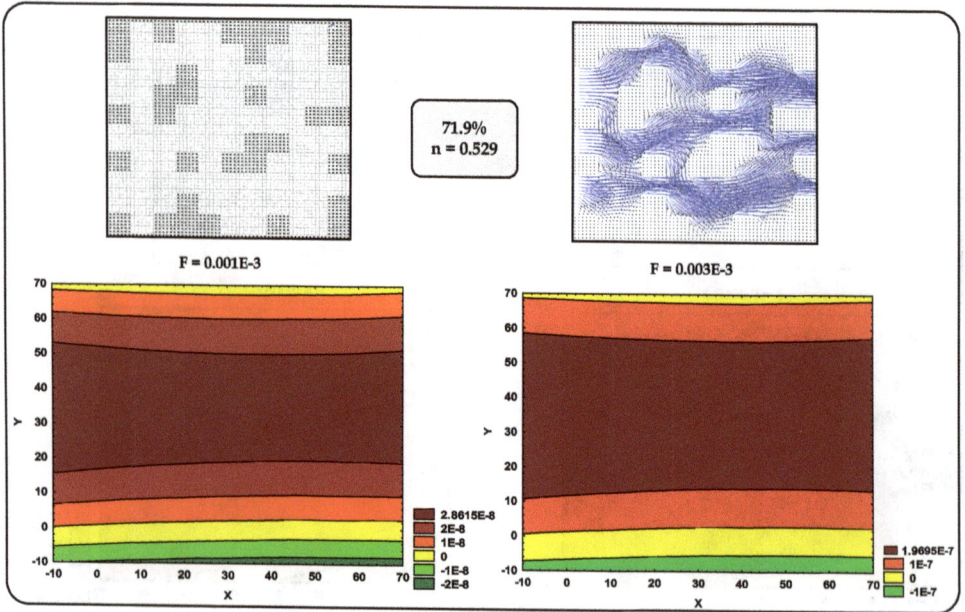

71.9%
n = 0.529

F = 0.001E-3   F = 0.003E-3

Fig. 5. (Continued)

Deterministic porous media		Ramdom porous media	
Porosity 68.75%	Probability distribution	Porosity 68.69%	Probability distribution
n = 0.529	Exponential	n = 0.529	Normal
n = 1.0	Normal	n = 1.0	Normal
n = 2.0	Gamma	n = 2.0	Gamma
Porosity 71.87%		Porosity 71.87%	
n = 0.529	Exponential	n = 0.529	Normal
n = 1.0	Weibull	n = 1.0	Normal
n = 2.0	Extreme	n = 2.0	Weibull
Ramdom porous media			
Porosity 47.37%	Probability distribution	Porosity 51.91%	Probability distribution
n = 0.529	Normal	n = 0.529	Gamma
n = 1.0	Normal	n = 1.0	Normal
n = 2.0	Normal	n = 2.0	Gamma

Table 1. Best fitting permeability probability distributions for number of experiments.

Porosity (%)	n Experimental	Average relative error (%)
10.18	n = 0.529	6.8359
	n = 1.0	0.0
	n = 2.0	1.3514
20.84	n = 0.529	3.8167
	n = 1.0	0.0
	n = 2.0	1.3514
30.82	n = 0.529	2.9907
	n = 1.0	0.0
	n = 2.0	1.0457
47.37	n = 0.529	2.3985
	n = 1.0	0.0
	n = 2.0	0.8065
51.91	n = 0.529	1.5123
	n = 1.0	0.0
	n = 2.0	1.0101

Table 2. Average relative error as a porosity function and the experimental value of n in our method.

5. Conclusions

The 2-D permeability variations considered in this work are applicable to real contaminated aquifers where the length and width of high-permeability inclusions are large relative to their height (or thickness). For example, fluvial deposits as "cut-and-fill" (e.g., scour pool and dissection elements) and accretionary (e.g., horizontally bedded gravel sheets) elements can have high length- and width-to-thickness ratios. The 2-D permeability variations considered in this work are not applicable to more complex contaminated aquifers, e.g., where high-permeability inclusions do not have large length- and width-to-thickness ratios, or where tortuous 3-D flows cause streamlines to twist.

Effective permeability distributions have been used to study geological phenomena for petroleum technology, among others. In this paper, a study of how different geometrical features and rheological parameters affect local permeability of a porous medium has been conducted in order to enhance the knowledge of the local permeability distribution and its importance when developing global permeability models (Lundström et al.,2004; Lundström & Norlund, 2005). We conclude that randomness, porosity, and the power law index for non-Newtonian fluids play an important role in the local effective permeability distribution obtained for each experiment (Normal, Gamma, or any other). Our detailed results are presented for a range of porosity from 47.4% to 71.9% in this paper. We are able of using our approach for porosities as low as 10% , but then the average relative error on n (comparing the experimental with the value obtained by LBM) can reach a maximum of 6.8% (for $n=0.529$). See Table 2.

We consider several heterogeneous porous media representations to illustrate flow performance; all of them are two-dimensional. A number of high-permeability inclusions within low-permeability zones are obtained from a variety of fluid characterizations (see Figures 2-5). The effect of porosity, randomness, fluid and flow parameters are correlated with permeability bands in a number of cases; original outputs shown in this paper.

It is intriguing to see how Newtonian fluids always produce symmetrical bands and high permeability bands included into small permeability bands with several layers provided porosity is large. We can also remark that for $n<1$, bands of low permeability are more numerous than in the case of $n>1$. Some throats can be formed if we introduce randomness and high porosity, see Figure 4, in the case of $n<1$ and high pressure gradients. Additional anisotropy appears for small porosity, see Figures 2.

Streamlines determined by numerical simulation are obtained and shown in Figures 2-5. From these, and our numerical results (rough input data to construct Figures 2-5), it is clear that flow is mainly parallel and uniformly distributed in the low-permeability zone. However, in the region containing the high-permeability inclusion there is very little flow in the low-permeability zone and almost all flow is focused in the high-permeability inclusion. In other cases, a number of high-permeability inclusions within a low-permeability zone are produced. In regions between high-permeability inclusions, most flow is uniformly distributed, see Figure 5. In regions containing high-permeability inclusions, most flow is focused in these inclusions and is parallel to the boundaries. From our numerical results and the path flows, vuggy zones can also be easily identified.

The effective permeability data (input for our figures), represented by path flow, show that the "vuggy" features are always in the neighborhood of the walls distributed around a central "non-vuggy" zone. These results show particularly low effective permeability near the entrance and towards the end of the tunnel. It is clear that a zone of high effective permeability can be found towards the end of the tunnel, especially for Non-Newtonian fluids and low effective permeability near the tunnel walls. Similar studies for effective permeability maps, as functions of the distance from the tunnel wall and the distance from the entrance, for sandstone perforated with underbalance have already been reported (Karacan and Halleck, 2000).

It is straightforward to see, from our experiments, that if n (fluid parameter, power law index) increases, the number of homogeneous effective permeability zones decreases; namely the local probability data take values in a smaller set of numbers. Besides, for Newtonian flows, the effective permeability field is always symmetric for large porosity, no matter the detailed structure of the porous media. For non-Newtonian fluids, this symmetry may be affected by an angle sweeping the horizontal axis, provided the porous medium is not randomly generated (here, symmetry in the local probability field is lost) and most flow is globally oriented at a clear angle to the horizontal direction.

Pseudoplastic fluids in randomly generated porous media generate effective permeability fields, where zones of constant values are oddly shaped (bottle necks, among others, that can be considered scale up, equivalent to geometrical features, result of our mesoscale experiments), totally abnormal behavior can also be obtained for a different set of parameters. There seems to be a critical value for porosity to keep symmetric effective permeability fields. Our results can be used as models for materials with several zones where different fine erosion and deposition processes occur and provides an original approach to learn from the effect of the fluid parameters (n, the power law index) in the flow, provided the remaining features are constant.

Finally, in what follows, we learn how our LBM (mesoscale information) can be used, by the inter-relationship of geometrical factors, randomness and fluid parameters, to produce permeability (as probability laws) distributions in order to predict scaled up subsurface permeability bands.

In our experiments, normal distribution for effective permeability is obtained mostly for randomly generated porous media provided the fluid is Newtonian or shear thinning, see Table 1. Eventhough this distribution is not expected for arbitrary generated porous media, it may appear provided the fluid is Newtonian. It should be remembered that the use of normal distribution can be theoretically justified in situations where a large number of effects act additively and independently together. The porous media, LBM and fluid factors used in these experiments can be a first approach to the effective permeability distribution of caffeine and testosterone solutions in silicone membrane (Khan et al., 2005).

Provided porosity increases, our results show that for Newtonian and shear thickening fluids, regardless the way the porous matrix (randomly or not) was generated, the effective permeability distribution tends to be Weibull. It is possible to conclude that the experiments shown in Table 1 as Weibull effective permeability distributions are describing systems involving a weakest link and/or non-linear effects derived from the fluid viscosity. Statistics of the mechanical and failure properties on the grain scale are often assumed to follow the Weibull distribution; here, the significant influence of microcrack length statistics has been emphasized (Wong et al., 2006). A Weibull distribution can accurately describe experimentally determined time trends of the infiltration rate (Faybishenko et al., 2003).

Infiltration experiments conducted on packs of rocks show that fluxes may stabilize into an exponential distribution (Tokunaga et al., 2005). Exponential distribution for effective permeability has also been reported in the analysis of heterogeneities in porous media (Savioli et al., 1996). In our simulation, the effective permeability distribution is exponential only for arbitrarily generated porous media and shear thinning fluids, no matter the porosity used in the simulation, see Table 1. They are fit to be used as models for petroleum processes. We believe that arbitrarily generated porous media may provide the deterministic background to obtain exponential effective permeability distributions since these are found in situations where certain events occur with a constant probability per unit distance. Clearly, for non-Newtonian fluids, non-linear effects are introduced; these are small if n tends to one.

The gamma distribution has been used as a statistical representation of surface roughness for flow in unsaturated fractured porous media and tortuosity distributions in porous media (Or and Tuller, 2000; Lindquist et al., 1996). In our simulation, the effective permeability distribution is gamma for shear thickening fluids and rather low porosity, regardless the way the porous matrix was produced, see Table 1. So the gamma effective permeability distributions found in our experiments can be considered applicable to particle tracking. This process has been modeled as the sum of elementary steps with independent random variables in the sand.

In our experiments, the effective permeability distribution is Generalized Extreme Value distribution for arbitrarily generated porous media, porosity 71.87%, $n=2$. This seems to be a very particular case, probably coming from a strong non-linearity. This is the limited distribution of the maxima of a sequence of independent and identically distributed random variables. Extreme hydrometereological phenomena follow an extreme distribution (Gutiérrez-López et al., 2005). At constant potential, the maximum peak currents in different time intervals during a potensiostatic test follow extreme distribution (Zuo et al., 2000).

Summarizing, this paper has presented a method to study scale up processes (mesoscale to core scale) of flow in porous media from geometry, fluid parameters, stochastic realization and LBM; namely, parameters such as tortuosity (arbitrarily generated porous media always have a clearly smaller tortuosity than randomly generated porous media), randomness in void space, geometrical distribution, porosity and consistency index of the non-Newtonian fluid, among others. This model is ready to predict as an output from thin small samples, effective permeability bands and the corresponding subsurface probability distributions scaled up for subsurface reservoirs, aquifers, etc.

6. Acknowledgment

The first author wishes to thank CONACYT and UNAM for financial support.

7. References

Ahronov, E. & Rothman, D.H. (1993). Non-Newtonian flow through porous media, a lattice Boltzmann method. *Geophys. Res. Lett.* 20, pp. 679.

Benzi R., Succi S. and Vergassola M. (1992). The lattice-Boltzmann equation: theory and applications. *Physics Report*, 222(3), pp. 145-197.

Bhatnagar P. L., Gross E. P. & Krook M. (1954). A model for collision processes in gases I: small amplitude processes in charged and neutral one-component system. *Phys. Rev.* 94(3), pp. 511-525.

Bird, R.B., Stewart, W.E. & Lightfoot, E.N. (1960). *Transport Phenomena*. Wiley, New York.

Boek, E.S. Chin, J. & Coveney, P.V. (2003). Lattice Boltzmann simulations of the flow non-Newtonian fluids in porous media. *Int. J. Mod. Phys.* B 17 (1), pp. 99–102.

Boghosian B. M., Coveney P. V. & Emerton A. N. (1996). A Lattice-Gas Model of Microemulsion. *Proc. R. Soc. Ser*, A 452, pp. 1221-1250.

Cercignani C. (1988). *The Boltzmann Equation and Its Applications,* Springer-Verlag New York Berlin Heidelberg.

Chapman S. and Cowling T. (1970). *The mathematical theory of non-uniform gases*. Cambridge University of Edinburg, Edinburg.

Chen S. & Doolen G. (1998). Lattice Boltzmann method for fluid flows. *Ann. Rev. Fluid Mech,* 30, pp. 329-364.

Chen S., Chen H., Martínez D. & Matthaeus W. (1991). Lattice Boltzmann model for simulation of magnetohydrodinamics, *Physical Review Letter*, 67(27), pp. 3776-3779.

Chen S., Dawson S. P., Doolen G. D., Janecky D. R. & Lawniczak A. (1995). Lattice methods and their applications to reacting systems. *Comput. Chem. Eng,* 19(6-7), pp. 617-646.

Chen S., Doolen G. & Eggert K. (1994). *Lattice Boltzmann Versatile Tool for multiphase, Fluid Dynamics and other complicated flows*, Los Alamos Science. 20, pp. 100-111.

Christakos, G. & Hristopulos, D.T. (1997). Stochastic indicator analysis of contaminated sites. *J. Appl. Probab.* 34, pp. 988-1008.

Cooke, R.A., Mostaghimi, S. & Woeste, F. (1995). Effect of hydraulic conductivity probability distribution on simulated solute leaching. *Water Environ. Res.* 67, pp. 159-168.

Doolen G. D. *Lattice Gas Methods for partial Differential Equations.* (1990). Addison-Wesley, Redwood City. CA.

Duderstadt J. J. & Martin W. R. (1979). *Transport theory.* Jonh Wiley & Sons, Inc.

Dullien, F.A.L., (1992). *Porous Media; Fluid Transport and Pore Structure.* Academic Press, New York. 1979. 396p.

Faybishenko, B., Bodvarsson, G.S., & Salve, R. (2003). On the physics of unstable infiltration seepage, and gravity drainage in partially saturated tuffs. *J. Contam. Hydrol.* 62 (3), pp. 63-87.

Frisch U., Hasslacher B., & Pomeau Y. (1986). Lattice-gas automata for the Navier-Stokes equations. *Physical Review Letter,* 56, pp. 1505-1508.

Geankoplis, Ch.J, (2000). *Transport Processes and Unit Operations.*

Gibson, L.J. & Ashby, M.F. (1988). Cellular Solids: Structure and Properties. Pergamon Press, Oxford.

Gutiérrez-López, A., Lebel, T., & Mejía-Zermeno, R. (2005). Space-time study of the pluviometric regime in southern Mexico. *Ing. Hidraul. Mex.* 20 (1), pp. 57-65.

Hardy J., Pomeau Y. & Pazzis O. de. (1973). Time evolution of a two-dimensional model system I: invariant states and time correlation functions. *Journal of Mathematical Physics,* 14(12), pp. 1746-1759.

He X. & Luo L. S. (1997). Theory of the lattice Boltzmann method: From the Boltzmann equation to the lattice Boltzmann equation. *Physical Review E.* 56(6), pp. 6811-6817.

Higuera F. J. and Jiménez J. (1989). Boltzmann approach to the lattice gas simulations. *Europhysics Letters,* 9(7), pp. 663-668.

James J. & William R. (1979).*Transport Theory,* Jonh Wiley & Sons, Inc.

Karacan, C.O. & Halleck, P.M. (2000). Mapping of permeability damage around perforation tunnels. In: Proceedings of ETCE/OMAE2000 Joint Conference, Energy for the New Millenium. February 14-17, New Orleans, LA.

Khan, G.M., Frum, Y., Sarheed, O., Eccleston, G.M., & Meidan, V.M. (2005). Assessment of drug permeability distributions in two different model skins. *Int. J. Pharm.* 303 (1-2), pp. 81-87.

Ladd A. J. C. (1994). Numerical simulations of particulate suspensions via a discretized Boltzmann equation Part 1: Theoretical Foundation. *J. Fluid. Mech,* 271, pp. 285-339.

Landaeta R., Herrera M. & Cerrolaza M. *(1997). Avances recientes en bioingeniería (Investigación y Tecnología Aplicada).* Sociedad Venezolana de Métodos Numéricos (SVMN).

Lebowitz J. L. & Montroll E. W. (1983). *Nonequilibrium phenomena I: The Boltzmann equation,* Amsterdam, North-Holland Publishing Company, New York.

Lindquist, W.B., Lee, S.M., Coker, D.A., Jones, K.W., & Spanne, P. (1996). Medial axis analysis of void structure in the three-dimensional tomographic images of porous media. *J. Geophys. Res. – Solid Earth* 101 (B4), pp. 8297-8310.

Lundström, T.S., & Norlund, M. (2005). Numerical study of the local permeability of noncrimp fabrics. *J. Compos. Mater.* 39, pp. 929–947.

Lundström, T.S., Frishfelds, V., & Jakovics, A. (2004). A statistical approach to the permeability of clustered fiber reinforcements. J.Compos.Mater.38,1137–1149.

Maxwell B. J. (1997). *Lattice Boltzmann methods in interfacial wave modelling.* PhD Thesis, University of Edinburgh.

Mc Namara G. R. & Zanneti G. (1998). Use of the Boltzmann equation to simulate lattice-gas automata. *Physical Review Letter,* 61(20), pp. 2332-2335.

Or, D., & Tuller, M. (2000). Flow in unsaturated fractured porous media: hydraulic conductivity of rough surfaces. *Water Resour. Res.* 36 (5), pp. 1165–1177.

Pai S-I. (1981). *Modern fluid mechanics.* Science Press; New York, Van Nostrand Reinhold Co.

Qian Y. H., d'Humieres D. and Lallemand P. (1992). Lattice BGK models for the Navier-Stokes equations. *Europhysics. Letters,* 17(6), pp. 479-484.

Rakotomala, N. Salin, D. & Watzky, P. (1996). Simulation of viscous flows of complex fluids with a Bhatnagar, Gross, and Krook lattice gas. *Phys. Fluids,* 8 (11), pp. 3200–3202.

Savioli, G.B., Bidner, M.S., & Jacovkis, P.M. (1996). Statistical analysis of heterogeneities and their effect on build-up and drawdown tests. *J. Pet. Sci. Eng.* 15 (1), pp. 45–55.

Schwabl F. (2002). *Statistical mechanics.* Springer-Verlag, Berlin Heidelberg, New York.

Shan X. & Chen H. (1993).Lattice Boltzmann model for simulating flow with multiple phases and components. *Physical Review. E,* 47(3), pp. 1815-1819.

Sitar, N., Cawlfield, J.D. & Kiureghian, A.D. (1987). First order reliability approach to stochastic analysis of subsurface flow and contaminant transport. *Water Resour. Res.* 23, pp. 794–804.

Struchtrup H. (2005). *Macroscopic Transport Equations for Rarefied Gas flows,* Springer–Verlag, Berlin Heidelberg.

Succi S. (2001). *The lattice Boltzmann equation for fluid dynamics and beyond.* Oxford University Press Inc., New York.

Succi S. (2002). *Colloquium:* Role of the H theorem in lattice Boltzmann hydrodynamic simulation. *Rev. Mod. Phys.* 74(4), pp. 1203-1220.

Telega, J.J. & Bielski, W.R. (2003). Flows in random porous media: effective models. Comp. *Geotech.* 30, pp. 271–288.

ThÄurey N. (2003). *A single-phase free-suface lattice Boltzmann.* PhD Thesis, Erlangen.

Tokunaga, T.K., Olson, K.R., & Wan, J.M. (2005). Infiltration flux distributions in unsaturated rock deposits and their potential implications for fractured rock formations. *Geophys. Res. Lett.* 32 (5), L05405.

Wilke J. (2003). *Cache optimizations for the lattice Boltzmann method in 2D,* PhD Thesis, Erlangen, March.

Wolf-Gladrow D. A. (2000). *Lattice Gas Cellular Automata and Lattice Boltzmann Models,* Springer-Verlag, Berlin Heidelberg.

Wong, T.F., Wong, R.H.C., Chau, K.T. & Tang, C.A. (2006). Microcrack statistics, Weibull distribution and micromechanical modeling of compressive failure in rock. Mech. Mater. 38 (7), pp. 664–681.

Zuo, Y., Du, H., Xiong & J.P. (2000). Statistical characteristics of metastable fitting of 316 stainless steel. *J. Mater. Sci. Technol.* 16 (3), pp. 286–290.

Part 3

Mechanism and Techniques

7

Radioisotope Technology as Applied to Petrochemical Industry

Rachad Alami and Abdeslam Bensitel
Centre National de l'Energie, des Sciences et des Techniques Nucléaires
Morocco

1. Introduction

Radioisotopes were first applied for industrial problem solving around the middle of the last century. Since then, their use has increased steadily. Today various applications of radioisotopes, as sealed sources and as radiotracers, are well established throughout the world for troubleshooting and optimization of industrial process plants. Petrochemical and chemical process industries are the main users and beneficiaries of the radioisotope technology.

In this chapter we present the three major techniques used in petrochemicals. Gamma-scanning is a very effective non-invasive technique used for on-line troubleshooting of distillation columns and pipes. Neutron backscattering is applied for level and interface detection in storage tanks and other reservoirs. Radiotracers are employed to establish the residence time distribution which is an important mean of analysis of the petrochemical units.

2. Gamma-scanning

Gamma Scanning is the best technique to carry out an internal inspection of distillation columns, vessels and pipes without interrupting production. The gamma-ray scanning technique is widely used for evaluating the operating characteristics of distillation columns considered as the most critical components in petrochemical plants. Gamma-ray scanning provides essential data to optimize the performance of the columns and to identify maintenance requirements.

Gamma Scanning is a very effective, well established, troubleshooting technique. This powerful diagnostic tool, which has been used for decades, has become popular as it allows inspection of the distillation column components without interrupting operation. In comparison to other non-destructive control techniques used in practice, gamma-ray scanning provides, in real time, the clearest vision of the production conditions inside a process reservoir. It is also cost-effective, particularly when compared to lost production (Hills, 2001).

Gamma scanning can help diagnose and solve approximately 70% of column problems encountered at refinery and petrochemical sites. For other cases, it allows resources to be focused on finding the true source of the problem by eliminating inadequate possible scenarios. It provides essential data to optimize the performance, track any deteriorating effects and identify the maintenance requirements of a distillation column so that it significantly reduces repair downtime (Alami, 2009).

2.1 Main problems met on distillation columns

The hydraulic capacity of trayed distillation columns is always limited by either a high liquid entrainment, or by an overload of downcomers. The technology of high capacity trays deals with these two fundamental limits by reducing the entrainment or by releasing the loads of downcomers. Both are generally connected (Hills, 2001).

2.1.1 Liquid entrainment

The entrainment can be defined as the physical rise of droplets, provoked by the ascending flow of vapour. The vapour tends to hoist droplets upward, while the gravity tends to pull them downward. If the vapour speed is relatively high, then the entrainment can surpass the gravity and some droplets can be transported from a tray to the tray above. The entrainment can be also seen as the excessive accumulation of liquid on trays, which is led by the ascending transport of liquid from a tray to the other one by the current of ascending gas.

A high liquid entrainment can lead directly to a column flooding called "flooding jet". It can also end in it indirectly by increasing the liquid load until reaching the limit of the downcomer, putting in danger the efficiency of separation. Consequently the reflux flow of the column is increased, increasing more the global liquid flow.

The devices of contact between liquid and vapour in the distillation and absorption columns which we meet most generally are trays with sieves and with valves and, in a more or less large extent, of trays with bubble caps. The capacity of retention of these trays is mostly limited by the phenomena of entrainment or flooding. In all the distillation columns of trays type, liquid entrainment and flooding can be present.

2.1.2 Flooding

The flooding of a distillation column is usually defined as the operating mode in which the entrainment is such as there is no downward flow or clear reflux. Conditions of flooding are useful for the engineer-designer, because they represent the maximal authorized operating mode which can serve as reference.

The flooding is usually caused by the accumulation of deposits (dirts) or a blockage on trays. It is also present when the feeding flow of reflux towards the column is upper to the flow coming down from the downcomer.

The flooding can also concern the capacity of downcomers to channel the liquid flow. Because downcomers are passages fitted out for the liquid flow coming down from a tray towards the other one, limitations in their capacity lead to a reduction of the efficiency of trays and, if the limitation is complete, the flooding of the column takes place.

The efficiency of the downcomer or its capacity is limited by:

- Its size
- The difference of pressure between trays, and
- The vapour pulled in the liquid passing by downward.

If the downcomer capacity is inadequate, the level of liquid in this one increases gradually until it limits the liquid flow on the tray above, or on its tray. This, however, increases the entrainment from one tray to the other one and cause finally the flooding. So conditions of flooding can result, either of an inadequate capacity of the downcomer, or of an excessive liquid entrainment in the vapour space.

In summary, the flooding can be seen as an indication of ineffectiveness due to an insufficient liquid-vapour contact (Hills, 2001).

2.1.3 Other encountred problems

Some other problems can be met. These are mainly:

- "Foaming": foam formation
- Destruction and collapse of trays
- "Weeping" or in other words presence of downward droplets in the vapour space between trays.

2.2 Measurement and physical principles of gamma scanning

In performing a scan of a distillation column or a similar reservoir, a small and adequately sealed source of gamma rays, with an appropriate collimator, is placed on one side of the column and a sensitive radiation detector is placed on the opposite side (Knoll, 2000). A collimated beam of gamma rays passes through the column wall, is affected by the column internals and hydraulic conditions, and passes through the other side. The source and the detector are simultaneously moved down along opposite sides of the column in small increments (Figure 2-1). Guide ropes for the source and detector are sometimes attached to the column to ensure scan orientation and as an additional factor of safety during the scan.

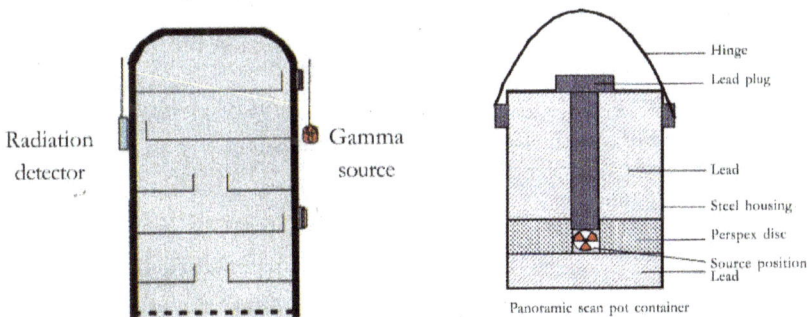

Fig. 2-1. Principle of column gamma scanning.

The scattering of the gamma radiation produces changes in the intensity of the beam that can be correlated to the density of the material inside the inspected column through which the radiation passed according to the following fundamental relationship:

$$I = I_0 e^{-\mu \rho x} \tag{1}$$

where:

- I_0 and I are the initial and transmitted intensities of the gamma beam,
- x is the thickness of the traversed material of density ρ,
- and μ is the coefficient of absorption which is constant for a given gamma-ray energy and material composition.

Equation (1) shows that an increase in material density will reduce the radiation signal and vice versa. The transmitted radiation intensity is measured and recorded via an interfaced portable computer at predetermined length intervals or positions along the side of the column. A radiation absorption profile (the "scan profile") is then produced.

By comparing the obtained scan profile with a mechanical drawing of the column, deductions can be pulled concerning, as well the possible mechanical damage of the trays as certain operating conditions of the column, such as flooding, blockages, weeping and other abnormalities of process.

Gamma scanning can be performed on almost any type and size of column (0.5 m to 10 m in diameter) and is not affected by pressure and temperature. It usually requires no preparation of the vessel or alteration to process and can be performed from existing platforms and through insulation on vessels.

The correlation of the changes in density recorded in the scan profile with the inner part of the column lead, therefore, to an accurate picture of performance and physical conditions (Hills, 2001).

Each tray and the space above it reflect its working state. For instance, a tray functioning correctly has a reasonable level of aerated liquid showing a fast decrease of the corresponding density until a clear steam space is reached just below the following tray.

- When gamma radiation goes through a medium containing a tray filled with aerated liquid, much of the incident beam is partially absorbed and the radiation quantity reaching the detector is relatively small.
- When a radiation beam goes through a non-aerated liquid, the majority of this radiation is absorbed and the detected signal is weak.
- When a radiation beam goes through steam, only a small amount of suspended material is present to absorb the radiation. This means that high intensities of radiation are transmitted to the detector.

To sum up, a scan using gamma radiation of a column can detect and localise regions of liquid and steam within the column. It can also discriminate between liquid aeration and can detect levels of foam and aerosol in steam regions (Urbanski, 1999) (Pless, 2002).

For distillation columns, without affecting processing unit, this reliable and accurate technique can be used to determine:

- the liquid level on trays,
- the presence or absence of internals, such as trays, demister pads, packing and distributors

- the extend and position of jet and liquid stack flooding
- the position and the density characteristics of foaming
- etc…

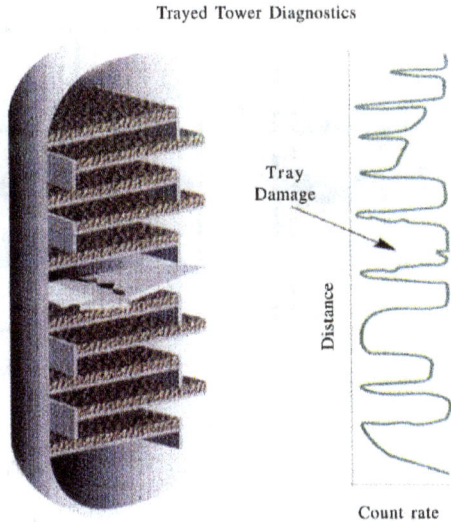

Fig. 2-2. Typical profile scan obtained (IAEA, 2002).

Baseline Gamma scans are performed after a startup or when a column is running efficiently to define an operational reference condition within the column. The baseline scan can be compared to future scans to determine how the column is responding over time or is responding to changes in operating conditions.

An important factor to take into account is that, as far as is possible, the operating conditions (such as feed rate, temperature and other process parameters) must remain constant especially during the scan investigation. It is very important to record any process changes during the time of the scan. This will facilitate the interpretation of the scan profile if anomalies are observed.

2.3 Scanning of trayed columns

To conduct a tray-column scan, it is advisable to execute a scan across the trays and to avoid scanning through the downcomers of the trays. Typical and recommended scan line orientations for trayed columns are shown in the following figure (IAEA, 2002).

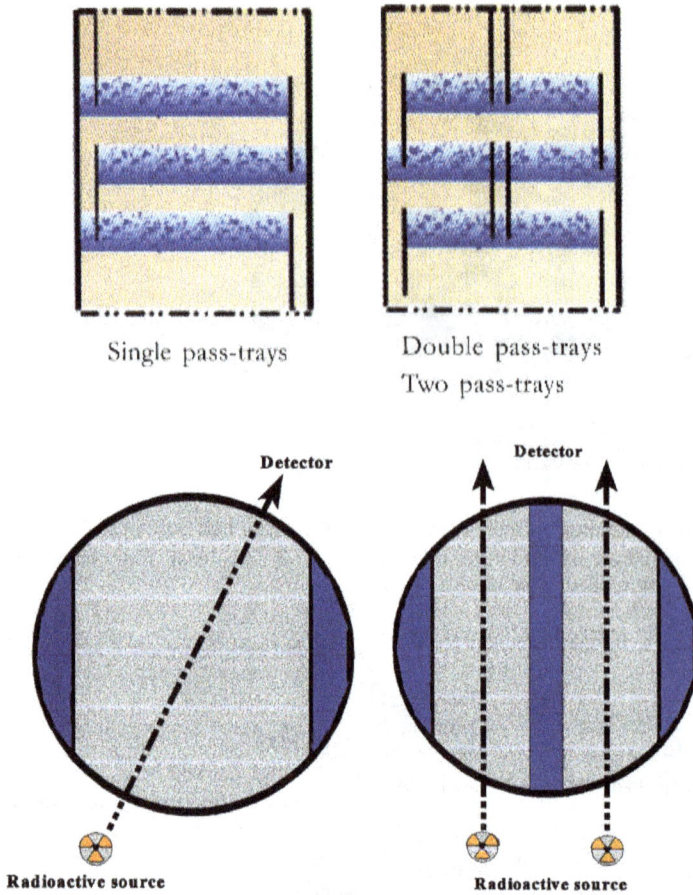

Fig. 2-3. Typical scan line orientation for single pass-trays' column (left) and double pass-tray column (right).

Examples of typical scan profiles obtained for various problems met on distillation columns are presented in Figure 2-4.

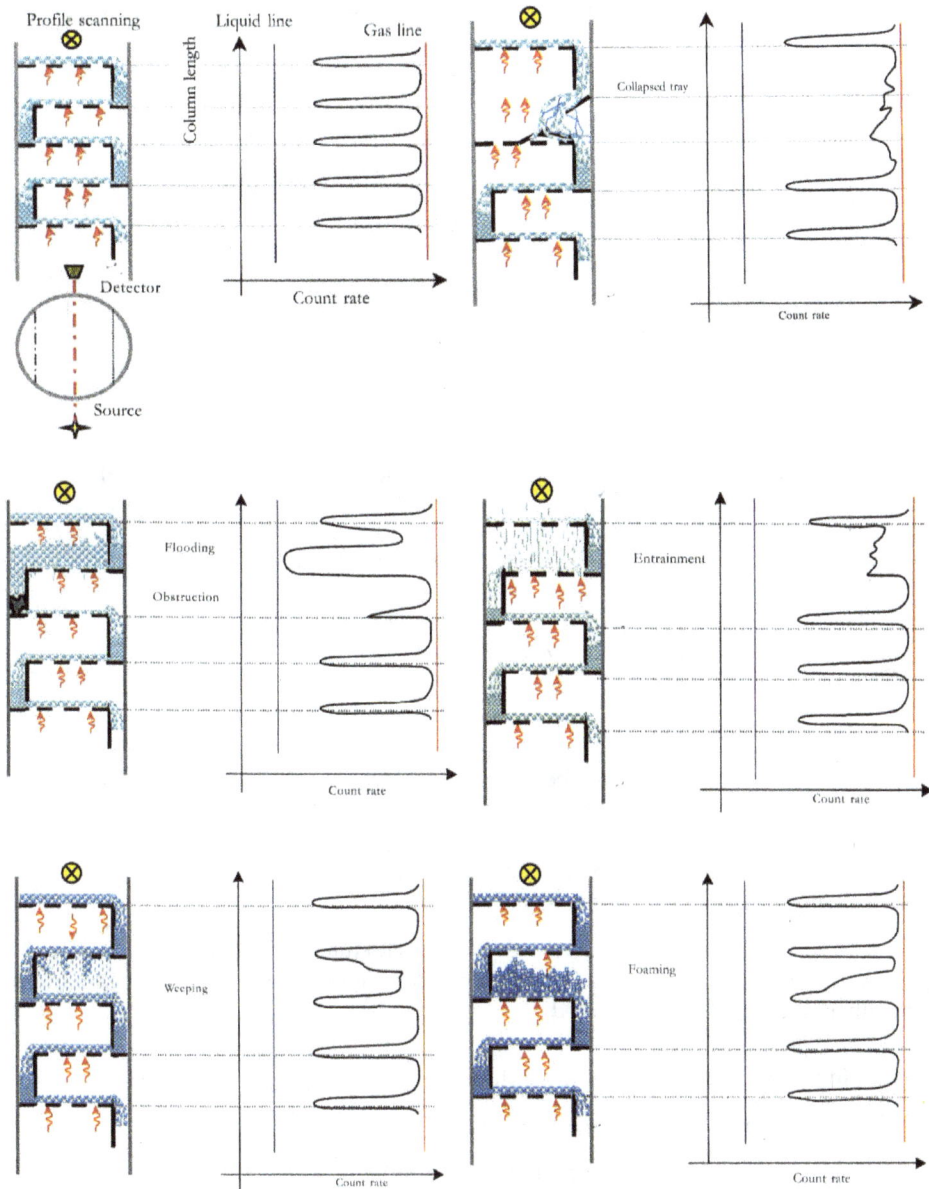

Fig. 2-4. Examples of typical scan profiles obtained for various problems met on distillation columns, respectively from up and left to bottom and right: normal column, collapsed tray, flooding, entrainment, weeping and foaming (IAEA, 2002).

2.4 Scanning of packed beds

Grid scanning is recommended for packed bed columns. A typical orientation of grid scan lines is shown in the figure 2-5. At least four scans are recommended to examine a packed bed column.

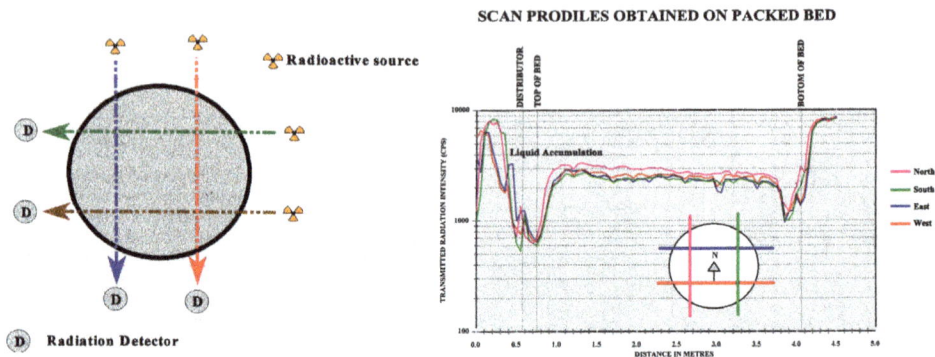

Fig. 2-5. Scan line orientations for packed beds (left) and example of real scan profiles obtained for packed bed (Hills, 2001).

Grid scans may be conducted to investigate process-related conditions such as:

- flooding or blockages
- entrainment or carry over of liquid or
- maldistribution of liquid flow through packed beds.

Grid scans can also be used to investigate mechanical construction problems such as collapsed packed beds or the correct installation of distributors as well as the correct distribution of incoming liquid feed. An irregular distributor can undermine the performance of the entire packed bed and column. Liquid distributors must spread liquid uniformly on top of a bed, resist plugging and fouling, and also provide free space for gas flow. An incorrectly water level installed distributor, that is a tilted distributor, could cause liquid to flow preferentially on one side of the column.

Grid scanning is recommended on packed columns with diameters up to approximately 3m. Larger diameter columns must be approached in a different way, since too large an area (especially in the centre) is not covered.

2.5 Planning of a gamma-ray scan investigation

The following data is required before a scan can be carried out:

- inside diameter and wall thickness of the column (mm)
- bulk density and type of packing material, for packed beds
- downcomer orientation and type of trays present (single, double pass trays)
- operating problems experienced, e.g. low or high pressure problems across the column, or temperature differences along the length of the column

- detailed mechanical drawings of the unit showing internal structure, such as elevations, tray or packing assemblies, nozzle and pipework locations as well as other special features.

Such information is vital for interpreting data from column scan profiles obtained and for identifying and visualising possible mechanical problems. The following additional information is useful:

- gamma scan profiles of an "empty" column (with all the internals but not in operation)
- a scan profile before a maintenance shutdown
- a scan profile after a maintenance shutdown when the column is under normal operating condition.

The activity required depends of the column diameter, ranging from 5-10 mCi for 1-2 m up to 60-70 mCi for 5-6 m and higher for greater diameters. Estimated source strength (activity) can be calculated as follows:

$$A = (D.(d)^2.(2)^{2wt/hl}) / T \qquad (2)$$

Where:

- D = dose rate required (mR/Hr)
- d = diameter of column (m)
- wt = total wall thickness of column (mm) + wall thickness of scan container
- hl = half layer thickness value of material (25 mm for steel for ^{60}Co)
- T = gamma-ray constant for a specific source (1.31 R/h on a distance of 1 meter for 1 Ci ^{60}Co source).

When using the above equation it is suggested that 200 mm be added to the diameter of the column to make provision for the source and detector container on the outside.

The above equation is an approximation, and build-up factors of the material are not taken into account. Shielding calculation software can be used to greater effect.

2.6 Factors Influencing the gamma-ray scanning technique

Major factors influencing the technique are:

- Accessibility to the distillation columns
- External construction of columns (piping, brackets, platforms, etc...)
- Variation of the operating conditions during the scan (feeding flow, temperature, reflux feeding, etc...)
- Weather conditions (wind and rain)

2.7 Pipe scanning

Pipe scanning technique is a derivation of gamma scanning technique for pipes. It can be used to detect:

- solids build-up
- refractory quality and losses

- slugging effects
- vapour and liquid presence in the line.

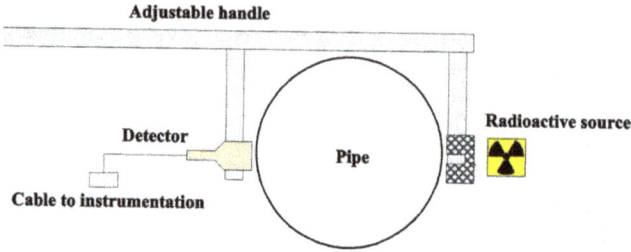

SCHEMATIC DRAWING OF THE PIPE SCANNER

Fig. 2-6. Pipe scanning principle (IAEA, 2002).

There are a number of radioisotope sources, which can be used; two of them are mostly used: ^{137}Cs with a gamma-ray energy of 662 keV (half-life 30 years), and ^{60}Co with gamma-ray energies of 1172 keV and 1332 keV (half-life 5.27 years). The source activity is calculated accepting a dose rate of approximately 1.0 – 1.5 mR/h, at the detector.

Before executing any pipe scanning, the following information is needed:

- the inside diameter and wall thickness of the pipe
- the medium in the pipe (gas, liquid or slurry).

A jack guide is used so that source and detector can be synchronised and always maintain the same distance. The source must be collimated with a collimator of 6 – 8 mm and 10 mm deep in order to obtain a narrow radiation beam. The detector also must be collimated for best results.

A reference scan is obtained on a representative area of the pipeline that is clean and deposit-free.

Figure 2-7 gives an example of a scan profile obtained for a pipe containing solid deposits. In figure 2-8 scan profiles for various conditions in pipes are presented (IAEA, 2002).

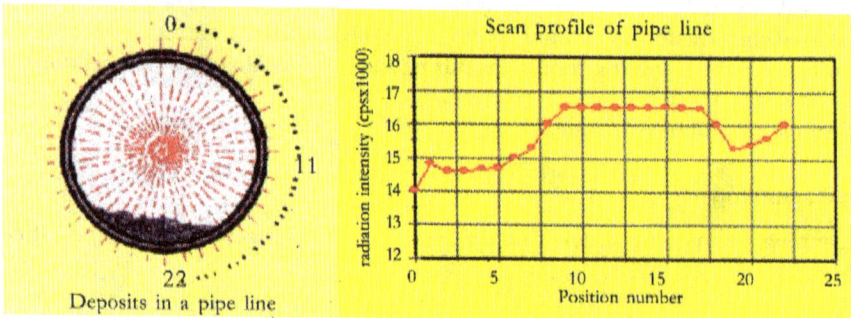

Fig. 2-7. Example of a scan profile obtained for a pipe containing deposits.

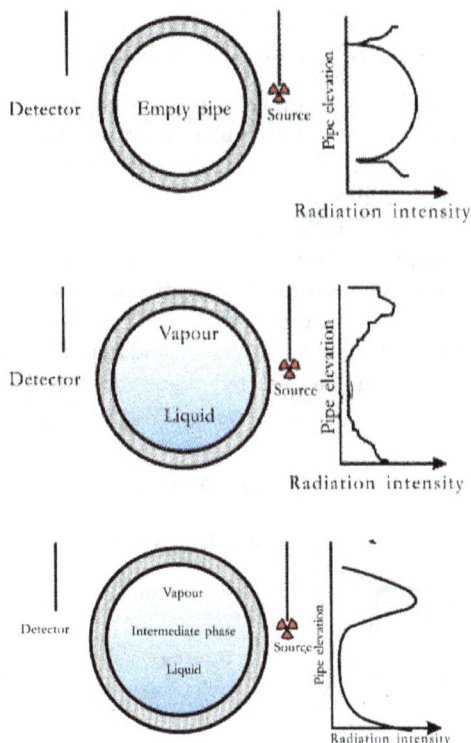

Fig. 2-8. Typical scan profiles obtained for respectively empty pipe (up), pipe with liquid and vapour (middle) and pipe with liquid and vapour separated by an intermediate phase (bottom).

3. Detection of level and interface by neutron back-scattering technique

Modern petrochemical plant operations often require accurate level measurements of process liquids in production and storage vessels. Although a variety of advanced level indicators are commercially available to meet the demand, these may not suit the specific needs of a situation.

3.1 Neutron backscatter principle

In a neutron backscatter gauge, fast neutrons (with energy 2 to 10 MeV) from a radioactive source (^{241}Am/Be or ^{252}Cf) are beamed onto the inspected vessel. Neutrons are particles with no charge and a relatively large mass. Because the distribution of nucleus is relatively sparse in the matter, it could be assumed, that the neutron radiation can pass deeply enough through the material. The effective measurement volume extends typically 100-150 mm into the vessel. By collision with the nucleus the neutrons lose their energy and change the direction of their movement. For ratio of energy of the neutron before collision E_1 and energy after collision E_2 the relation holds (Thyn, 2002):

$$E_2/E_1 = (A-1)^2/(1+A)^2 \tag{3}$$

Where A is the mass number of atom whose nuclei participated in the collision.

Thus the elements with small atom number have the greatest ability to slow down the neutrons. So, fast neutrons are slowed down mainly by collisions with hydrogen atoms of material inside the vessel (50% of the neutron energy in average is transferred to the hydrogen nucleus in a single collision). The neutron flow has substantially lower energy (thermal neutrons) and different direction after collision with material containing hydrogen. A part of thermal neutrons are then bounced back towards the source. By placing a thermal neutron detector next to the source, these backscattered neutrons can be measured. The number of backscatter neutrons is directly proportional to the concentration of hydrogen atoms in front of the neutron detector. As the source and detector move down the side of the vessel (Figure 3-1), interfaces can be detected thanks to the change in hydrogen atom concentration (Charlton, 1986).

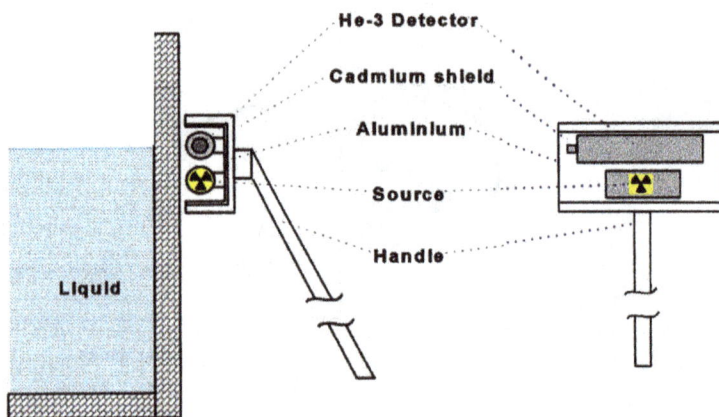

Fig. 3-1. Neutron backscatter level measurement principle (Hills, 2001).

Neutron backscatter gauge clearly indicates solid/liquid and liquid/liquid boundaries and, with careful interpretation of the data, foam levels.

The inspection of the interface between water and oil, as well as among hydrocarbon fractions is the major application of this technique.

Typical profile using the neutron backscattering technique, obtained for a storage tank filled with various liquid and vapour phases, is given in Figure 3-2 (IAEA, 2002).

In practice this method is seldom applied to vessels with wall thickness above 40mm. Although this technique is not suitable in units with thick walls (over 40mm) and although the detector "sees" not deeper than 10 to 15cm into the system, the portable measuring equipment is a valuable tool often used in analysis and diagnostics of chemical processes.

As long as the vessel has a wall thickness less than 100mm, the use of neutrons is a quick and versatile technique, ideally suited if access to both sides of the vessel is not possible.

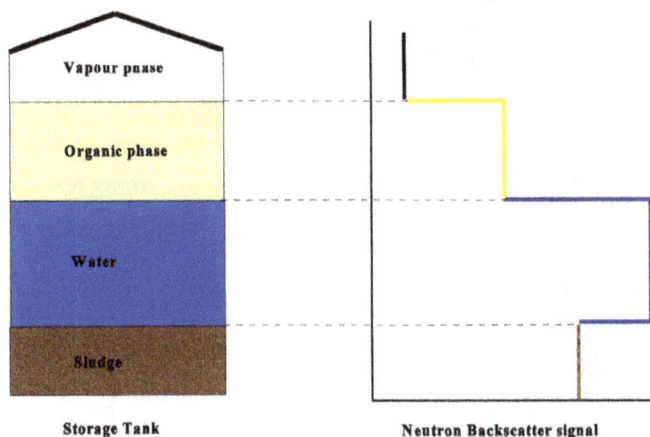

Fig. 3-2. Typical neutron backscattering profile obtained for a storage tank with various liquid and vapour phases.

3.2 Source and detector used

Helium (He-3) or BF3 neutron detectors can be used. He-3 detector has a higher efficiency and is mostly utilized in recent neutron gauges. The neutron source mostly used is 1 Ci ^{241}Am/Be neutron source, which produces a flux of 2.2 x 10^6 n/s with energies from 0.1 MeV to 11.2 MeV, and average energy of approximately 5 MeV. Cf-252 neutron source is used as well, but it is more expensive.

Neutron source	Reaction	Half-life	Neutron Average Energy (MeV)	Flux of fast neutrons
^{241}Am- Be	^{9}Be$(\alpha,n)^{12}$C	433 y	4.46	2.6 x 10^6 n/Ci/s (0.3 g/Ci)
^{252}Cf	Spontaneous fission	2.645 y	2.1	2.314 x 10^{12} n/(s.g)

Table 1. commonly used neutron sealed sources (Charlton, 1986) (Johansen, 2004) (Martin, 1999).

3.3 Applications and limits

Applications of the neutron backscattering technique include:

- Inventory in oil storage tanks without gauges
- Level determination of liquid petroleum gas storage vessels
- Calibration of permanent conventional level gauges
- Determination of sludge or water layers in tanks
- Measurement of packing levels in absorption towers
- Detecting collapsed beds in packed columns
- Monitoring levels of toxic or corrosive liquids in tank cars and railway tankers

- Identifying build-up and blockages in pipes and reactor coils
- Measurement of catalyst levels in reactors
- Detecting ice formation in flate/vent systems.

Neutron backscatter gauge can be used to measure level and interface of transported liquids in pipes as well.

The following factors may influence the measurement and give wrong results:

- Hydrogen-rich materials in the vicinity of the source and detector but outside the process
- Moisture in insulation
- Non-uniform insulation thickness
- Angle and curved surfaces.

4. Radiotracers applications

The use of the radioactive tracers in industry started about fifty years ago, with the arrival on the market of diverse radioisotopes, and did not, since then, stop knowing a continuous extension. The success of the radiotracers applications is mainly due to the possibility, offered by the unique properties of the radioactive materials, to collect data which cannot be obtained by other techniques of investigation (Margrita, 1983).

The radiotracers allow a diagnosis of the functioning of manufacturing units to reach one of the following objectives:

- to determine the characteristics of circuits: flow rate measurements, leaks on heat exchangers...
- to detect the abnormalities of functioning of reactors: short circuit, dead volume, defective mixture
- Detailed knowledge of the conditions of flow: identification of a mathematical model of the residence time distribution and determination of the characteristic parameters (e.g. Peclet number);
- Define the data to be introduced into the automatic circuits: determination of the transfer functions of phases, substances in solution or in suspension
- Complete the kinetics information by hydrodynamics data: obtaining the residence time distributions of the various phases in reaction.

The importance of such studies is underlined by the fact that the competition to which are subjected the diverse chemical industries brings these to optimize their units' efficiency to make more competitive the made product. A better knowledge of the flow of fluids passing through devices contributes to reach this objective: we are led in certain cases to internal mechanical modifications of the devices which induce an improvement of their performances.

The implementation of radiotracers methods so allows to answer a certain number of questions concerning the characteristics of material transfer in the units without disrupting the operation. The main applications concern very diverse industrial domains such as the oil, the cement, the inorganic chemistry, the chlorine, etc....

Radiotracers play an important role in troubleshooting of processes in petrochemical industry.

4.1 Methodology

4.1.1 What is a tracer?

A tracer is by definition a substance which can become identified with a product, the characteristics of flow of which we want to know. This tracer will have to have a behaviour identical to the product, while being able to be revealed and measured by an appropriate technology.

The used radioactive tracers are generally gamma-ray emitters to be detectable through the walls of the installations by external probes.

These radioactive tracers, so called radiotracer, line up in two categories:

- The tracer is an isotope of the product to be studied; in that case it constitutes the intrinsic tracer so giving the confidence of a physico-chemical behaviour and a hydrodynamics strictly identical to that of the product;
- The tracer is an isotope of an element other than the constituent of the product, when this one can not be or badly activated; in that case we have to make sure that its behaviour is the same than the one of the product to be studied.

The radioactive tracer, obtained by irradiation of some grams of the product in the neutronic flow of a nuclear reactor, is mainly characterized by:

- The period or half-life T (time at the end of which the initial activity is divided by 2),
- The energy of the emitted gamma-rays (E keV),
- The physico-chemical form (gaseous, liquid or solid pulverulent as the case may be).

4.1.2 Why a radioactive tracer? Advantages of the radiotracers

The used tracers are, as the case may be, coloured, fluorescent, radioactive chemicals, etc...

The radioactive tracers present various advantages on the other types of tracers:

- Their qualitative and quantitative detection can be made through walls;
- The ease of qualitative analysis (presence, absence of tracer) or of quantitative of the radioactive tracers, analyze which are independent from the matrix. We shall obtain mostly a maximal quantity of information with a radioactive tracer.
- The radioactive methods offer the widest range of tracers, what is often major when we examine a problem of labelling.
- Another characteristic of the radioactive measures is their high sensitivity. This interesting property finds its origin in the nature of the used detectors, in the fact that in spite of an omnipresent natural activity the signal on noise ratio is favourable, and finally, in the possibility that we have to improve the precision of measure of a sample by increasing the counting time.

The results obtained by this method are interesting during all the duration of evolution of a chemical process since the conception until the exploitation, including the improvement of the installation and the optimization of the operation.

4.1.3 Choice of tracer

The best labelling consists in using radioactive isotopes identical to those of the medium to be marked and under the same chemical form. We then have to deal with an internal tracer. If we do not arrange such isotopes, we shall use an isotope of different nature, so called external tracer. This tracer will be chosen so that we can plan that it will have a dynamic, chemical, physico-chemical behaviour, identical to that of the product to be marked.

- For compounds, mostly under the macro-aggregates form such as powders, the direct irradiation in a flow of thermal neutrons gives rise to radioactive isotopes of often present elements in the state of traces. The sodium, for example, is often used in this purpose.
- If the substance to be marked is made of rather big blocks, we shall make for it holes in which are included the tracers (powder, wire).
- The marking of solid substances is still made by soaking them with solutions of a radioactive substance or by putting down on the surface a radioactive substance.

We are sometimes led to realize, in laboratory, tests the object of which is to show that the proposed tracer is valid.

The gamma rays must have enough energy to get through the wall and be exactly recorded. Its period must be long enough to take into account deadlines between the dates of the irradiation and the use.

Let us indicate finally that the radioactive tracers have, for the used activities, very low masses, of the order of fractions of micro-grams. We often use them, if there are risks of adsorption, in parallel with important quantities (several grams) of similar but not radioactive substances. The last ones are so called "trainers".

4.1.4 General principle of an industrial application of radiotracer

The principle of a tracer experiment is the one of any common method impulse-response (Figure 4-1): injection of a tracer in the entrance of a system and a recording of the concentration-time curve at the exit.

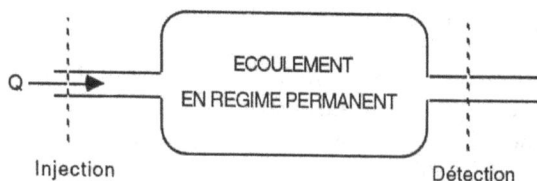

Fig. 4-1. General principle of radiotracer application in industry.

The operation consists in marking a slice of material, in the entrance of the device to be studied and to observe, in various characteristic points of this one, the response curve of the concentration of the tracer versus time, C(t).

1. Injection: the injection of tracers can be realized by various manners: very fast injection, injection with constant flow, injection in a very high-pressure circuit and under diverse

physical forms: gas, liquid, powder. The injection will have to be the most brief possible so that the recorded response in the chosen points of measure can be considered as a Residence Time Distribution (RTD) of the marked phase. The tracer, packaged e.g. in a quartz bulb for gases, is placed in an injection device surrounded with a lead shielding. The injection is made instantaneously in the flow of material in the entrance of the device, by a push of nitrogen the pressure of which is adapted to the operating conditions of the unit.

2. Detection: Detectors allowing to measure through the walls of the installations the gamma rays, emitted by the tracer, are generally scintillation detectors type. The signals delivered by these probes are registered in a data acquisition system, monitored by a computer allowing the storage of the information of about ten sensors (or more) collected at high sampling rate. So, we can have, for every test, of the order of 5 000 information for each measuring channel. Later the data so stored are treated in batch mode by a computer in the laboratory.

4.1.5 What is a Residence Time Distribution (RTD)?

Since its introduction in the chemical engineering by Danckwerts in 1953, the concept of the Residence Time Distribution (RTD) became an important means of analysis of the industrial units. We know that particles crossing a device stay in this one more or less for a long time according to the route which they go through. The Residence Time Distribution (RTD) represents the density of probability which has a particle entering a system, at a given moment, to go out of this system during time. The RTD of the marked phase is calculated from the response curves of the concentration in tracer versus time, C(t), measured in a point of the installation (Guizerix, 1970) (CEA, 1990).

The experimental residence time distribution E(t) is calculated:

$$E(t) = \frac{C(t)}{\int_0^\infty C(t)dt} \tag{4}$$

4.2 Results which can be obtained

The Residence Time Distribution allows to calculate:

- The time of arrival (t_a) which corresponds to the time of transit of the fastest particles between the injection and the point of measure.
- The time (t_m), the abscissa of the maximum of the curve, which indicates the time of transfer of the maximal concentration of tracer.
- The mean residence time, (τ), of the marked product in the system, defined by the difference of the abscissas of the centre of gravity of entrance-exit curves.

The physical parameters directly deductible from the RTD are:

- Speed of the flow of material
- flow of material
- Density

- Dead volumes
- By-pass

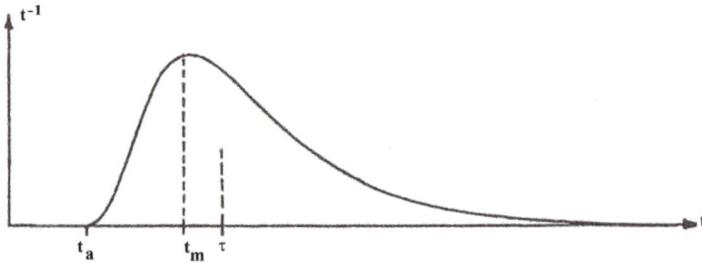

Fig. 4-2. Residence Time Distribution (RTD) curve.

4.3 Applications

4.3.1 Flow measurements

The flow measurement represents one of the most ancient applications of the radioactive tracers. According to the cases, two methods are used:

1. Allen's method (or method of "two peaks"), for the pipes of known geometry (figure 4-3): it is the simplest technique where a small quantity of radiotracer is injected in the process and its passage downstream is monitored by two radiation detectors placed outside of the pipe. The used radioisotope should be a Gamma emitter allowing external detection. If the first detector is at a sufficient distance from the injection point, to assure a complete mixture of the tracer, the time measure of transit of the impulse between both detectors allows the calculation of the average speed of flow (speed of the fluid). Knowing the surface of the pipe section, the measured speed can be converted in volumetric flow. This method allows generally measures of average flows with better than ±1 %, when the internal diameter of the pipe is exactly known.

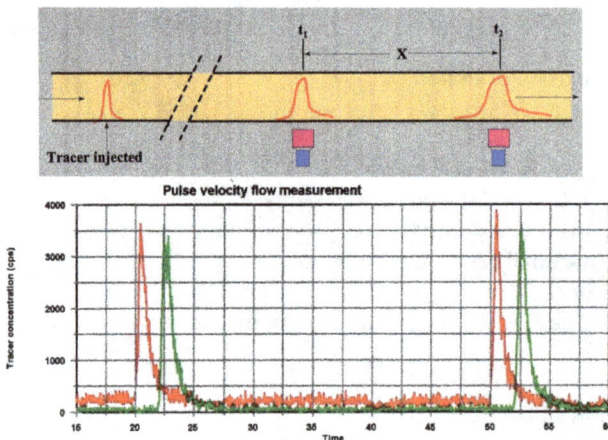

Fig. 4-3. Flow rate measurement by the "two Peaks" or transit time tracer method (IAEA, 2008).

2. The method of dilution, with its two variants: integration method and injection method with constant flow. Here an appropriate radiotracer is injected in the process, and samples are collected in a point enough downstream to assure a complete mixture. The injection of the tracer and its concentration, both in the injected solution and in the diluted samples, are measured. From these data, the flow can be easily calculated.

a. Integration Method: the tracer (activity A) is injected during a short interval of time. In a section, situated downstream, we measure in a point the function C(t) representing its passage; the flow is given by:

$$Q = \frac{A}{\int C(t)dt} \qquad (5)$$

The good mixing condition is: $\int$ C(t)dt = constant, whatever is the point of sampling in the section of measure.

b. Injection method with constant flow: the tracer is injected in a section at constant flow, q, and at a concentration C_0. In a section situated downstream we determine, after reaching a constant regime of concentration, the concentration C of the tracer. If Q is the flow to be measured, an equation of balance of the tracer gives:

$$Q = \frac{C_0}{C \times q} \qquad (6)$$

This relation can be written only if the distance between the section of injection and the section of measure is rather big so that in this last one there is good mixture of the tracer, condition which we represent by: C = constant, whatever is the point of sampling in the section. This technique is ideal for the measure of flow in open channels such as sewers and mouths of rivers (the range of measure is very wide: flows of rivers of several thousand cubic meters per second can be measured). The precision is among the best which we can obtain (±1 % or better). A precision of ± 0.5 % can be reached for regular flows with adequate distances of mixture.

These methods of dilution apply (cf. equation of balance) for the measure of constant flows. An analysis can be however made for the application of the method of injection with constant flow in case Q varies. It allows to measure continuously the flow of industrial waste over reference periods, for example of 24 hours.

These methods, apply as well to gases as to liquids, at high or low pressure. The techniques of measure of flow are immediate and are not thus useful for continuous measures. We apply them for the calibration of classic devices (ratemeters), with a 1 % precision, or in the studies of circuits with tapping.

The flows of products (liquid, solid and muddy) or of cooling water are measured in a 1-2 % precision in situations where ratemeters are not installed or are not reliable any more because of problems of deposits or corrosion.

Finally, the implementation of these techniques is very simple and induces only a minimal disturbance of the network; it is need, for example, only of a "pricking" for the injection of the tracer.

4.3.2 Volume measurements: Determination of dead (or stagnant) volumes

The capacity of a device is not always fully used, and it can exist zones, so called dead, which participate little or not in the flow.

The volume of the dead zones is determined by exploiting the obtained residence time distribution (RTD) using the method of dilution which consists in injecting an activity A and in measuring, after homogenization, the resultant concentration, what allows to reach the volume in which diluted the tracer.

The RTD allows to calculate t, the residence time of the material in the device. If τ is the mean residence time, determined by:

$$\tau = \frac{V}{Q} \tag{7}$$

Where V is the volume of the device and Q the flow.

In case the volume is completely occupied, t is equal to the mean residence time τ. If $t < \tau$ we deduct generally the existence of a dead volume V_m given by:

$$t = \frac{V - V_m}{Q} \tag{8}$$

Then:

$$V_m = V \times (1 - \frac{\tau}{t}) \tag{9}$$

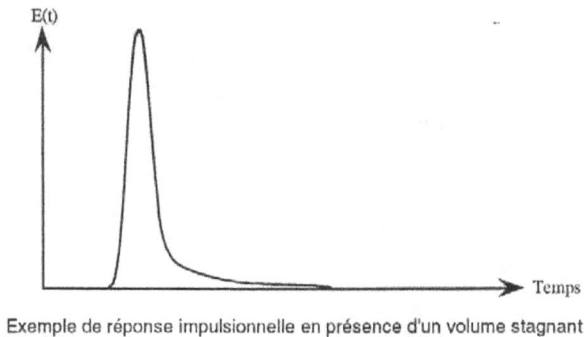

Exemple de réponse impulsionnelle en présence d'un volume stagnant

Fig. 4-4. Typical RTD curve obtained in case of dead volume.

The stagnating volume, which corresponds to a zone of fluid little accessible to the main flow, is represented on the impulse response by a trail of the curve.

4.3.3 Leak detection

Leak detection and leak location using radiotracer techniques are probably the most widespread applications of radiotracers in industrial troubleshooting. Leaks create problems in process plants or in pipelines, spoil the quality of the final product or reduce the capacity of oil and gas pipelines and contamination of surface or ground water and soil could also happen (IAEA, 2009) (IAEA, 2004).

The sensitivity inherent to the radiotracers techniques makes them extremely precious for the detection of leaks in pipes, heat exchangers, condensers and valves. The measures are realized while the units stay on operation and without interfering with the process.

Leaks in heat exchangers are more frequent problem in many processing plants of petrochemical industry. Radiotracers are very efficient and most competitive for detecting small leaks inside the heat exchangers. Detection limits of 0.5% of stream flow can be achieved. Radiotracers methods are also very effective when the pipes are buried.

The experiments of leak determination are two types: qualitative or quantitative.

- In the first case, it is only a question of showing that there is a leak between two media. The answer is given by marking the fluid of the first medium and by showing the appearance of the tracer in the second medium.
- In the second case, we try to determine the flow rate of the leak. An injection of tracer is realized in the primary circuit and the impulse responses are measured downstream to both circuits. If the leak exists, some of the tracer will pass into the lower pressure stream. There it will be detected either by sampling the process fluid downstream or by monitoring the movement of the tracer using detectors mounted externally. The equation of balance of the tracer allows easily to calculate the flow rate of leak. Leaks so low as 0.1 % can be measured with online external detection, while sampling can identify leaks as small as 0.01% of the main fluid flow rate.

For leak detection and leak location in buried pipes the so called "pig technique" is used. Here the pig means a sensor for detecting and recording radiation signals from inside the pipeline. Radiation detection pig consists of a detector and data logger assembled together inside a compact watertight cylinder. Pig moves inside the pipelines.

The method is a two step procedure. In the first step, radiotracer solution is pumped into the pipeline as a tracer plug without any interrupting operation. Where the "plug" meets a hole or fracture, a small amount of the tracer will penetrate and be trapped outside the pipe wall. In the second step, the pig is launched into the pipeline for leak detection run (Fig. 4-5).

This method has a higher sensitivity for leakage detection in underground pipelines as the pig moves in close contact with the leak surrounds. Leaks of the order of 0.1 litter per minute can pinpointed (IAEA, 2009).

4.3.4 Determination of rate of by-pass (short circuit)

We can identify short circuits, i.e. the existence of fluid flows quickly circulating in the reactor. In that case, following an instantaneous injection of tracer, the residence time distribution (RTD) presents abnormalities of shape due to the presence of a narrow peak coming to overlap in the main response. This means the existence of a preferential passage inside the unit (Fig. 4-6).

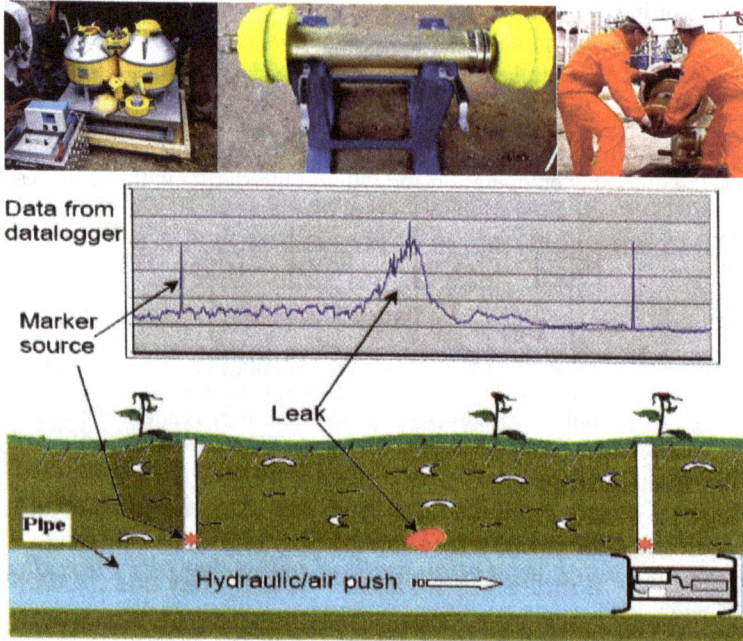

Fig. 4-5. Principle of radiotracer pig method: radiotracer injection device (up, left), pig container (up, middle), pig introduction in pipe (up, right), schematic principle of leak detection in buried pipe (bottom) and typical signal obtained from datalogger (middle).

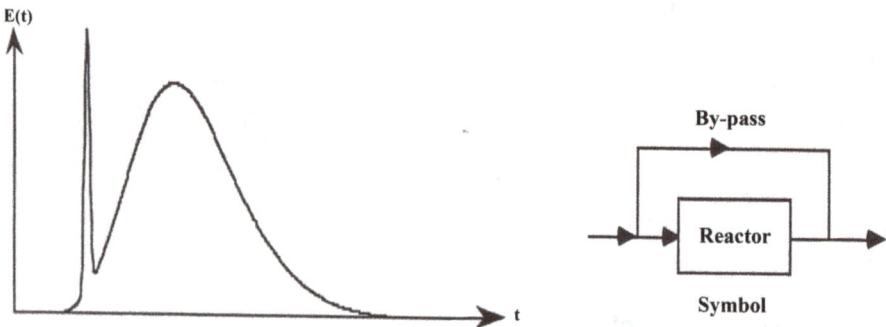

Fig. 4-6. Typical RTD curve obtained in case of by-pass or short circuiting.

The importance of the by-pass can be quantified from the ratio of the area of the peak over the total area of the curve: by interpolating under this peak the main response, we can resolve the impulse response into two responses corresponding to both modes of flow, with respective areas A and B. The rate α of by-pass is (Fig. 4-7):

$$\alpha = \frac{A}{A+B} \tag{10}$$

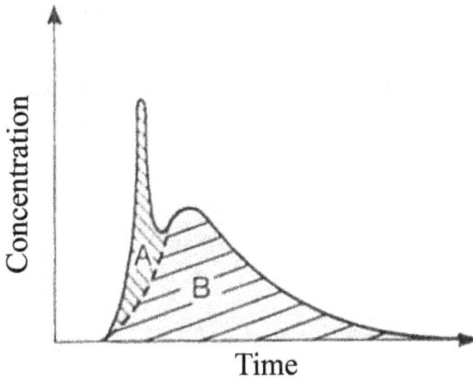

Fig. 4-7. By-pass rate calculation method.

4.3.5 Determination of recycling

The current of recycling corresponds to the reincorporation of a fraction of the outgoing flow, to the entrance of the reactor, and which we can detect on the impulse response by successive, but weakened waves, at regular intervals.

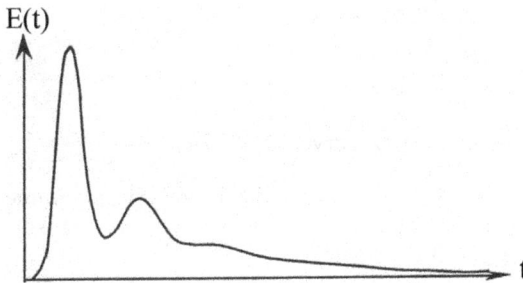

Fig. 4-8. Typical RTD curve obtained in case of recycling.

4.4 Construction, use of physico-mathematical models

One of the most interesting and the most promising modern applications of the tracers, generally, and the radioactive tracers, in particular, concern either the use, or the construction of physico-mathematical models. This is in a general tactics which is always adopted to optimize a situation. Better we know how to describe a phenomenon, more we master it (Margrita, 1983).

4.4.1 Models of flow

The behaviour of the material passing through manufacturing units is very often similar to a type of flow being situated between two extreme behaviours of theoretical flow: "Piston" and "Perfect Mixer".

1. "Piston" model: the fluid moves altogether according to the image of a piston in a cylinder, in parallel slices without any exchange between them. The impulse response of such a system is h (t) = δ (t - τ), δ being a Dirac function (Fig. 4-9).

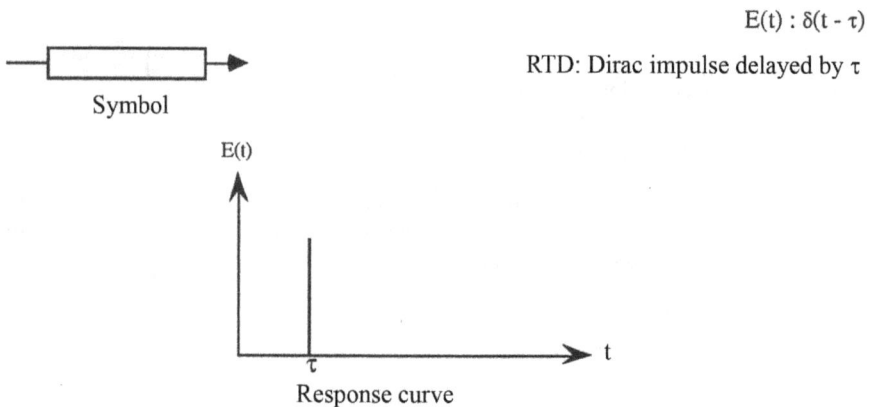

$$E(t) : \delta(t - \tau)$$

RTD: Dirac impulse delayed by τ

Symbol

E(t)

τ

t

Response curve

Fig. 4-9. Piston model and its response curve.

2. "Perfect Mixer" model: this type of model represents an immediate and uniform mixture of the entering fluid. The impulsive response of such a system is the exponential Shape (Fig. 4-10) and spells:

$$E(t) = \frac{1}{\tau}e^{(-t/\tau)} \tag{11}$$

Symbol

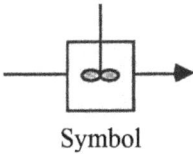

$$E(t) : \frac{1}{\tau} \exp(-t/\tau)$$

RTD : Decreasing exponential

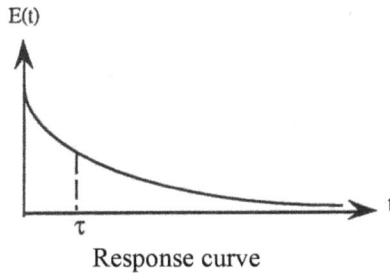

Response curve

Fig. 4-10. Perfect Mixer model and its response curve.

3. "Piston-Dispersion" model (Fig. 4-11): it is the type of flow that we meet frequently. It is situated between the previous two models. Slices quoted in the "Piston" model exchange some material, leading to a phenomenon of dispersion of this material, quantified by a so called dispersion coefficient. This dispersion is also characterized by a without dimension parameter P called "Peclet Number" and which is calculated by the relation:

$$P = \frac{u \times L}{D} \tag{12}$$

A high value of Peclet number represents a low dispersion. When $D \to 0$, this model tends towards the Piston model; on the contrary when $D \to \infty$ (big dispersion, P low) this model is comparable to a Perfect Mixer.

The impulsive response has the form:

$$h(t) = \frac{u}{\sqrt{4\pi \times Dt}} e^{-\frac{(L-ut)^2}{4Dt}} \tag{13}$$

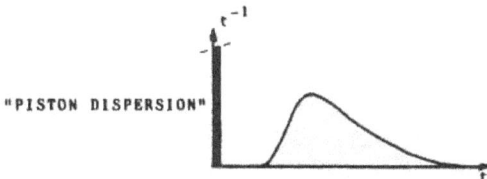

"PISTON DISPERSION"

$$h(t) = \frac{u}{\sqrt{4\pi \, Dt}} \cdot EXP - \frac{(L-ut)^2}{4 \, Dt}$$

Fig. 4-11. Piston-Dispersion model and its response curve.

4.4.2 Identification to a model

The relation between the entrance function and that of the exit of some system is given by a so called Convolution Operation. This operation allows to determine the exit function **s(t)** when we know the entrance function **e(t)** and the impulse response **h(t)** (Fig. 4-12).

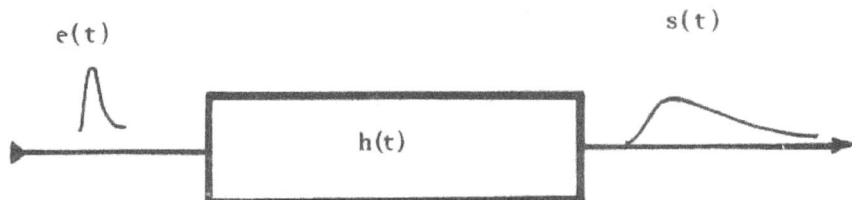

e(t)   s(t)

h(t)

Fig. 4-12.

The fit of a "Piston-Dispersion" model to an experimental curve in exit of a device, consists in determining u and D parameters of the impulse response h(t), in a way that the product of convolution of the entrance function by the impulse response of the considered system, overlaps at best with the experimental curve registered in exit.

The general approach is the following one:

- A physical model of the flow is imagined; its mathematical formulation allows to clarify, according to certain parameters, the residence time distribution.
- A tracer experiment on the unit gives the real residence time distribution.
- A fit between both distributions (experimental and theoretical) by the methods of mathematical optimization allows to evaluate the validity of the proposed model on one hand and to fix the values of the parameters on the other hand.

A good agreement between both distributions means that the proposed model is useful, either to optimize the system, or, still to extrapolate its dimensions to those of another device in project.

The values of the found parameters can be used to the establishment of tables which will facilitate, later, the extrapolations which we have mentioned.

The purpose of fit of a mathematical model can also be the automatic monitoring of the units by computer.

5. Conclusion

The economic benefits that may be derived from the use of radioisotope technology in petrochemical industry are large. In this chapter we tried to present the state-of-the-art in major techniques used in petrochemicals such as gamma-scanning as a diagnostic tool for distillation columns and pipes, neutron backscattering for level and interface detection in storage tanks and other reservoirs and finally radiotracer applications for troubleshooting and optimizing processes (residence time distribution establishment, flow measurement, dead volume determination, leak detection, short-circuiting or by-pass calculation, etc...). It aims to provide a comprehensive description of what can be achieved by the application of such techniques and to promote their benefits to industrial end-users.

6. References

Alami, R.; Ouardi, A.; Laghyam, H.; Lhaiba, S.; Bensitel, A. & Benchekroun, D. (2009). Developments on software tools for data acquisition and numerical simulation for Gamma-ray scanning, *IEEE Proceedings of the 1st ANIMMA (International Conference on Advancements in Nuclear Instrumentation, Measurement Methods and their Applications) Conference*, Marseille, France, June 07-10, 2009

CEA-Damri-SAR (1990). Les Traceurs en Génie des Procédés, In: Guide des Applications des Radioéléments N°1, Avril, 1990, Saclay, France

Charlton, J.S. (1986). *Radioisotope Techniques for Problem-solving in Industrial Process Plants*, Leonard Hill, Glascow

Guizerix, J. (1970). Application des Traceurs Radioactifs dans le Génie Chimique, In: *La Technique Moderne*, October, 1970

Hills, A.E. (2001). *Practical guidebook for radioisotope-based technology in industry*, Technical Report, IAEA/RCA/8/078, Vienna, Austria

Hills, A.E. (2001). Radioisotope Tracer Applications in Industry, In: *Practical Guidebook for Radioisotope-Based Technology in Industry (2nd Ed.)*, Technical Report, IAEA/RCA RAS/8/078, Vienna, Austria

IAEA (2002). *Radioisotope Applications for Troubleshooting and Optimizing Industrial Processes*, RCA in India, March, 2002

IAEA (2004). Radiotracer Applications in Industry – A Guidebook, *Safety Report Series N°423*, ISBN 92-0-114503-9, ISSN 0074-1914, Vienna, Austria

IAEA (2008). Radiotracer Residence Time Distribution Method for Industrial and Environmental Applications, *Training Course Series 31*, ISSN 1018-5518, Vienna, Austria

IAEA (2009). Leak Detection in Heat Exchangers and Underground Pipelines Using Radiotracers, *Training Course Series 38*, ISSN 1018-5518, Vienna, Austria

Johansen, G.A. & Jackson, P. (2004). *Radioisotope Gauges for Industrial Process Measurements*, John Wiley & Sons Ltd, ISBN 0-471-48999-9, Chichester, West Sussex, England

Knoll, G.F. (2000). *Radiation Detection and Measurement*, John Wiley & Sons

Margrita, R. & Santos-Cottin, H. (1983). Etude par Traceurs Radioactifs des Ecoulements de Matière dans les Unités de Cimenteries, In: Ciments, Bétons,Plâtres, Chaux, N° 743-4/83

Martin, R.C.; Knauer, J.B. & Balo Choi, P.A. (1999). Production, Distribution and Applications of Californium-252 Neutron Sources, *4th Topical Meeting on Industrial Radiation and Radioisotope Measurement Applications*, Raleigh, North Carolina, USA, October 3-7, 1999

Pless, L.; Asseln, B. (2002). *Using gamma scans to plan maintenance of columns*, Mechanical Engineering Book, PTQ, pp 115-123, Spring 2002

Thyn, J. & Zitny, R. (2002). *Analysis and Diagnostics of Industrial Processes by Radiotracers and Radioisotope Sealed Sources*, vol. II, CTU Faculty of Mechanical Engineering, Department of Process Engineering, ISBN 80-01-02643-4, Prague

Urbanski, N.F.; Resetarits, M.R.; Shakur, M.S.M. & Monkelbaan, D.R. (1999). Applying Gamma scanning a column containing closely spaced trays, *Annual Meeting AIChE*, Dallas, Texas, USA, November 4, 1999

8

Resonance in Electrical Power Systems of Petrochemical Plants

Job García and Gabriel García
Instituto de Investigaciones Eléctricas
Mexico

1. Introduction

One of the most serious problems in the operation of petrochemical plants is the breakdown of the electrical energy in the production processes. This interruption could be due to the presence of failures that inhibit the continuity and reliability of the electrical energy service. A possible cause of fault in the electrical power system with permanent damages in the primary electrical equipment is the occurrence of the electromagnetic resonance phenomenon. The presence of this phenomenon is directly associated with the grounding scheme of the electrical power equipments. At the moment, in some petrochemical plants only one neutral is connected to ground. When a fault appears, the protection relays operate in order to release the fault and leaving the system temporarily without ground reference. During this time the electrical system remains floated the resonance phenomenon can appear. In petrochemical plants, it is a common practice to have a neutral bus where all the power equipments are connected through an energy cable with a length up to 1 km, with a parasitic capacitance in the order of 50 to 1,000 nF.

This chapter provides an analysis of research on the occurrence of resonance overvoltages in the electrical power system of petrochemical plants. Using digital simulation by *Graphic Alternative Transients Program (ATPDraw)* from *EMTP* was determined that this phenomenon appears when the electrical system does not have reference to ground. It was found that the occurrence frequency of this phenomenon is within a range near to 240 kHz and the magnitude of the overvoltages is in the order of 10 to 20 p.u. Likewise, the evidence of practical cases are discussed where the results were catastrophic for the plants. Also, an analysis of overvoltages by connecting potential transformers windings in open delta is presented.

As a practical solution to keep away from the presence of overvoltages by series resonance, a hybrid neutral grounding scheme is used with the aim of avoiding the power system could be floated during stationary and fault operation of the electrical power system in petrochemical plants. Hybrid neutral grounding consists on an implementation of high resistance grounding in parallel with the low resistance scheme on each of the electrical generators.

The solution for avoid series resonance in potential transformer with open delta windings is the used of wye connection to ground.

2. Grounding methods in petrochemical plants

The electromagnetic resonance phenomenon, will be described in section 3, it is directly associated with the method of grounding the electrical system of the petrochemical plants. Protection schemes and practical grounding of industrial power systems of medium voltage, such as industrial petrochemical plants, are based on IEEE standards, the National Electrical Code (NEC), IEC standards and technical publications [IEEE Standard 141-1993; IEEE Standard 142-1982; IEEE Standard 242-1986; IEEE Standard 241-1983; ANSI C62.92-1987; National Electrical Code National; IEC 61936-1 2002; IEC 60364-1 2001; Dunki-Jacobs, 1977].

The grounding of the electrical system is physically connecting a point of the system (usually the neutral transformer or generator) to ground and create a reference voltage for the electrical system [IEEE Standard 141-1993; IEEE Standard 142-1982; IEEE Standard 242-1986; IEEE Standard 241-1983; ANSI C62.92-1987; National Electrical Code National; IEC 61936-1 2002; IEC 60364-1 2001; Dunki-Jacobs, 1977].

In practice, the electrical power system of a petrochemical plant on 13.8 kV is grounded using "low resistance" method, grounding only one or maximum two points of the system, as the neutral of star winding neutral transformer or generator. Also, on 4.16 kV transformers are used in delta-star connection so the neutral is grounded using the "low resistance" method. At the level of 480 V as used by the method of "solid" grounding by the neutrals transformers.

On the other hand, when a fault to ground occurs on 13.8 kV, the system can be "floating" as described in section 2.3, in this case, the system is "ungrounded". The following describes each of the grounding methods used in the electrical system of the petrochemical pants.

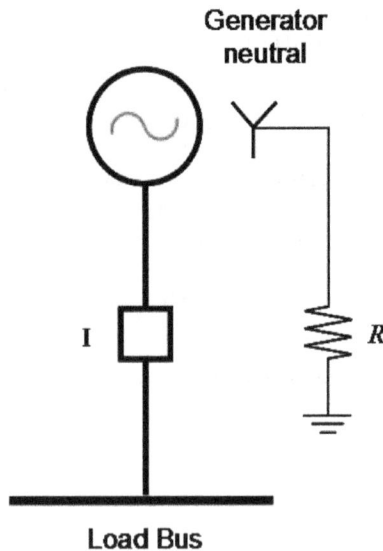

Fig. 2.1.1. Resistance with grounding.

2.1 Low resistance grounding

In large systems where we have a high capital investment in equipment or losses which have a significant economic impact, we use the low impedance grounding [Manning, 1964; Shipp & Angelini, 1991]. In this scheme, a resistor is connected, normally a value that allows only a ground fault current of 200 to 800 A, through the neutral grounding system, see Fig. 2.1.1. As the grounding impedance is a resistance, any transient overvoltage is quickly damped and the overvoltage does not stay for a long time [Shipp & Angelini, 1991].

2.2 Solidly grounding

Solidly grounding systems are the most commonly used. These schemes can be used in systems operating with a single phase loads, tending to a safe operation, see Fig. 2.2.1. If the system is not solidly grounded, the neutral system remains floating to ground, and remains as function of the load line to neutral and exposed to unbalanced voltages and instability [Shipp & Angelini, 1991].

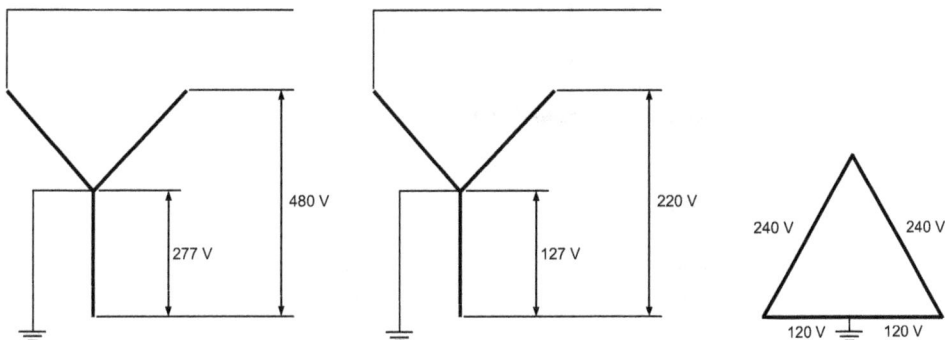

Fig. 2.2.1. Neutral solidly grounding.

2.3 Isolated from ground

A floating system is used in those systems where service continuity is the main concern, see Fig. 2.3.1. The perception is that the floating systems provide greater continuity on the service. This is based on an argument that the ground fault current is smaller and therefore, the heating produced by the presence of such failure is negligible [Shipp & Angelini, 1991]. That way, the line to ground faults can remain on the system until it's convenient to remove them. This idea has some validity only if snapshots or "solids" failures are considered. However, in practice, most faults begin as ground faults by arcs of low energy. When starting a ground fault arc, we can have the following problems:

- Multiple ground faults
- Resonance
- Transient overvoltages

Multiple failures to ground may occur in floating systems. While a ground fault on an isolated system does not cause an exit, if the fault continues, increases the possibility of occurrence of a second fault in another phase. That is because the ground insulation of the unfaulted phases acquires the phase to phase voltage. In other words, the insulation is subjected to stress 73% above the normal operating condition.

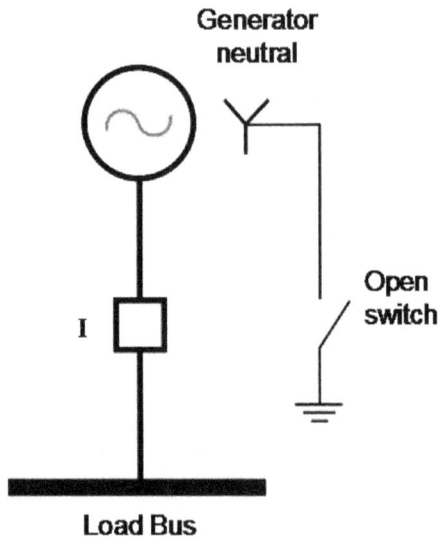

Fig. 2.3.1. Ungrounded neutral.

Even though it isn't common, the phenomenon of resonance can occur when one phase is grounded through an inductance, for example, a ground between the winding of the instrument transformer. When this happens, the high current flow causes high voltages across the unfaulted phases. Transient events due to switching or intermittent faults to the ground can produce overvoltages on floating systems [Shipp & Angelini, 1991].

2.4 Advantages and disadvantages of grounding methods

Table 1. summarizes the most relevant characteristics of grounding schemes, available in the petrochemical plant electric system [Shipp & Angelini, 1991; Reyes Martínez, 2001].

3. Series resonance

Series resonance is the tuning of the inductive and capacitive reactance of the elements of an RLC series circuit. During the occurrence of this phenomenon dangerous overvoltages may occur depending on quality factor Q of the circuit [Rüdenberg, 1950; Boylestad, 1998; Boylestad, 1998; Fuentes Rosado, 1988], that way damage to associated equipment. This effect is observed in the digital simulation results presented in section 3.3. The resonance can also occur by the presence of harmonic frequencies [Lemerande, 1998; Currence, et al., 1995; Girgis, et al., 1993; Fujita, 2000]. This sections contains a description of the resonance phenomenon, the causes may brings as present in the electrical system of an industrial petrochemical plant during a fault.

	floating (ungrounding)	Solidly grounding	Low resistance grounding
Single phase fault current in percent of the three phase fault current	≤ 1%	Variable, can be 100 % or greater	5 a 20%
Transient overvoltage	≤ 6 p.u.	≤ 1.2 p.u.	< 2.5 p.u.
Segregation automatic fault zone	No	Yes	Yes
Surge arresters	Floating neutral type	Grounding neutral type	Floating neutral type
Comments	Not recommended due to the overvoltages and the impossibility of segregation of the fault	Typically used in systems of 600 V or less and up to 15 kV	Generally used in industrial systems from 2.4 to 15 kV

Table 1. Characteristics of the electrical system with several grounding methods currently used in petrochemical plants.

3.1 Description of grounding practices that induce the presence of this problem

It is extremely necessary to maintain the electrical power supply of petrochemical power plants; therefore most of them have their own electrical energy generation. In some petrochemical plants there is a common practice to associate an electrical generator with a load connected directly to the electrical generator terminals. Most of the plants use 30 MW generators at 13.8 kV with a load on the order of 15 to 20 MW associated to their terminals. The generators are interconnected trough a synchronizing bus with short circuit current limiting reactors as shown in Fig. 3.1.1. The use of these limiting reactors is associated with voltage regulation problems and with the necessity to overexcite the generators, causing problems with generator stator core. However, the use of this scheme maintains the continuity of energy supply to the load sustained through the public network when generators are out of service.

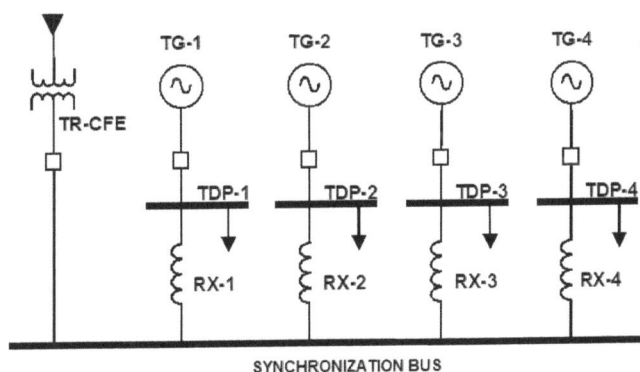

Fig. 3.1.1. Typical generation scheme of a petrochemical plant.

3.2 Factors affecting series resonance

Petrochemical plants with 13.8 kV generators; use low-resistance neutral grounding method at just one generator. The ground connection through a resistance bank has a value of 8 to 13.2 Ω which limits the ground fault current between 600 and 1,000 A. In other hand, if all generators are connected to ground, all protection devices will operate to leave the entire generators out of service. Furthermore, the short circuit ground fault current increases at a point that it will easily exceed the 1,000 A.

The neutral grounding method used in a petrochemical plant is shown in Fig. 3.2.1. The system has two power units, the first unit consists in two generators of 32 MVA (TG-1 and TG-2) and the second one with 2 generators of 40 MVA (TG-3 and TG-4). The four neutral generators are connected to a neutral bus through an interrupter. The neutral bus has a low resistance bank connected to ground. Therefore, only one generator is connected to ground.

When a ground fault occurs in a cable or in any other electric power equipment and this does not release appropriate, the protection system acts over the generator with the neutral connected to ground. The generator outs of service and the system operate temporality "ungrounded". Even though, it is in fact connected to ground through the ground system capacitance.

3.3 Overvoltages caused by series resonance

Considering an ungrounded generator, that feeds an electrical load, as shown in Fig. 3.3.1. The generator model is represented by a source and an inductance in series. The cable and the load inductance are illustrated too. Also it represents the cables capacitance, the capacitors banks and the surge arresters.

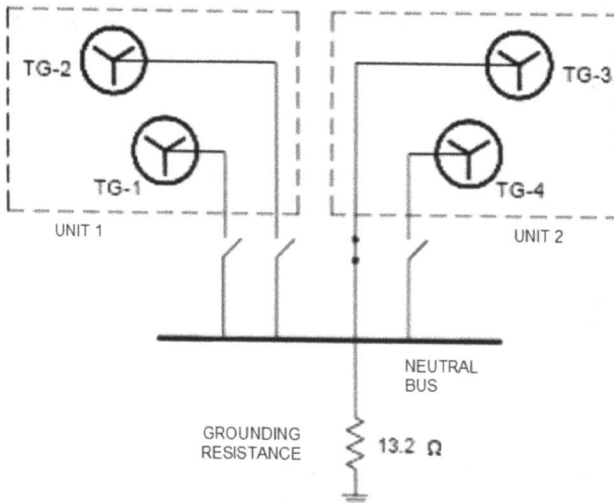

Fig. 3.2.1. Actual generator neutral grounding scheme.

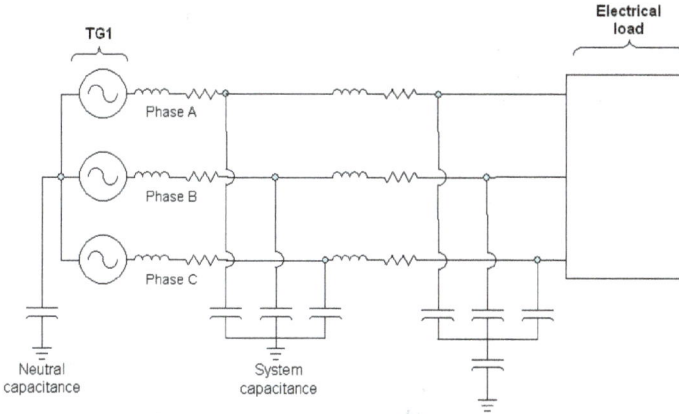

Fig. 3.3.1. Diagram of one generator neutral grounding.

The system has a steady performance. The load is feeding by the generator; the neutral current is almost zero. The generator is grounded by its own capacitance. There is not current through the neutral generator and there is a third harmonic voltage of 1 kV peak to peak approximated. When a ground fault occurs in the system, a return current through the generator capacitance and a voltage appears through the neutral, which affects the voltage applied in the un-faulted phases, as shown in Fig. 3.3.2.

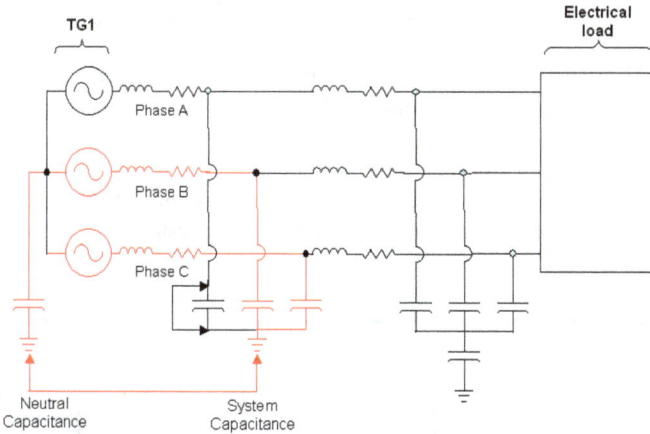

Fig. 3.3.2. Fault in phase A and current return path.

When a current is injected through the neutral capacitance, the neutral potential increases, in such a way that ground potential of the phases change to voltage between phases due to neutral shift. Although it is negative, this situation is considered in the electrical systems designs. For example, power cables are designed with 173% of insulated level. As shown in Fig. 3.3.3, an increase in the magnitude for the next phase in sequence to the ground fault phase is observed.

This voltage increase is not a problem for electrical equipments due to it is considered in design. The problem turns pessimistic due to short circuit current of ground fault phase to neutral, affecting the waveform of un-faulted phases, which stay with a no sinusoidal waveform. As a result of the interaction between fault phase and the others phases, it is obtained a waveform, as shown in Fig. 3.3.4.

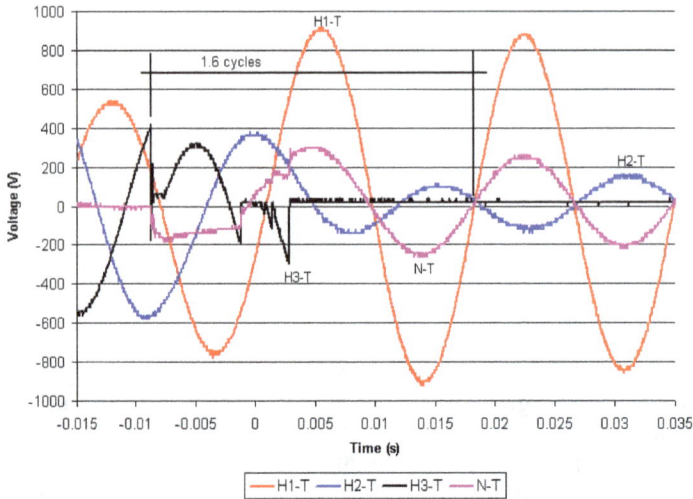

Fig. 3.3.3. Voltage increase in the no fault phase A into an ungrounded circuit.

During a short circuit the phase-neutral voltage is affected, and it is attributable to harmonics components, whose frequency depends on the generator inductance and capacitance equivalent values, furthermore, the equipments and accessories that are connected with the generator to a distribution switchboard. If a Fourier transform is applied to the voltage waveform illustrated in Fig. 3.3.4, it is possible to find the frequency of harmonic components. Fig. 3.3.5 exposes the Fourier analysis of generator phase-neutral voltage.

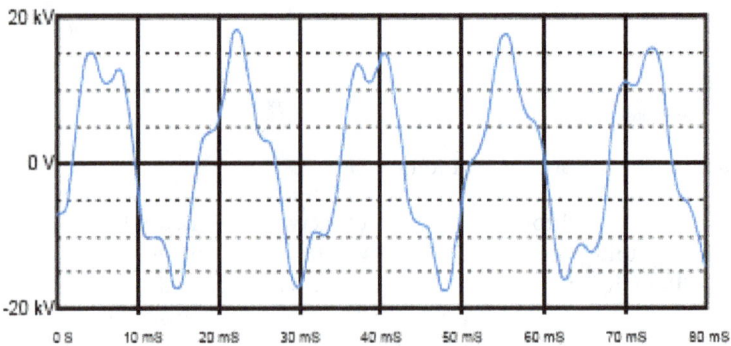

Fig. 3.3.4. Waveform of phase neutral voltage generated during a short circuit.

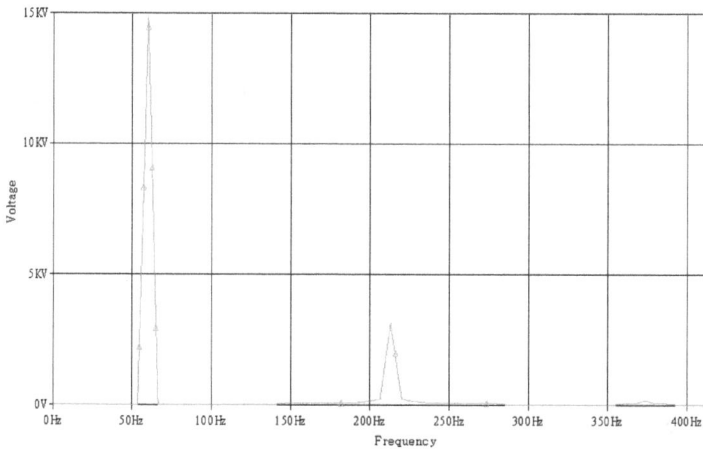

Fig. 3.3.5. Frequency spectrum of the voltage waveform illustrated in Fig 3.3.4.

During the ground fault short circuit, the un-faulted phases have a distorted waveform. With this distorted wave, the generator supplies the demanded electrical energy in conjunction with the electrical public network. At this moment, the question is if all network with capacitance and inductance distributed values of all component, are in resonance or it is represented in somewhere of the circuit where it has appropriate tuned values. Founded evidences in some faults, demonstrate that resonance is located specifically and the damage takes place in the nearby equipments to the overvoltage site. The ground fault occurs in a different phase from that had failed previously. In practice, the second fault could occur at a different site where it occurred at the first time, the crucial condition is that the inductance and capacitance are synchronized and forming a series resonant circuit as shown in Fig. 3.3.2.

3.4 Detailed analysis of equations of series resonance

The series resonance phenomenon can occur when a disturbance occurs as a phase to ground fault in power system operation. Resonance occurs when the inductive reactance equals the capacitive reactance [Rüdenberg, 1950; Enríquez Harper, 1978; Boylestad, 1998; Fuentes Rosado, 1988; Halladay and Shih, 1985; Hayt & Kemmerly, 1990; Strum & Ward, 1973; Bodger & Norriss, 1988; Kraft & Heydt, 1984].

In a series RLC circuit as shown in Fig. 3.4.1, the current $i(t)$ is given by equation (1).

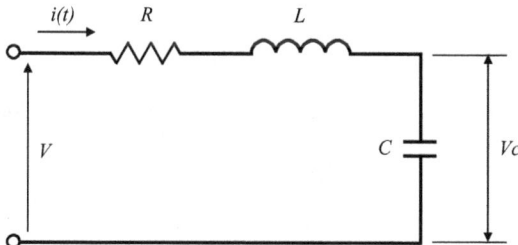

Fig. 3.4.1. Series RLC circuit.

$$i(t) = \frac{v(t)}{Z} \tag{1}$$

Where Z is the impedance of the circuit by Kirchhoff's voltage law is equation (2).

$$Z = R + j\left(\omega L - \frac{1}{\omega C}\right) \tag{2}$$

Where ω is the angular frequency of the circuit and its value is $\omega = 2\pi f$; the magnitude of Z is given by equation (3).

$$Z = \sqrt{R^2 + \left(\omega L - \frac{1}{\omega C}\right)^2} \tag{3}$$

When $X_L = X_C$ or $\omega L = \frac{1}{\omega C}$, the circuit parameters are limited mainly by the resistance R and it says that the circuit is in resonance. Under these conditions the resonance frequency is given by

$$f = \frac{1}{2\pi \sqrt{LC}} \tag{4}$$

For the particular value of the frequency of equation 4, the current I and the voltage V are in phase (P.F = 1.0) and current-voltage relationship agree with Ohm's law. Analyzing the resonant circuit in Fig. 11 is obtaining resonance overvoltage.

$$Vc = \frac{-jX_C V}{R + j(X_L - X_C)} \tag{5}$$

in resonance $X_L = X_C$ and

$$V_C = -j\frac{X_C}{R}V \tag{6}$$

or also

$$|V_C| = Q\ V \tag{7}$$

Where

$$Q = \frac{\omega L}{R} = \frac{1}{\omega RC} \tag{8}$$

The parameter Q is the quality factor of the resonant circuit and is defined as 2π times the storage energy ratio between the energy dissipated in half cycle [Rüdenberg, 1950; Boylestad, 1998; Hayt & Kemmerly, 1990; Strum & Ward 1973].

At resonance, the energy is stored in the inductor and the capacitor and only dissipated in the resistor. Also, the current through the resonant circuit is in phase with the voltage V of power source.

3.5 Practical example (showing photos of catastrophic effects in electrical equipment)

In some plants have occurred serious problems related to actual generator neutral grounding. This section describes three events in petrochemical plants.

3.5.1 Fault description in Plant "A"

The fault began with ground fault in the TDP-2 – RX-2 feeder (cable), see Fig. 3.5.1.1, when the ground fault occurred in this feeder, the negative sequence current circulated through the TG-3 neutral and TR-CFE transformer, condition that triggered the generator ground protection. After the trip from TG-3 also triggered transformer protection, therefore the power electrical system was temporarily ungrounded. Subsequently, TG-2 differential protection was operated. This protection operated because there was a ground fault in a coil of R phase into the TG-2 generator. Based on the system conditions at the time of TG-2 fault occurred, and the position of poles damaged, it is considered that this fault was caused, by the series resonance phenomena as a result of ungrounded system.

The resonance phenomenon occurs at the tuned frequency with the capacitive and inductive reactance of the circuit. In case of industrial system could be defined by 1) system capacitance, it is includes the cable, the windings and parasitic capacitance; and 2) the inductive reactance of the cable and equipment windings.

The generator core damage by electro-erosion, produce by arc flash, is shown in Fig. 3.5.1.2. The arc flash must begin by a magnitude voltage that broke the dielectric distance. For that distance, it is estimated an overvoltage that exceeding 100 kV.

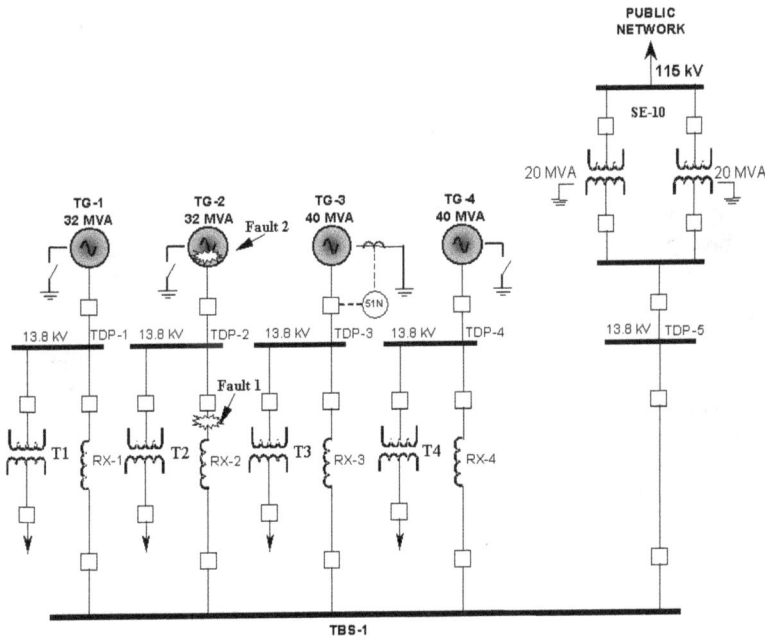

Fig. 3.5.1.1. Localization of faulted cable in the power electrical system.

Fig. 3.5.1.2. Damages evidences in the stator core laminations of TG-2.

3.5.2 Fault description in Plant "B"

This plant was operated with three electrical generators: TG-2, TG-3 y TG-4. The power electrical system was synchronized to the public network through a "link" principal transformer. The electric network of this plant was ungrounded during a fault or there was not ground reference in system by a previous fault. Subsequently, in another site a fault occurs in a switchboard called "extension bus bar of TG-2" at voltage level of a 13.8 kV. The fault consists of an arc flash between bus bar and switchboard walls when the network was steady. Transient oscillations of the frequencies derived of the arc were tuned to the circuit, and it was in resonance. The evidence shows that intermittent arc flash occurs, promoted electro-erosion as is shown in Fig. 3.5.2.1.

The distance between bus bars and switchboard walls connected to ground was 14 cm; therefore, it must be at least an overvoltage of 160 kV to break down the air dielectric strength. Then, another arc flash occurred in the un-faulted phases, after it there was produced a three-phase fault at bottom of the bus bar. Due to three-phase fault the copper material was fused in bars until it was detected by the network protections.

Another switchboard fault was presented in this plant. The evidence showed that there was an arc flash between CT´s connection bars to the switchboard wall. A few minutes before the fault occurred in the 115 kV transmission line in the public network which was feeding the load plant, then it was produced an overvoltage in the 13.8 kV network of the plant. Due to the principal transformer is connected through a reactor to the synchronizing bus; the transient was spread to the feeder circuits connected to the bus. Due to the disturber in the 115 kV line service, the interrupter was operated. Immediately after, the load segregation system was operated.

The voltage regulator of the TG-4 generator tried to absorb the load, however, the generator left of operation by overexcitation protection. The TG-1 and TG-2 generators continued operating and feeding the not segregated load. Fig. 3.5.2.2 shows the evidence of arc flash between CT´s connection bars and the switchboard wall.

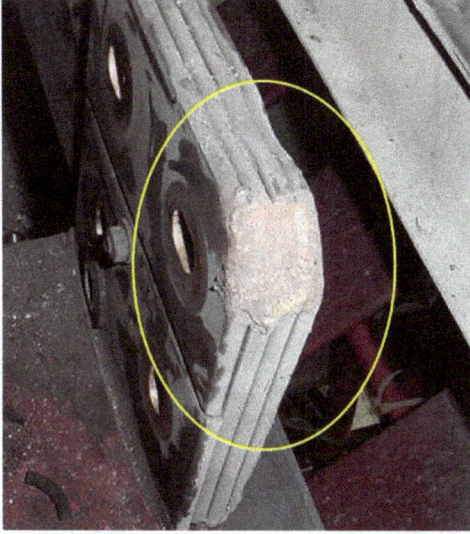

Fig. 3.5.2.1. Electroerosion damages by repetitive or intermittent ground fault arc flash.

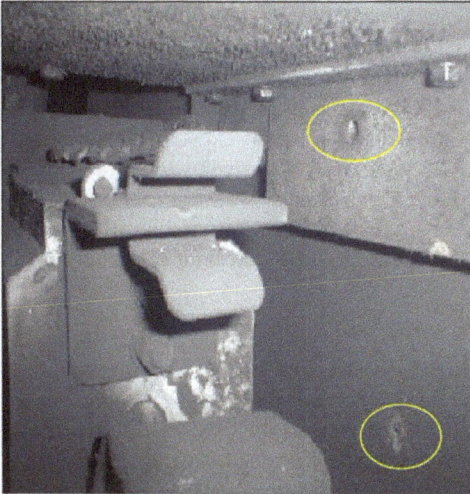

Fig. 3.5.2.2. Arc flash evidences between CT's and metallic walls.

3.5.3 Fault description in Plant "C"

In the cases above mentioned there was not an evidence of previous fault. In this case, the first fault occurred in a 13.8 kV cable feeder connected to TDP-1 switchboard about 200 meters away from where, 15 seconds after, occurred the second one. The second fault occurs in the called "cell fundamentals" where de protection and measure device's are located. The evidence shows that the generator surge protection capacitors were severely damaged, the connection bars were fused and the two potential open delta connecting transformers, showed on theirs surfaces, a evidence of three-phase short circuit (see Fig. 3.5.3.1).

Fig. 3.5.3.1. Capacitor and capacitor bars connections damages.

The fault began in a phase 2 feeder as a ground fault, it had produced an overvoltage in phase 3 with sufficient magnitude to cause the capacitor ground fault. The ground reference during the fault was in TG-3 generator; the current must be circulated through the ground resistance in TG-3. The fault performance was apparently like there was not a ground reference; the zero sequence current did not found ground return, it produced overvoltage in the not fault phases. This fact caused the generator operated like an ungrounded system.

3.6 Problem solution

It has been observed trough the years, that in some instances, when there is a ground fault in the 13.8 kV system, the generator that has a reference to ground sees the failure and the protection system takes it out of service. In this conditions, the electrical power system remains without a ground reference without being able to release the failure. The short circuit currents is limited by the capacitance of the generator, that is in the order of 0.5 A. The injected failure current trough the neutral produces a non sinusoidal voltage in the generator output. This voltage excites the failed system and a series resonant circuit can be excited. In this conditions a very dangerous overvoltages are generated. It practice, it has been observed arc flashover between electrodes that are 15 cm apart; and then the overvoltages are estimated in the order of 10 to 12 p.u. So, in order to avoid resonance overvoltages is necessary to have grounded systems to keep and intentional reference to ground. Fig. 3.6.1 shows a high resistance grounded system. When a ground fault occurs the short circuit magnitude is given by the high resistance value, in this diagram is 5 A. Simulations showed there are not maximums overvoltages more than 1.73 p.u. Finally, Fig. 3.6.2 shows a high resistance transformer installed in a petrochemical plant.

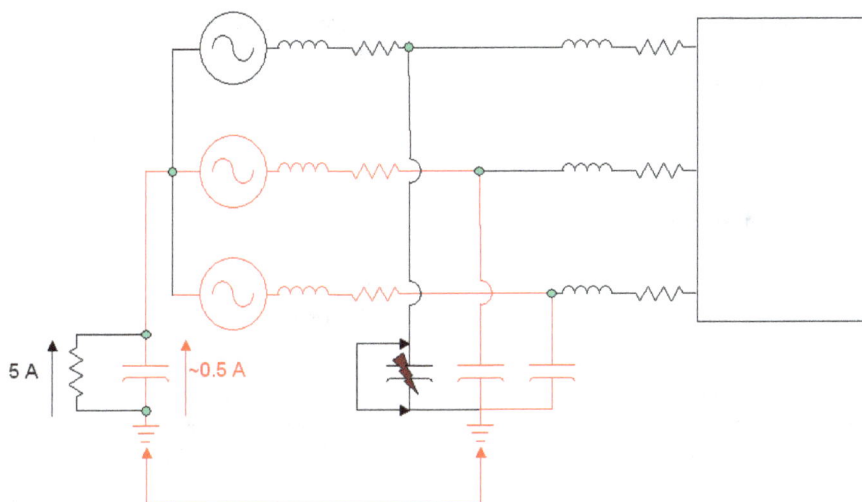

Fig. 3.6.1. High resistance grounded scheme to avoid resonance overvoltages.

Fig. 3.6.2. High resistance transformer installed in a petrochemical plant.

4. Resonance in potential transformers

It has been observed that the series resonance can occur in potential transformers (PT). In this case overvoltages were also presented by the series resonance phenomenon.

4.1 Description potential transformers in electrical switchgears

The PT is the transformer designed to provide the proper voltage to the measuring instruments such as voltmeters, frequency meters, wattmeters, watt-hour meters, etc., as well as protective devices such as relays, in which the secondary voltage is proportional to the primary and phase shift with respect to it an angle close to zero.

The terminals of the primary winding of a potential transformer are connected to the two lines of the system where is necessary to measure the high voltage and measuring instruments are connected in parallel to the secondary terminals. Its function is to provide an image proportional in magnitude with the same angle of tension in the power circuit connected. There are 2 types of PT´s, one inductive (PT) and a capacitive type (PD).

A common practice in electrical energy distribution switchgears is the open delta connection of PT's used for measurement and protection. However, it is shown that failures occur between phases with relative ease [Magallanes Ramírez & Alba Medina, 1991]. In three-phase industrial systems, the use of this connection involves using only two PT's, so that economically there is a saving in purchasing one PT for each panel. It also requires less space to accommodate two PT's within the switchgears.

The statistics regarding the incidence of failures in PT's, in some industrial plants indicate that as time passes, the number of failures in these devices increases substantially. For example, Fig. 4.1.1 shows the annual failure rate recorded in industrial plants until March 2001.

As shown in Fig. 4.1.1, the failure rate will increase. This is due to growth, sometimes excessive, of the installed capacity carried out without updating studies of the fault current levels and margins of safety switches and feeders. Also it is not considered necessary to make modifications to the grounding system. On the other hand, is a common practice in 13.8 kV to ground only one power equipment, either a generator or a transformer, so that the electrical system is exposed to surges by close-open events on switches, etc., which can be very large.

According to the records that have in some petrochemical plants, PT's that have failed, are usually located near a power source (generator or transformer), which is isolated from ground.

Therefore, considering all the above factors, it is feasible to consider the submission of PT's failures by phenomena of resonance.

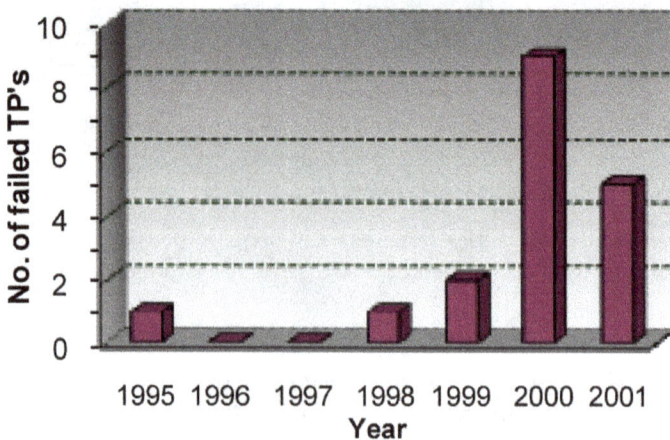

Fig. 4.1.1. Annual rate of PT's failures in petrochemical industrial plants.

4.2 Description of grounding practices that induce the presence of this problem

It is estimated that the possible cause of failure of the PT's, is due to series resonance effect caused by the open delta connection of the primary winding of the PT.

The Fig. 4.2.1 shows that when a ground fault occurs, at the end of one of the feeders there is an RLC circuit formed by the resistor (R11 and R12), inductance (L11 and L12) and capacitance (C12) of the feeders. Under fault conditions the current flows through these elements, this stream may contain a component of a frequency (1 to 10 kHz) that could generate a series resonant circuit.

The typical primary winding inductance of a PT is into the order of 28.5 H at 60 Hz, if the power cable would have a length of 1 m, this would present a capacitance of 1 nF. Therefore, in order to present the series resonance phenomenon requires a 990.8 Hz frequency. On the other hand, in the industrial petrochemical plants surges were detected in 5 to 7 kHz. These surges may be caused by maneuvers or system failure, see Fig. 4.2.2. Therefore, it is concluded that the open delta connection of PT's makes possible the resonance effect.

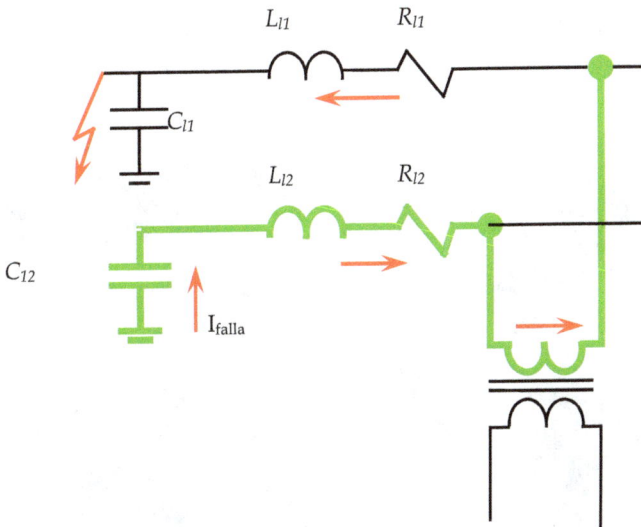

Fig. 4.2.1. Diagram of a PT in open delta connection.

4.3 Practical example (showing photos of catastrophic effects in electrical equipment)

Consider the following case, there was a failure of one of the PT's installed on the 13.8 kV level. The PT was located into a medium voltage switchgear connected to the output of a generator in an industrial petrochemical plant in Mexico. PT failure completely destroyed the switchgear that contained the PT, causing in turn the shot of the generators connected to the system and total loss of electrical power plant. As a result, the industrial petrochemical plant was out of operation for several hours, causing significant economic losses. Fig. 4.3.1 shows the damage to the PT, and Fig. 4.3.2, the damage on the switchgear.

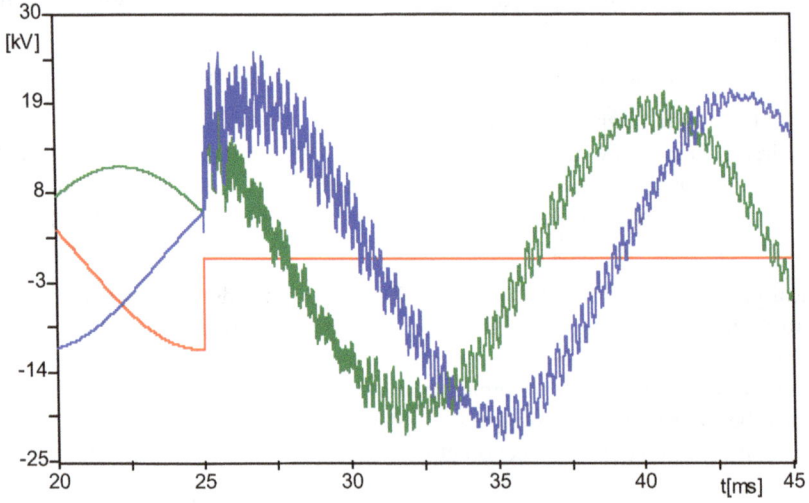

Fig. 4.2.2. Overvoltages registered in an industrial system due to maneuvers in the system itself.

Fig. 4.3.1. Damage to the PT and panel where PT was connected.

Fig. 4.3.2. Damage to the switchgear where the TP was connected.

The switchgear where PT was connected was of kind of metal-enclosed and was composed of 11 panels of 13.8 kV with a current capacity of 2,000 A. Nine of the panels were vacuum switches with their auxiliary equipment. One of the panels was used exclusively for measurement. The remaining panels, was empty because it could not be restored after the failure of one of the PT's occurred three years ago. One of the feeders of the switchgear was connected to one of the generators of the plant.

4.3.1 The process of deterioration that causes the failure of PT is slow

The industrial petrochemical plant where the TP failure occurred has a monitoring and recording system for events and variables. Fig. 4.3.1.1 shows the data of voltage, current and power recorded 20 minutes before the failure occurred.

The system logs indicate that the failure of PT began 5'30" before the three-phase fault occurred on the switchgear, see Fig. 4.3.1.2.

According to the records of power, there is an increase of active power on the order of 25% between 3:52:30 and 3:58.00. The PT had a two sections winding, each of 55 layers. Before the fault, the voltage was increased from 14.1 kV to 17.6 kV (25%). According to the transformer design, this means that one of the sections is affected at a rate of 12 seconds per layer. The initial conversion ratio was 120 and changed to 150 prior to the occurrence of the fault. That is, the transformer explosion happened after that was damaged 25% of the primary winding. The energy stored in the internal failure of the PT was large enough to produce the explosion of the transformer and the spread of the arc.

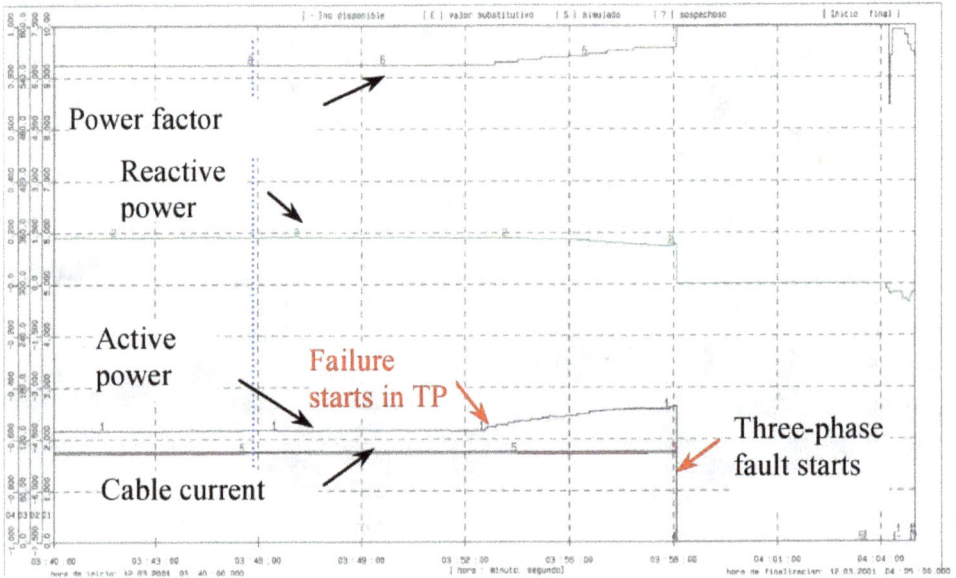

Fig. 4.3.1.1. Signals recorded of the feeder in the switchgear where the PT was installed.

Due to the open delta connection of PT's, it follows that the switchgear fire was started by the failure phase (between phases 2 and 3) of PT. According to the records of the fault currents, it is estimated that this event had duration of about 2 cycles (see Fig. 25). As shown in the figure, because the fault is a fault-phase, unbalanced residual current is generated through the common point of the current transformers. From the oscillograms it is concluded that the evolution of biphasic failure to three-phase fault was about 30 to 40 ms.

The electric arc between phases began in the TP that connects stages 2 and 3 of the switchgear. Subsequently, the arc evolves to cause the failure of phase 1 to generate a three-phase fault. Under these conditions, the arc length is estimated in the order of 60 cm. The arc is very unstable so a fire can easily spread through the fiberglass bulkheads separating the panels and the different switchgear cubicles.

Thus, according to the sequence of events leading to the occurrence of the fault, it is determined that the process of failure of a PT is a slow process, in the order of 5 to 10 minutes. However, the energy stored during this time causes serious consequences.

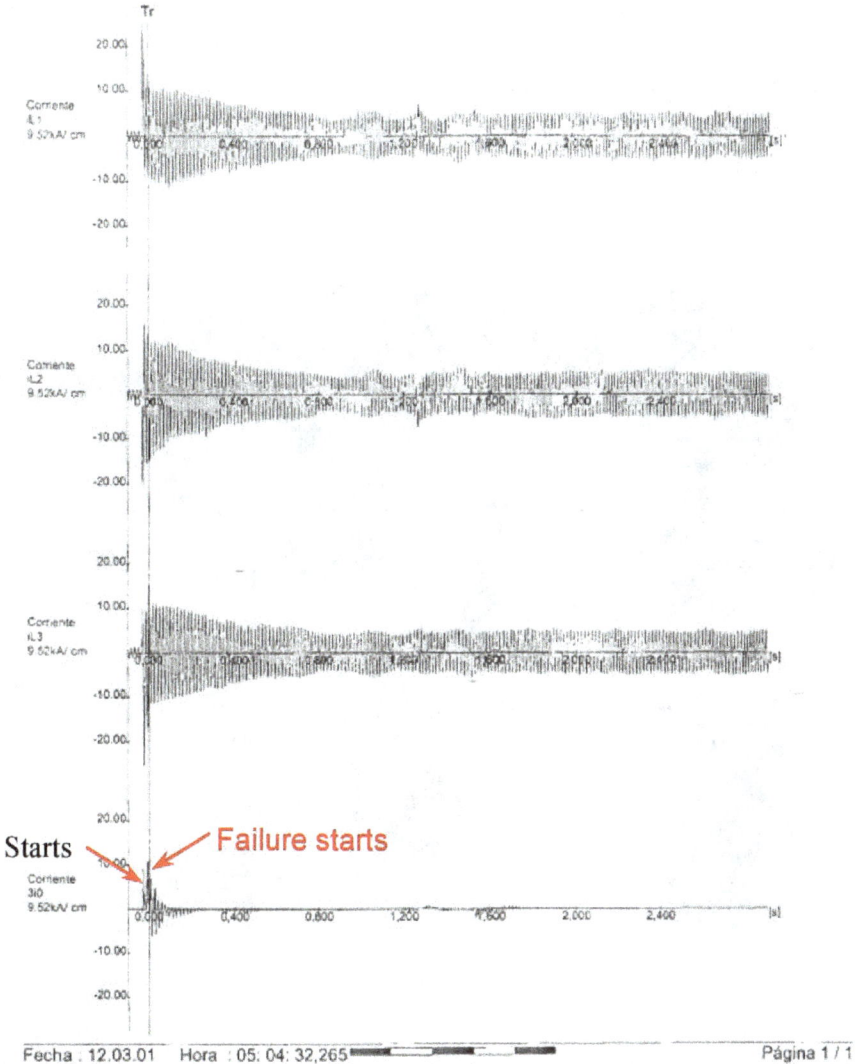

Fig. 4.3.1.2. Current signals recorded in the generator during the fault of the switchgear.

4.3.2 Design of PT

The failed PT had a capacity of 275 VA, therefore, the rated current of high voltage winding is limited to about 30 mA. To be connected in open delta, the level of its primary winding voltage is 14,400 V and the secondary winding is 120 V.

The failure of a PT can be started as a short circuit between turns or between rounds in the section of the primary winding of the transformer (see Fig. 4.3.2.1). The fault between turns can be maintained for a period of 5 to 10 minutes, until the pressure built up by the insulation decomposition gases cause the explosion of the PT. This event causes mainly the carbonization of insulation and the generation of an arc between phases.

An inspection at the failed PT indicated some manufacturing defects. Due to the arrangement used for the construction of the windings do not have adequate penetration of the resin between layers. This produces cavities between layers and therefore high levels of partial discharges.

Fig. 4.3.2.1. Potential transformer that caused the failure in switchgear.

The arrangement of the windings of the PT is shown in Fig. 4.3.2.2. Each different sections of the primary winding is made of 31 AWG wire, whose capacity for maximum current is 0.160 A, with a resistance of 435 Ω / km. Each section consists of 55 layers and 200 turns per layer so that it has an electrical gradient of about 130 V per layer. Between each layer exists one or two layers of insulating paper impregnated with varnish. It has 3 metal screens (brass) placed in different areas, which are not only used to obtain a uniform distribution of the electric field inside the PT, but also serve as a connection between the different windings.

The secondary winding is isolated from the primary winding with a section of paper based on insulation and resin, with a cross section of about 7 mm. The secondary winding is formed with 16 AWG wire, whose current capacity is 6 A. This winding is made of two layers, the upper with 93 turns and the lower with 99. Between these layers are placed a paper based on insulation and varnish.

The core of the PT is covered with cardboard. It is estimated that this arrangement is used to absorb the mechanical stresses (expansion and contraction of materials) produced during the curing of the resin. This will reduce the possibility of fracture of the encapsulation of PT.

It was detected a manufacturing defect in the high voltage windings of PT. In the section that connects the terminal P2, shows a surplus of winding and insulating material in a way peak. That is, the coil is not parallel to the core, which causes the protruding part of the coil winding.

On the other hand, as a mechanism of protection for the PT's are generally used fuses of 2 A in the high voltage side. However, this protection is only suitable for the case of full ground faults in PT, and prevents internal short circuits in the windings of the transformer.

As noted in the records of power system switchgear the failure was caused by the fault between turns of the primary winding of the PT. The most critical condition occurred at the instant it had damaged 25% of the primary winding of the PT, in which case the maximum current that could flow through this winding would be about 45 mA. It is noted that this magnitude of current is very small, and therefore the fault cannot be detected by the fuses installed in the PT.

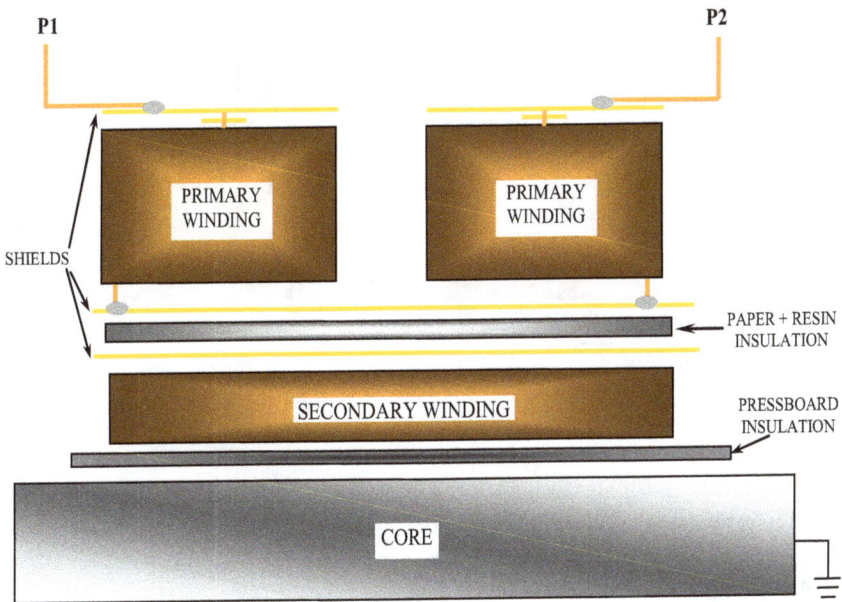

Fig. 4.3.2.2. Schematic design of PT.

4.3.3 Destruction of the switchgear as a result of the failure of PT

One of the causes that led the destruction of the switchgear was the materials currently used in it, which contributed to the rapid spread of fire. The switchgear was metal-enclosed type, in which the manufacturer used fiberglass spacers bonded to the separation between the phases of the switches and the separation between panels. This material (fiberglass) is highly flammable, so that the arc fault produced by PT caused instantaneously the fire in panel separators.

The temperature produced by the arc is very high (around 10,000 - 20,000 K). Therefore it only takes a few milliseconds to melt the materials installed on switchgear. If the switchgear does not have a fire fighting system and the failure continues to be powered by the generator, the fire may spread to the rest of the panels [ABB].

Fig. 4.3.3.1 shows the relationship between arc duration and damage in different materials of the switchgear. This figure shows that from the 500 ms the damage caused by the arc can be considered catastrophic.

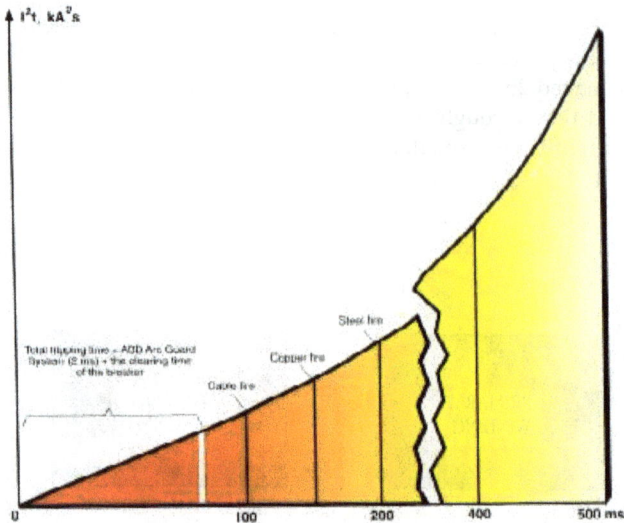

Arcing time	Result
< 100 ms	Minimal damage to equipment
> 100 ms	The personnel and equipment are at risk
> 500 ms	Catastrophic damage to equipment and personnel

Fig. 4.3.3.1. Graph damage caused by an arc of power versus time.

4.4 Solution to the problem

To avoid the resonance phenomenon in the PT's there are two alternative solutions that are mentioned below.

4.4.1 Winding wye connection of PT

To solve the problem of series resonance presented in open delta connection the PT windings can be connected in grounded wye. This connection limits the possibility of occurrence of the phenomenon of resonance, because only the parasitic capacitance of the transformer is presented. Fig. 4.4.1.1, shows the arrangement of PT grounded wye connection. As shown in Fig. 4.4.1.1, if a fault occurs at point F in one phase, it is not possible that series resonance occur in the circuit, because one end of the primary winding of the PT is grounded.

When a failure occurs in the grounded wye connection of PT's, the fuse will act protecting the PT. It is necessary to prevent failures that may arise between phases to prevent arcing that could ignite the switchgear. It is necessary to install non-flammable separators between the panels of the switchgears to prevent fires from spreading to other panels.

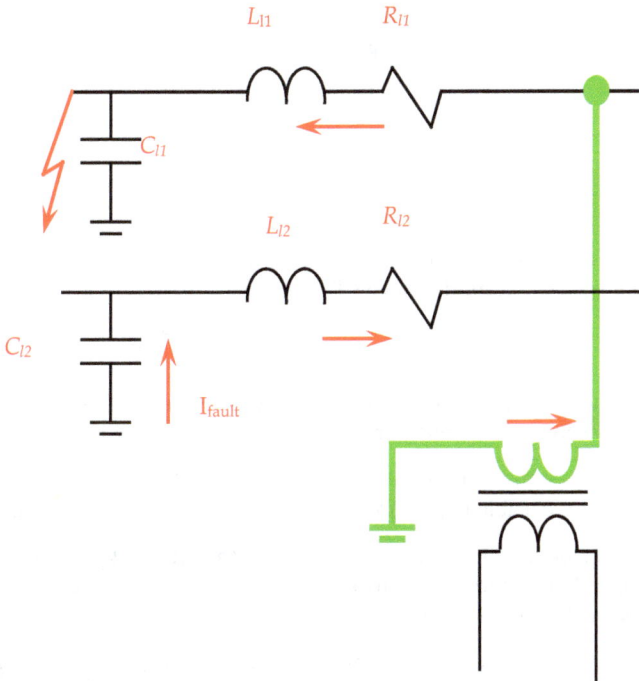

Fig. 4.4.1.1. Circuit connecting a grounded wye PT.

4.4.2 Implementation of disconnector link

Another way to avoid damage in switchgears due to faulty PT's is installing disconnects under load conditions in each of the phases of the open delta, see Fig. 4.4.2.1. Since the failure process is slow it can be detected through a relay to operate as a differential voltage and sent a signal to open a disconnect. This change is easy to implement in new switchgears and requires a specific analysis for each switchgear due to the fault currents that may arise in each.

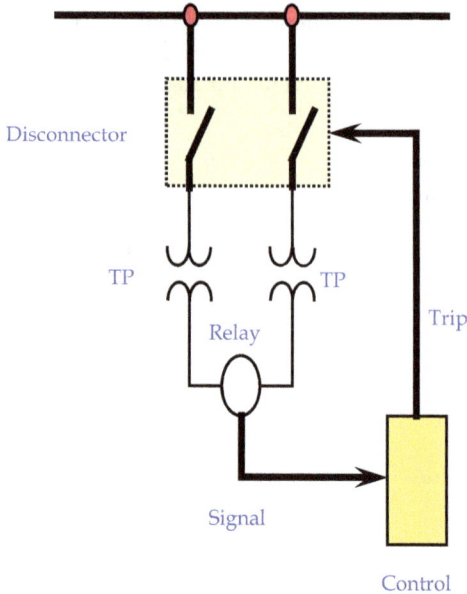

Fig. 4.4.2.1. Connection of disconnector link under load conditions in open delta PT's winding ircuit connecting a grounded wye PT.

5. Conclusion

The low resistance neutral grounding scheme at just one source of generation in industrial power systems creates favorable conditions for resonance overvoltages.

The TP's are probably the most susceptible to failure when they are operating in open delta connection, because it is feasible to generate faults between phases with relative ease.

The resonance faults occurred in specific places of the network where the concentrate parameters of capacitance and inductance are tuning at resonance frequency and not precisely the weak point of the system.

Resonance overvoltages can be avoided by the use of a proper grounding method. All energy sources, generator and transformer, must be grounded. It is necessary to consider the protection scheme and the interconnection´s characteristics of the electrical power equipments to avoid failures.

6. References

A. Girgis, C. M. Fallon, J. C. P. Rubino and R. C. Catoe. (1993). Harmonic and Transient Overvoltages Due to Capacitor Switching. IEEE Transactions on Industry Applications, pp. 1184-1188, Vol. 29, No. 6, November-December, 1993

ABB. Technical Information and Application Guide, Medium Voltage Metal-Clad Switchgear, *Bulletin* No. TB3.2.5-1A

ANSI C62.92-1987, IEEE Guide for the Application of Neutral Grounding in Electrical Utility Systems, Part I – Introduction, American National Standards Institute

B. Bridger, Jr. (1983). High Resistance Grounding, *IEEE Transactions on Industry Applications*, pp. 15-21, Vol. IA-19, No. 1, Jan.-Feb. 1983

C. J. Lemerande. (1998). Harmonic Distortion: Definitions and Countermeasures, Part 2. *EC&M Electrical Construction and Maintenance*, pp. 56-60 & 114 , Vol. 97, No. 5, May, 1998

Charles J Mozina, P.E. (2004) Upgrading the Protection and Grounding of Generators at Petroleum and Chemical Facilities", IEEE, Paper No. PCIC-2004-6

D. D. Shipp & F.J. Angelini. (1991). Characteristics of Different Power Systems Grounding Techniques: Facts and Fiction. *IEEE 1991 Textile, Fiber and Film Industry Technical Conference*, pp. 1535 – 1544, Greenville, SC, USA, May 8-9, 1991

Douglas Moody et al. (2004). Medium Voltage Generator Hybrid Grounding System. *IEEE Industry Applications Magazine*. May – June 2004

E. J. Currence, J. E. Plizga and H. N. Nelson. (1995). Harmonic Resonance at a Medium-Sized Industrial Plant. *IEEE Transactions on Industry Applications*, pp. 682-690, Vol. 31, No. 4, July-August, 1995

F. Crespo. (1975). *Sobretensiones en las Redes de Alta Tensión*. Asinel. España, 1975

G. Enríquez Harper. (1978). *Técnica de las Altas Tensiones Vol. II, Estudio de Sobretensiones Transitorias en Sistemas Eléctricos y Coordinación de Aislamiento*. Ed. Limusa, México, 1978

H. Fujita, T. Yamasaki and H. Akagi. (2000). A Hybrid Active Filter for Damping of Harmonic Resonance in Industrial Power Systems. *IEEE Transactions on Power Electronics*, pp. 215-222, Vol. 15, No. 2, March, 2000

IEC 60364-1 (2001), Electrical Installations of Buildings – Part 1: Fundamental Principles, Assessment of General Characteristics, Definitions, 2001

IEC 61936-1 (2002), Power Installations Exceeding 1 kV a.c. – Part 1: Common Rules, 2002

IEEE Standard 141-1993, IEEE Recommended Practice for Electric Power Distribution for Industrial Plants

IEEE Standard 142-1982, IEEE Recommended Practice for Grounding of Industrial and Commercial Power Systems

IEEE Standard 241-1983, IEEE Recommended Practice for Electric Systems in Commercial Buildings

IEEE Standard 242-1986, IEEE Recommended Practice for Protection and Coordination of Industrial and Commercial Power Systems

J. A. Halladay and C.H. Shih. (1985). Resonant Overvoltage Phenomena Caused by Transmission Line Faults. *IEEE Transactions on Power Apparatus and Systems*, pp. 2531-2539, Vol. PAS-104, No. 9, September 1985

J. Fuentes Rosado. (1988) Sobretensiones Resonantes en Sistemas de Transmisión. *Tesis de grado de maestría. SEPI-ESIME Instituto Politécnico Nacional*, México, 1988

J. R. Dunki-Jacobs, (1977) State of the Art of Grounding and Ground Fault Protection. *IEEE 24th Annual Petroleum and Chemical Industry Conference*, Dallas, TX, September 12-14, 1977

J.P. Nelson. (2002). System Grounding and Ground-Fault Protection in the Petrochemical Industry: A Need For a Better Understanding. *IEEE Transactions on Industry Applications*, pp. 1633 – 1640, Vol. 38, No. 6, November-December, 2002

John P. Nelson et al. (2004). The Grounding of Marine Power Systems: Problems and Solutions. Fifty-first Annual Technical Conference of the Petroleum and Chemical Industry Committee. San Francisco, California, USA. September 13-15, 2004

L. A. Kraft y G. T. Heydt. (1984). A Method to Analyze Voltage Resonance in Power Systems. *IEEE Transactions on Power Apparatus an Systems*, pp. 1033-1037, Vol. PAS-103, No. 5, May 1984

L. Magallanes Ramírez & J. Alba Medina. (1991). La Confiabilidad en Transformadores de Instrumento, Evaluación, Mejoras y Requerimientos. *Tecnolab*, agosto, 1991

L. W. Manning, (1964) Industry Power Systems Grounding Practices". *Industrial and Commercial Power Systems Technical Conference*, Philadelphia, PA, October, 1964

National Electrical Code National – National Fire Protection Association 70

O. A. Reyes Martínez. (2001). Desarrollo de un Modelo de Generador Síncrono con Transformador Saturable para el Análisis de Sobrevoltajes Transitorios en Sistemas de Mediana Tensión. *Tesis de grado de maestría. Universidad Autónoma de Nuevo León*, México, 2001

P. S. Bodger y T. B. Norriss. (1988). Ripple Control Interference at Irrigations Installations. *IEE Proceedings*, pp. 494-500, Vol. 135, Pt. C, No. 6, November, 1988

R. D. Strum and J. R. Ward. (1973). *Electric Circuits and Networks*. Ed. Quantum. USA, 1973

R. L Boylestad. (1998). *Análisis Introductorio de Circuitos*. Octava edition. Ed. Prentice Hall, México, 1998

R. Rüdenberg. (1950). *Transient Performance of Electric Power Systems*. Ed. Mc Graw Hill. USA, 1950

W. H. Hayt and J. E. Kemmerly. (1990). *Análisis de Circuitos en Ingeniería*. Cuarta edición. Ed. Mc Graw Hill, Colombia, 1990

9

Attenuation of Guided Wave Propagation by the Insulation Pipe

Jyin-Wen Cheng[1], Shiuh-Kuang Yang[2],
Ping-Hung Lee[3] and Chi-Jen Huang[3]
[1]General Manager, Cepstrum Technology Corp.
Ling-Ya Dist. Kaohsiung,
[2]Dept. Mechanical and Electro-Mechanical Engr.,
Natl. Sun Yat-Sen Univ.; Kaohsiung,
[3]Taiwan Metal Quality Control Co., Ltd.
Taiwan, ROC

1. Introduction

Pipeline systems are widely used in gas, refinery, chemical and petro-chemical industries, which usually carry high pressure, high temperature or even highly corrosive fluids. Cracks and corrosion are often found at the outer or inner surface of pipeline and can lead to a serious thinning of wall thickness. Leaks or sudden failures of pipes can cause injuries, fatalities and environmental damage. Ultrasonic nondestructive techniques are available for the detection of wall loss associated with defects in the pipe. Unfortunately, a high proportion in pipelines of these industrial are insulated, so that even external corrosion cannot readily be detected by the conventional ultrasonic testing (single position measurement) without the removal of the insulation, which in most case is time-consuming and cost expensive. Especially in typically industrial plants, there are hundreds of kilometers of pipelines can be in operation. Making inspection of full pipelines is virtually impossible in industrial plants. There is therefore a quick reliable method for the detection of corrosion under insulation (CUI). This technique, called guided wave, employs a pulse-echo system applied at a single location of a pipe where only a small section of insulation need to be removed, using waves propagation along the pipe wall. The changes in the response signal indicate the presence of an impedance change in the pipe. The shape and axial location of defects and features in the pipe are also determined by reflected signals and their arrival times. Propagation distance of many tens of meters can readily be obtained in steel pipes [1-6]. Since these guided waves are cylindrical Lamb waves along the pipe, no lateral spreading can occur and the propagation is essentially one-dimensional. In a uniform pipe, their amplitude with propagation distance is therefore only reduced by the material attenuation of the steel [7].

However, many pipes of interest are necessary wrapped with the coated materials for manufacturing process in refinery, chemical and petro-chemical plants. When the pipe is coated, some energy from the guided wave in the pipe wall can leak into the insulation material and the wave attenuates as it goes along the pipe, thus reducing the wave propagation distance. The rate of energy leakage depends on the wave propagating mode in the pipe, on the acoustic impedance and attenuation properties of the insulation material [8]. The insulation material like a mineral wool has very low acoustic impedance and is not strongly adhered to the pipe. So it has virtually no effect on either the torsional or longitudinal wave. However, if the pipe is coated by a heavy viscous substance such as bitumen, epoxy, then both shear and longitudinal modes can leak from the pipe. The rate of leakage is then controlled by the properties and thickness of the coating, the displacement of the pipe surface in all direction (radial, circumferential and axial) and the frequency. This paper describes an investigation of the attenuation of a guided wave propagating on the pipe due to various coated materials such as mineral wool, polyethylene, and bitumen which are typically used to insulate the pipeline in industrial plants. The effect of reflected signals and the traveling distance of guided waves are demonstrated both on analytical predictions and experimental measurements for a variety of the coated materials on the pipe.

There have been few studied on this subject. However, Lowe and Cawly [9] considered the plate of adhesive and diffusion bonded joints to study the leakage of energy by both longitudinal and shear wave which leads to very high attenuation rates, especially when the acoustic impedance of the materials mismatch highly between another. This case is similar to the case of a pipe embedded in a solid. Wave propagation studies in structural composites are of relatively similar [10,11]. Since the composite laminates used in aircraft and aerospace structures are usually thin plate. They have been developed in an effort to gain model-based understanding of the nature of the guided waves that can be transmitted in the composite thin plates. Jones and Laperre [12,13] also considered the propagation of Lamb waves in bi-layered elastic plates, but they did not explore the effects of internal losses. More recently, the internal losses in the coating are modeled according to the theory of linear viscoelasticity by Simonetti [14]. He found that at low frequency the guided wave attenuation is only slightly affected by the longitudinal bulk attenuation, while the contribution of the shear bulk attenuation is substantial. However, while internal losses in metals are negligible at the low frequency used, the presence of attenuate media in combination with the propagation wave can dramatically attenuate the energy of guided wave. Since a significant proportion of industrial structures are coated or embedded in other material, the assessment of the attenuation characteristics of the propagating wave modes becomes a major issue. The attenuation of the torsional guided wave in a Coal-tar-enamel-Coated pipe was measured by Kwun [15]. He used torsional guided wave to measure the attenuation of buried pipe which is coated the Coat-tar-enamel material. However, no information is currently available on the effect of the coated material on the pipe when a guided wave is employed for pipeline inspection. The aim of this paper is to evaluate the attenuation of the industrial designation coated material on the pipe when the guided wave is traveling in the pipe.

2. Materials

Since the pipe is coated with a viscous material and the energy can be carried away from the pipe by both shear and compressional waves, there is a need to analyze the wave structure under energy attenuation. The wave structure describes the displacement pattern of the propagation mode for different frequencies across the thickness of the pipe. However, the wave structure of the torsional T(0,1) mode is not frequency dependent due to its completely non-dispersive characteristics at all frequencies. The wave structure of T(0,1) mode in a 6 inch steel pipe is shown in Fig.1. It shows the profile of the tangential displacement through the thickness of the pipe wall. The axial and radial displacements are zero in this mode. It can be seen that the tangential displacement are approximately constant through the wall thickness. It is also indicated that defects can be detectable anywhere in the cross section of the pipe as illustrated at 28 kHz in Fig.1, since the T(0,1) mode is a constant with frequency. An example of the wave structure for various coated material at frequency 28 kHz are also shown in Fig.2. These plots represent the amplitudes of the circumferential displacement from the centre line of pipe of the outer half space, i.e., through the inner of pipe, then through the pipe wall, and finally through the coated material layer. For the T(0,1) mode, the circumferential displacement through the coated material layer is dominant, since the inner of pipe is empty [16]. When the pipe is coated with the strong adhesive material, such as bitumen (Fig.2a), and with the weak adhesive material, such as polyethylene (Fig.2b), the energy of guided wave will leak into the coated material from the pipe wall. The rate of leakage is controlled by the strength of adhesion on the pipe and the frequency of propagation mode. If the pipe is coated with the insulation material, such as mineral wool (Fig.2c), it is not adhered to the pipe, and then the amplitude of circumferential displacement decreases in the pipe with distance due to the material properties of pipe. This T(0,1) mode and 6 inch steel pipe are also used in the experimental measurements for the results comparison.

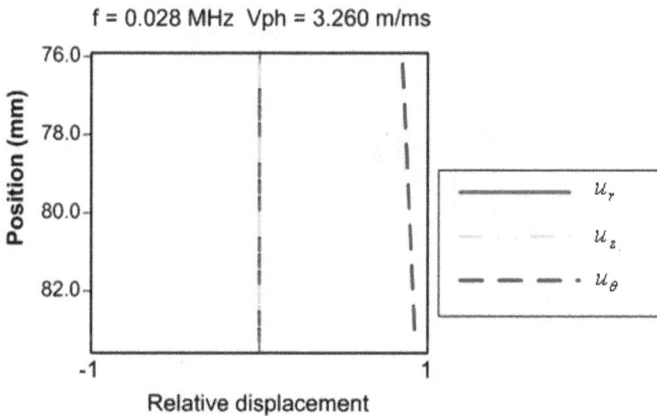

Fig. 1. The wave structure of T(0,1) mode in a 6 inch pipe at 28 kHz. This profile shows the relative displacements from inside wall 76 mm to the outside wall 83.2 mm. The radial displacement u_r and axial displacement u_z are zero.

(a) bitumen

(b) polyethylene

(c) mineral wool

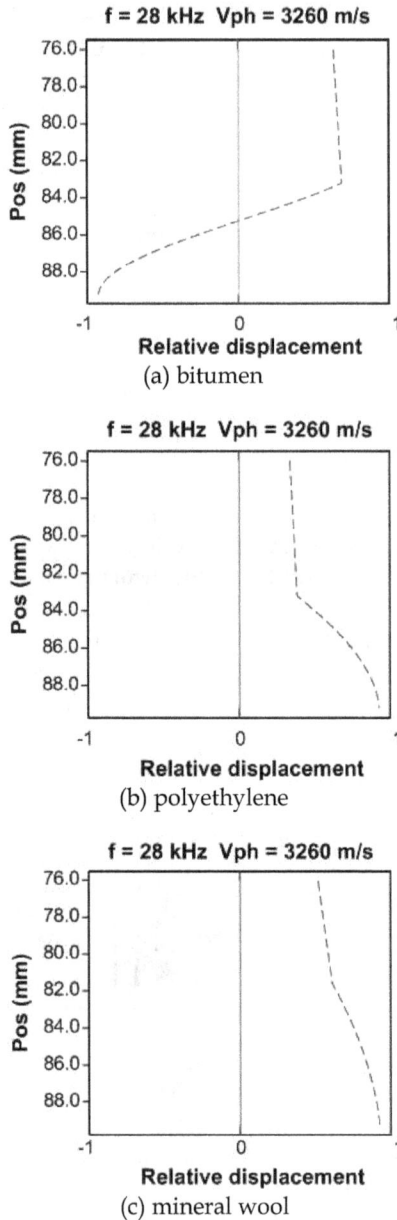

Fig. 2. The wave structures of various coating materials at frequency 28 kHz; (a) for bitumen coating, the amplitude of circumferential displacement is vary from the inside wall (position 76 mm) to outside wall (position 83.2 mm) then through the coating material. The relative displacement changes significantly high in the viscous layer; (b) for polyethylene coating, the variation of the relative displacement is slightly high in the viscous layer. (c) for mineral wool coating, its relative displacement is near the same as in the viscous layer.

The coated material used on the pipe is usually a viscoelastic layer adhered on the pipe. The internal losses in the coated material are modeled according to the theory of linear viscoelasticity, which is also the model implemented in the software DISPERSE [17] used for the wave structure analysis. The shear velocity and shear attenuation of bitumen are obtained from the result of Simonetti measurement [16] for the software used to predict the attenuation of guided wave. The material properties of the other two coated materials are found from the data bank in the DISPERSE software. The theory of linear viscoelasticity for isotropic and homogenous media is modeled in the frequency domain, which leads to linear equation of motion [18]. Thus

$$(\tilde{\lambda} + \tilde{\mu})\nabla(\nabla \bullet \vec{u}) + \tilde{\mu}\nabla^2\vec{u} + \rho\omega^2\vec{u} = 0 \tag{1}$$

Where $\vec{u}$ is the Fourier transformed displacement vector, ρ is the density of material, ∇ is the three dimensional differential operator and ω is the angular frequency. The complex frequency dependent functions $\tilde{\lambda}$ and $\tilde{\mu}$ are related to the relaxation functions of the material $\lambda(t)$ and $\mu(t)$.

$$\tilde{\lambda} = \lambda_0 + i\omega\int_0^\infty \lambda(t)e^{i\omega t}dt \tag{2}$$

$$\tilde{\mu} = \mu_0 + i\omega\int_0^\infty \mu(t)e^{i\omega t}dt \tag{3}$$

Where λ_0 and μ_0 are the asymptotic values of the relaxation curves.

Simonetti [14] studied the relationship among the guided wave attenuation, the strain energy in the viscoelastic layer, and the acoustic properties of the coated layer. For the case of torsional mode, it has already been shown [19] that by considering a volume, V, of one cycle in the circumferential direction (2π r), and with height equal to the thickness of the coated layer, the guided wave attenuation, α, can be related to the average dissipated power, P_d, in the volume and the average in-plane power flow, $< P >$, [14]

$$\alpha = \frac{dP_d / dz}{2 < P >} \tag{4}$$

Where z is the axial direction of pipe.

Generally speaking, at the considered portion of a viscoelastic medium, the average dissipated power per unit volume can be related to the peak strain energy per unit volume [20]. As a result, the total attenuation in volume, V, for the guided wave can be expressed as [14]

$$\alpha = \frac{1}{2}\omega\frac{\tilde{u}_i}{\tilde{u}_r}\frac{dE / dz}{< P >} \tag{5}$$

Where the subscript i refers to the imaginary part of the quantity, and r refers to the real part of the quantity. E is the peak strain energy of the portion of the viscoelastic layer contained in V. There are two different material attenuation regimes can be defined in

Equation (5). When the material attenuation is low which is accompanied by small imaginary parts of the Lame constant, the viscoelastic layer embraces both the non-propagating and propagating branches of the corresponding elastic layer. It will disperse in the phase velocity and guided wave attenuation spectra. For the material attenuation is high which is representative of steel pipe coated with bitumen, the energy of guided wave is employed into two parts. The first one travels primarily in the elastic layer and the second one is trapped in the viscoelastic layer. While the energy of the second family have little practical interest, as they are highly attenuated with distance, the energy of the first family can be employed in suitable frequency ranges.

3. Experimental setup

All the experiments have been performed on a 6-inch schedule 40 steel pipe, which is usually used in refinery and petro-chemical factories, to measure the reflected signal of the torsional T(0,1) mode. The pipe has three flanges, two bends, one branch, one welded support, one simple support, one patch, and ten welds of features on it. Fig.3 shows the positions of various features on the experimental pipe. The flange is assumed to be a 100% reflector, since it breaks all of the energy in the guided waves and this signal is the level of the outgoing guided wave. At a bend the geometry of the pipe changes and the pipe is no longer symmetric about the plane normal to its axis. The major characteristics of the bends are the reflected signal from the two closely spaced welds, but there will some mode conversion caused by the change in geometry of the pipeline at the bend. The branch has a side outlet hole at right angles to the axial run of the pipeline. This hole, which changes in geometry, produces direct and mode converted reflected signals. The response from this branch is similar to a bend with two closely spaced weld reflected signals. The welded support is a bracket welded to the pipe over a significant axial length. The response from this welded support is complicated due to mode conversion. The amplitude of reflected signal from welded support depends on the length of the support. The clamped support is a U type of clamp. The reflection from clamped support depends on the tightness of clamp and the frequency of excitations. The patch is similar to a defect that extends in the axial direction order more than 1/4 of a wavelength. It generates reflections from the front and back of the patch. The reflected signals are the combination of the front and back of the reflections. In general, the weld is a symmetric geometry of pipe. The weld caps cause a geometric change in the pipe wall section, which also reflects the guided wave. The amplitude of the reflected signal is related to the percentage increase in the wall cross-section at the weld. The reflected signals from various features on the experiment at frequency regime 1.0 are shown in Fig.4. The frequency regime is introduced instead of frequency in the system since the behavior of the guided waves propagation in the pipe varies with pipe dimensions. It is defined as a number which normalizes the wavelength to pipe diameter. In Fig.4, the horizontal axis is the distance from the position of the transducer mounted on the test pipe. The vertical scale is the reflected amplitude in millivolts. The gray area in Fig.4 is the near field of transducer. The amplitude of the reflected signal in this near field cannot be measured for calculation. The green area in Fig.4 is the dead zone of the measuring distance. This dead zone is near the transducer that cannot be inspected. The − W1 and +W1 sign shown in Fig.4, represent the location of the welds where the distance are

2.5-m and 3.5-m from the transducer, respectively. The amplitude and shape of the reflected signals of the weld do not change with frequency. That is why the amplitude of the reflected signal of the weld is measured for comparison standard in this paper. The differences in the amplitudes of the individual welds, -W1 and +W1 sign, indicate the variation in size of the weld caps. The –N1 and +N1 signatures in Fig.4 are the location of artificial defect which are 3.6-m and 4.9-m from the transducer, respectively. Most defects are non-symmetric. The position of these reflected signals are generally suspected as indicating anomalies, with the shape of defect signature distorted from peak shape. The amplitude of the reflected signal from defect is dependent on the depth and circumferential extent of the defect. The position of –P1 in Fig.4 is the patch feature which locates at 4.7-m from the transducer. The –F1 sign shown in Fig.4 represents the location of the flange where the distance is 6.0-m from the transducer. The shape of the flange signature depends on the duration of the excitation signal and the geometry of the flange. No energy is propagated across flanges, and testing is only possible between flange breaks. The +E1 signature in Fig.4 is the location of elbow of pipe which is 6.35-m from the transducer. The dashed lines, which indicate exponential decay, represent the distance amplitude correction curves from the features of the flange, weld and noise, respectively.

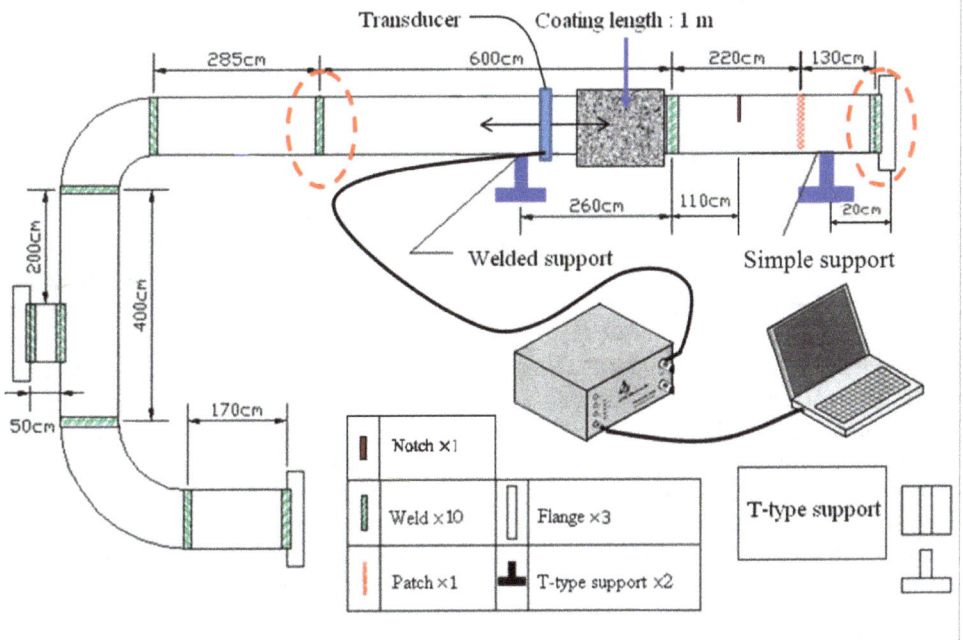

Fig. 3. Profile of the test pipe and the measuring system.

A commercial instrument is used to generate a 10 cycles Hanning-winded tone burst signal to excite the transducers. The frequency regime of guided wave is excited from -1.2 increased equally 0.2 of frequency band to 3.4 to propagate the torsional T(0,1) mode on the pipe for the measurement, respectively. When the transducer ring generates the T(0,1) mode

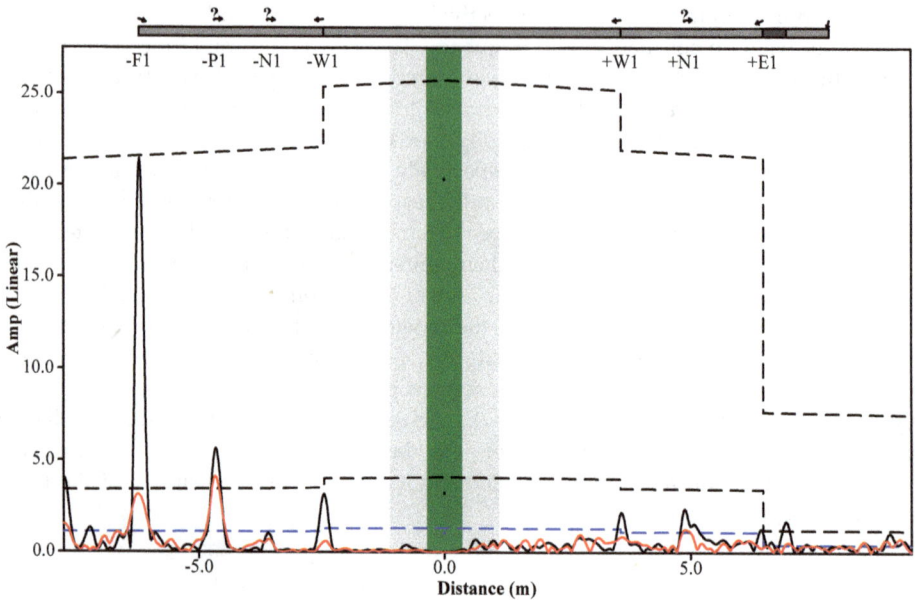

bare pipe

Fig. 4. The reflected signals of the test pipe at frequency regime 1.0 (unwrapped).

on the pipe, the guided wave can be propagated to both ends of pipe. The coated material is embedded on the pipe between the position of transducer ring mounted and flange for the experiments. Various coated materials which are used commonly in industrial plants, such as mineral wool (1-m length, 5-cm thickness, no adhesive material). Polyethylene (1-m length, 1-cm thickness, adhesive strength: 1.56 kg/25 mm), and bitumen (1-m length, 2-cm thickness, adhesive strength: 2 kg/25 mm), are used in experiments to measure the reflected signal for attenuation investigation (see Fig.5). In Fig.5, the amplitude of the reflected signal could be measured at the same position for the attenuation evaluation. At the flange feature –F1 for bitumen coating shown in Fig.5(a), the amplitude of reflection is 8.51 mV at incident frequency regime 1.0, the value is 13.6 mV for polyethylene coating shown in Fig.5(b), and the measured amplitude is 22.6 mV for mineral wood coating shown in Fig.5(c). Comparison of the amplitude value of the flange feature among Fig.5(a), 5(b) and 5(c) shows that the attenuation is significantly high for bitumen coating. In order to evaluate the effect of the coated materials for guided wave T(0,1) mode, the attenuation of reflected signal is measured by comparing the amplitude of the guided wave when the waveguide is in air to the amplitude as the waveguide is embedded by coated material over a given length, L, on the pipe. As mentioned before, measuring the reflected amplitude of the weld feature before and after embedding is measured as the reference reflector. The guided wave attenuation α_e over the embedded length L in the experiment is then calculated as

$$\alpha_e = \frac{1}{2L} 20 \log[\frac{AF_a / Aw_a}{AF_b / Aw_b}] \qquad [dB/m] \qquad (6)$$

Here L is 1-m since the embedded length of coated material is united in the experiments. At a given frequency regime, the amplitude of the Fourier transforms of the flange reflection and the weld reflection before embedding are denoted by AF_b and Aw_b, respectively, and after embedding by AF_a and Aw_a, respectively. This attenuation α_e corresponds to Equation (5) of α for the excited frequency propagating in the pipe.

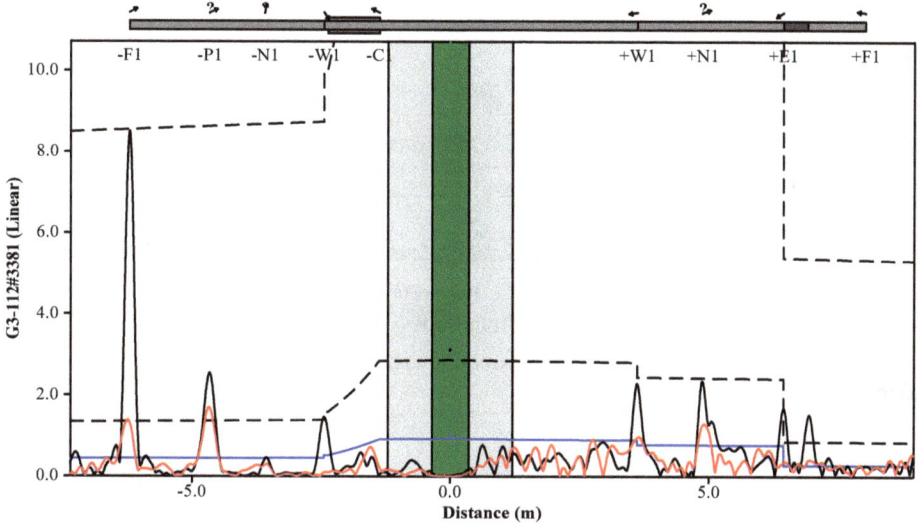

(a) bitumen

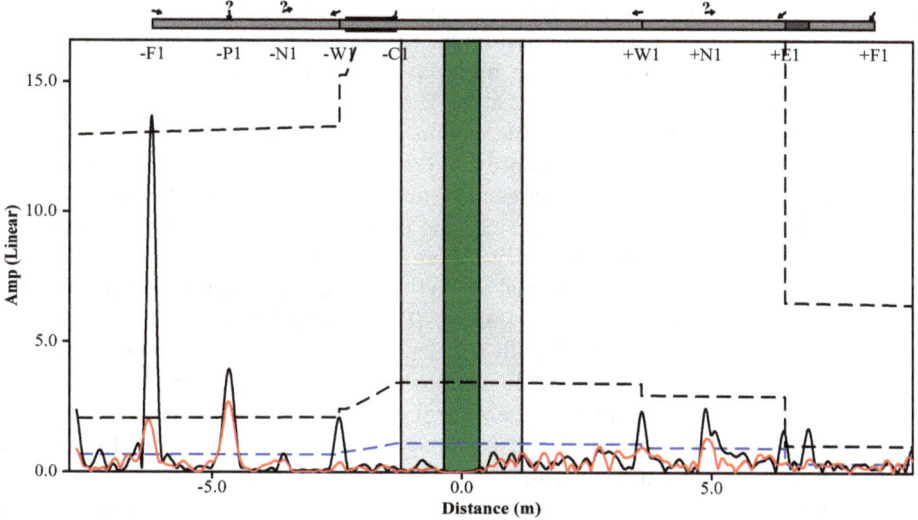

(b) polyethylene

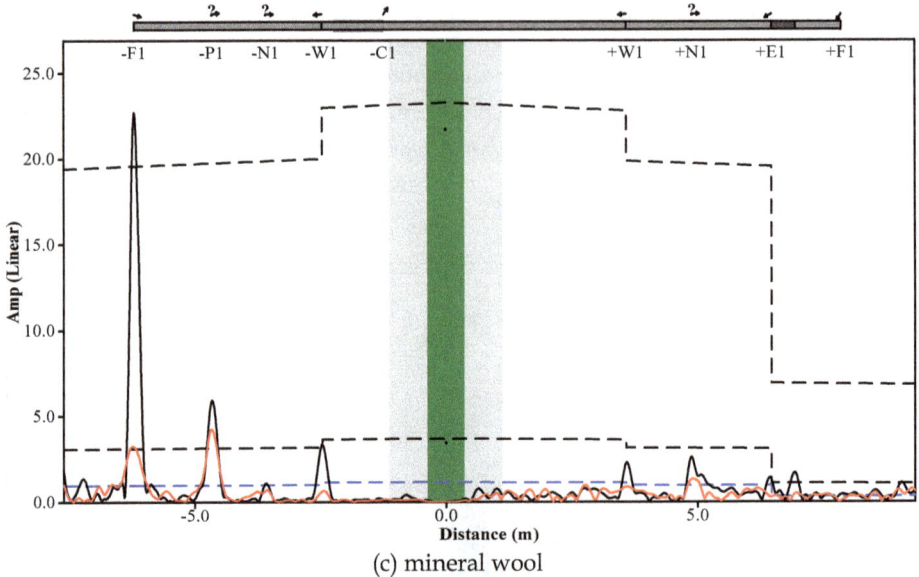

(c) mineral wool

Fig. 5. The reflected signals of the experimental measurement; (a) bitumen, (b) polyethylene, and (c) mineral wool.

4. Results and discussion

4.1 Predictions

All the predictions presented in this paper were produced using the DISPERSE software. The software is based on the global matrix method for the analysis of multi-layered structures that over-comes the problem of instability at high frequency-thickness products commonly associated with the Thomson-Haskell transfer function technique [21]. When predicting the attenuation due to the material damping of the coated material, the DISPERSE software is used in this research. The acoustic properties of various coated materials are shown in Table 1. The attenuation of the prediction discussed for various coated materials are shown in Fig.6. Notice that the attenuation of guided wave for mineral wool is nearly zero at the frequency range of 10 to 40 kHz (there is no attenuation diagram can be shown in DISPERSE software), since the coated material of mineral wool has no adhesives strength on the pipe. It means that there is zero traction at the surface of the pipe, and the energy of guided wave has no leakage to the coated material. It is obviously that the attenuation of guided wave for polyethylene (Fig.6a) and bitumen (Fig.6b) of coated material have significantly decayed with frequency. This attenuation results due to the energy dissipation of guided wave by the viscous of viscoelastic material. The behavior of the viscoelastic material combines the energy-storing feature of elastic media, the dissipating feature of viscous layer, and frequency dependent in elastic moduli [22]. The traveling distance of guided wave $T(0,1)$ mode in the pipe will be reduced by the higher attenuation. In Fig.6, the circumferential displacement through the pipe wall is dominant by the $T(0,1)$ mode. The amplitude of displacement is reduced apparently for the cases of

Material	Density (Kg/m³)	Shear Velocity (m/ms)	Shear attenuation (dB/m)	Thickness (mm)
bitumen	970	0.46	1516	6
polyethylene	950	0.95	1371	6
mineral wool	2500	2.0	0	10

Table 1. The acoustic properties of various coating materials.

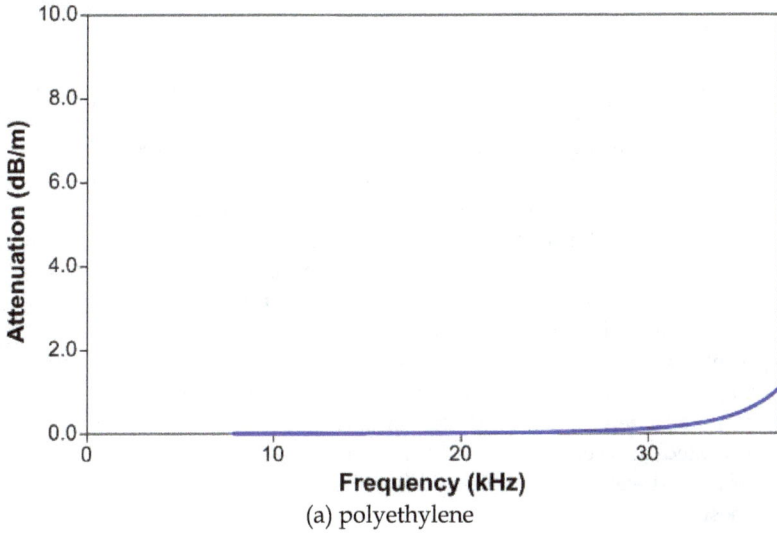

(a) polyethylene

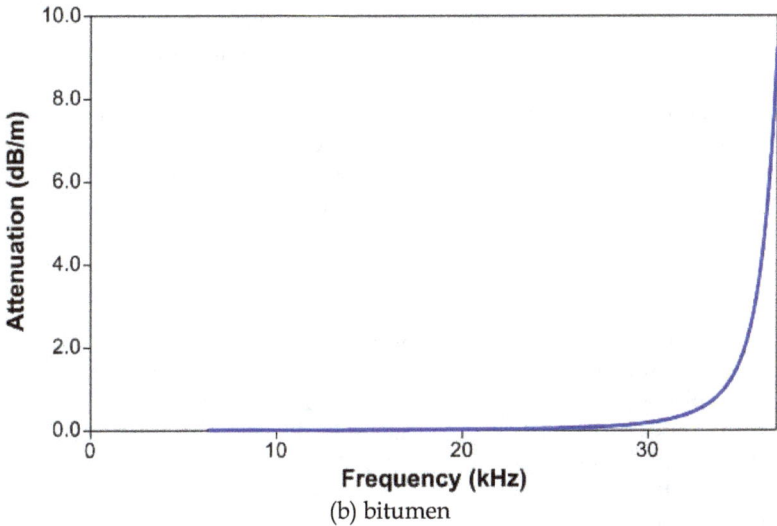

(b) bitumen

Fig. 6. Prediction of attenuation of guided wave for (a) polyethylene and (b) bitumen. The attenuation of guided wave of mineral wool cannot be shown in the DISPERSE software since the attenuation of guided wave is zero.

bitumen and polyethylene coated, while the material damping of them is heavier than mineral wool is. Another interesting result in Fig.6 is the higher frequency, the heavier attenuation of the guided wave. The reason is the high frequency has small wavelength, and can be easily dissipated the energy of guided wave into the viscous layer.

4.2 Experimental results

The experimental data is obtained using the experimental set-up shown in Fig.3. The amplitude of the reflected signals for the weld reflector and flange feature is measured to calculate the attenuation of guided wave before and after embedding the coated materials. Since the viscoelastic materials are very sensitive to temperature changes [22]. The measurements use the same set-up for various coated materials measurements and control the same temperature 28°C during the experiments in this paper. The results of measurements are shown in Table 2 for various insulation materials at selective frequency regime (0.2 to 3.0). The attenuation of guided wave is also shown in Table 2 which is calculated using Equation (6). It is found, as expected, that the attenuation of guided wave in the mineral wool coated material is near zero. This is because of the adhesive strength for mineral wool coated has no viscosity. Therefore, neither the amplitude of reflected signal nor the attenuation could be changed before and after embedding the coated material in the experimental measurements. Other interesting results are the cases of bitumen and polyethylene coated, since they have large adhesive strength. The amplitude value of the attenuation for guided wave is from 4.046 to 4.668 dB/m for bitumen coated, and the value range is from 2.235 to 2.453 dB/m for polyethylene coated at selective frequency regime (0.2 to 3.0). It became evident that the attenuation of guided wave is due to the leakage of energy into the viscous layer. On the other hand, the adhesive strength of bitumen coating and polyethylene coating is large enough that the $T(0,1)$ wave leaks the energy into the viscous layer. The results of the experiment for various coated materials at selective frequency range are shown in Fig.7. Comparison of the results in Fig.7 with the results of predictions in Fig.6 shows that the attenuation of guided wave is significantly high for bitumen coating, slightly high for polyethylene coating, and there is almost no effect for mineral wool coating. The phenomenon of the predicted curves is clearly in agreement with the experiments for all case of various insulation materials on the pipe. Hence the agreement between the predictions and experiments for the attenuation of guided wave is excellent. In addition, the behavior of the higher attenuation of guided wave in the higher frequency has been observed. The reason for this variation with frequency is the small wavelength at high frequency cannot carry incident energy largely and easy to leak the energy into surrounding media.

For the traveling distance evaluation, the average attenuation of guided wave is used to calculate the propagating range for pipe inspection at selective frequency regime (0.2 to 3.0). The signal to noise ratio used in this experiment is 32 dB for the measuring instrument. It means that the noise or baseline level is set to 32 dB below the flange level, the 100% reflector, of the reflected signal. The traveling distance on pipe for the bitumen coating is about 7.53-m since the decay rate of guided wave is 4.250 dB/ m, take frequency regime 1.0 for example. For the polyethylene coating, the traveling distance is near 14.03-m for the decay rate of guided wave at frequency regime 1.0 is 2.281 dB/m. There is no effect on the traveling distance for the mineral wool coating. Hence, the amplitude of guided wave decreases exponentially with distance for $T(0,1)$ mode propagating in the pipe. The value of

Coating: mineral wood

Frequency regime	0.2	0.4	0.6	0.8	1.0	1.2	1.4	1.6	1.8	2.0	2.2	2.4	2.6	2.8	3.0
AW0	2.47	2.39	2.32	2.24	2.17	2.11	2.05	2.01	1.98	1.94	1.93	1.91	1.90	1.87	1.85
AF0	24.9	24.0	23.1	22.3	21.4	20.6	19.7	18.9	18.1	17.3	16.5	15.7	14.9	14.0	13.2
AW1	2.63	2.54	2.45	2.37	2.29	2.23	2.17	2.12	2.09	2.07	2.04	2.03	2.01	1.99	1.97
AF1	26.4	25.4	24.5	23.5	22.6	21.7	20.9	20.0	19.1	18.3	17.4	16.6	15.7	14.8	13.9
dB / m	-0.02686	-0.01062	-0.00765	-0.000321	0.00626	0.00050	-0.00055	-0.00099	-0.00203	-0.03558	-0.01132	-0.01583	-0.01904	-0.02155	-0.02371

Coating: polyethylene

Frequency regime	0.2	0.4	0.6	0.8	1.0	1.2	1.4	1.6	1.8	2.0	2.2	2.4	2.6	2.8	3.0
AW0	2.47	2.39	2.32	2.24	2.17	2.11	2.05	2.01	1.98	1.94	1.93	1.91	1.90	1.87	1.85
AF0	24.9	24.0	23.1	22.3	21.4	20.6	19.7	18.9	18.1	17.3	16.5	15.7	14.9	14.0	13.2
AW1	2.67	2.58	2.50	2.41	2.33	2.27	2.21	2.17	2.14	2.12	2.10	2.08	2.07	2.05	2.02
AF1	16.1	15.5	14.8	14.2	13.6	13.0	12.5	11.9	11.4	10.8	10.3	9.81	9.29	8.75	8.20
dB / m	-2.23518	-2.24017	-2.2562	-2.27239	-2.28074	-2.30523	-2.32209	-2.33864	-2.35843	-2.40678	-2.39589	-2.41350	-2.42880	-2.44249	-2.45262

Coating: bitumen

Frequency regime	0.2	0.4	0.6	0.8	1.0	1.2	1.4	1.6	1.8	2.0	2.2	2.4	2.6	2.8	3.0
AW0	2.47	2.39	2.32	2.24	2.17	2.11	2.05	2.01	1.98	1.94	1.93	1.91	1.90	1.87	1.85
AF0	24.9	24.0	23.1	22.3	21.4	20.6	19.7	18.9	18.1	17.3	16.5	15.7	14.9	14.0	13.2
AW1	2.65	2.55	2.46	2.37	2.29	2.22	2.16	2.11	2.07	2.05	2.03	2.01	2.00	1.98	1.96
AF1	10.5	9.99	9.46	8.98	8.51	8.06	7.63	7.23	6.85	6.48	6.13	5.78	5.44	5.11	4.77
dB / m	-4.04576	-4.09656	-4.15152	-4.19787	-4.24978	-4.29583	-4.34772	-4.39731	-4.43647	-4.50485	-4.52555	-4.56658	-4.60539	-4.63903	-4.66763

Table 2. The measurement results for various coating materials.

attenuation of guided wave for each frequency regime shown in Table 2 can also be calculated. Take bitumen coating at incident frequency regime 2.2 for example, the attenuation is 4.523 dB/m listed in Table 2. Therefore, the traveling distance of guided wave is 7.07-m for incident frequency regime 2.2 propagating on the pipe of the bitumen coating when the T(0,1) mode is excited by 10 cycles Hanning-windowed tone burst signal.

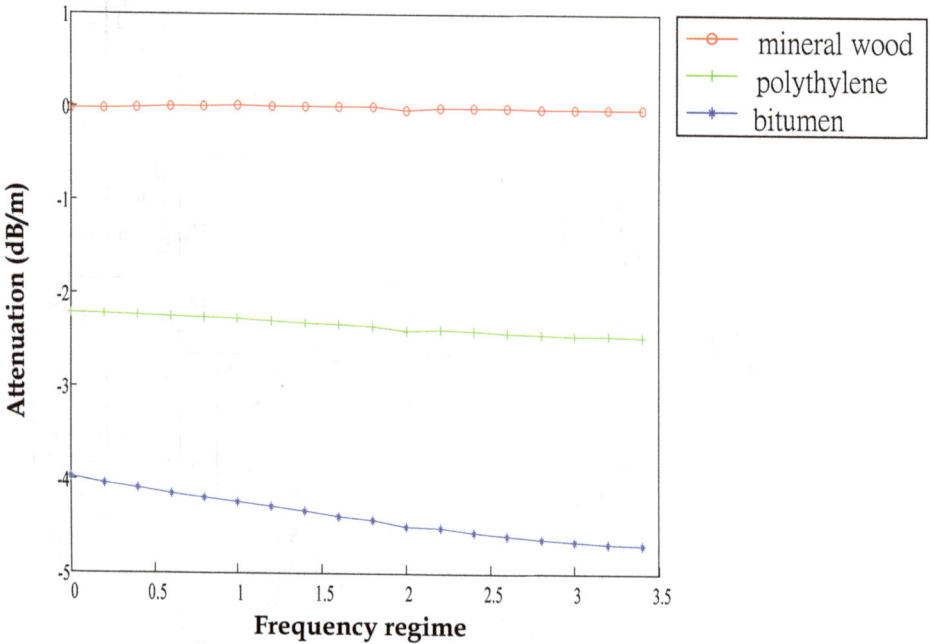

Fig. 7. The results of the experiment for various insulation materials.

5. Conclusions

Good agreement has obtained between the experiments and predictions for the attenuation of guided wave study on the insulation pipe. The effect of the adhesive strength on the pipe is investigated for various coated materials. The results show that the bitumen coating on the pipe has 4.046 to 4.668 dB/m attenuation of guided wave, the polyethylene coating has 2.235 to 2.453 dB/m, since the adhesive strength of the bitumen coating is larger than the polyethylene coating is. There is no effect on the torsional T(0,1) mode for mineral wool coating because it has no adhesive strength applied on the pipe. In addition, the traveling distance of the torsional T(0,1) mode on the pipe is evaluated in this paper. It can travel about 7.53-m for bitumen coating, and the traveling distance is near 14-m for polyethylene coating, while the guided wave is excited to propagate the pipe for pipeline inspection. For mineral wool coating, the propagating range depends only on distance due to the natural decay of material properties of the pipe. The results of this study are of interest refinery, chemical and petro-chemical industrial plants for pipeline inspection, since a high proportion of these industrials pipelines are wrapped with the coated material for the necessary of their work.

6. References

[1] D.N. Alleyne, M.J.S. Lowe. and P. Cawley, "The Reflection of Guided Waves from Circumferential Notches in Pipes," ASME Journal of Applied Mechanics, Vol.65, 1998, pp. 635-641.

[2] D.N. Alleyne and P. Cawley, "The Effect of Discontinuities on the Long Range Propagation of Lamb wave in Pipes," Proc. I Mech E, Part E: J. Process Mech. Engng. Vol.210, 1996, pp. 217-226.

[3] W. Mohr and Holler "On Inspection of Thin Walled Tubes for Transverse and Longitudinal Flaws by Guided Ultrasonic Waves," IEEE Trans. Sonics Ultrason. SU-23, 1976, pp. 369-374.

[4] W. Bottger, H. Schneider and W. Weingarten, "Prototype EMAT System for Tube Inspection with Guided Ultrasonic Waves," Nuclear Engng. Des. Vol.102, 1987, pp. 356-376.

[5] J.L. Rose, J.J. Ditri, A. Pilarski, K. Rajana and F.T. Carr, "A Guided Wave Inspection Technique for Nuclear Steam Generator Tubing," NDT&E. Int. Vol.27, 1994, pp.307- 330.

[6] G.A. Alers and L.R. Burns. "EMAT Designs for Special Applications," Material Eval. 45, 1987, pp. 1184-1189.

[7] D.N. Alleyne and P. Cawley, "The excitation of Lamb waves in pipes using dry-coupled piezoelectric transducers," Journal of Nondestructive Evaluation, Vol.15, No.1, 1996, pp. 11-20.

[8] D. Gazis, "Three Dimensional Investigation of the Propagation of Wave in Hollow Circular Cylinders," J. Acoust. Soc. Am. 31(5), 1959, pp. 568-578.

[9] M.J.S. Lowe and P. Cawley,"The Applicability of Plate Wave Techniques for the Inspection of Adhesive and Diffusion Bonded Joints," J. NDE 13, 1994, pp. 185-200.

[10] A.H. Nayfeh, *Wave Propagation in Layered Anisotropic Media with Applications to composite* Elsevier, Amsterdam, 1995.

[11] D.E. Chimenti, "Guided Waves in plates and their use in Material Characterization", Appl. Mech. Rev. 50, 1997, pp. 247-284.

[12] J.P. Jones, "Wave Propagation in a Two-Layered Medium," J. Appl. Mech. 31(2), 1964, pp. 213-222.

[13] J. Laperre and W. Thys, "Experimental and Theoretical Study of Lamb Wave Dispersion in Aluminium/ Polymer Bilayers," J. Acoust. Soc. Am. 94, 1993, pp. 268-278.

[14] F. Simonetti, "Lamb Wave Propagation in Elastic Plates Coated with Viscoelastic Materials," J. Acoust. Soc. Am. 115, 2004, pp. 2041-2053.

[15] H. Kwan, S.Y. Kim, M.S. Choi. And S.M. Walker, "Torsional Guided Wave Attenuation in Coal-tar-enamel Coated, Buried Piping," NDT&E Int. Vol.37, 2004, pp. 663-665.

[16] F. Simonetti and P. Cawley, "A Guided Wave Technique for the Characterization of Highly Attenuative Viscoelastic Materials," J. Acoust. Soc. Am. Vol.114 (1), 2003, pp. 158-165.

[17] B. Pavlakovic, M.J.S. Lowe and P. Cawley, DISPERSE: a general purpose program for creating dispersion curves. In: Thompson D, Chimenti D, editors. Review of progress in quantitative NDE, Vol. 16, New York, Plenum Press, 1997.

[18] R.M. Christensen, *Theory of Viscoelasticity: An Introduction*, Academic, New York, 1971.

[19] F. Simonetti and P. Cawley, *On the Nature of Shear Horizontal Wave Propagation in Elastic Plates Coated with Viscoelastic Materials*, Proc. R. Soc. in press, 2004.

[20] B.A. Auld, *Acoustic Fields and Waves in Solids*, Volume 1, Kriger, Malabar, FL, 1990.

[21] M.J.S. Lowe, "Matrix Techniques for Modeling Ultrasonic Wave in Multilayered Media," IEEE Trans. Ultrason. Ferroelectr. Freq. Contr. 42, 1995, pp. 525-542.

[22] J.L. Rose, *Ultrasonic Waves in Solid Media*, Cambridge University Press, 1999.

10

Causes of Floccules Formation in Hydro-Treated Lubricating Base Oil

Tao Huang[1], Hong Gao[2] and Xingguo Cheng[3]
[1]College of Resource and Environmental Science, Lanzhou University, Lanzhou,
[2]Key Laboratory of Western China's Environmental Systems (Ministry of Education),
College of Resource and Environmental Science, Lanzhou University, Lanzhou,
[3]Lanzhou Lubricating Oil R&D Institute of PetroChina, Lanzhou,
P. R. China

1. Introduction

150BS bright oil is an important kind of high viscosity lubricating base oil. Its main used field is to modulate some oils, such as gas engine oil, gear oil, compressor oil and aircraft hydraulic oil et al. The amount of thickening oil can be reduced greatly using this component. However, only a few refineries can produce this kind of product in China because of the special requirements of raw material and its high investment etc.

150BS bright oil produced by Karamay Refinery is a new series of health-friendly products. In comparison with traditional solvent-refined base oil, it has many excellent characteristics, such as high viscosity index, low volatility loss, and low aromatic content (Galiano-Roth and Page, 1994). However, the hydro-treated base oil became hazy and emerged floccules in low temperature after the high-pressure hydrogenation unit running for a period. This disadvantage seriously impact on the practical application and marketable credit of this product. Thus, it is necessary to find out the reason for the forming of floccules in KH150BS and to find an effective means to overcome this problem.

In recent years, many studies have discussed the problem of deposition emerging in base oils at low temperatures. Singh et al. (1999) pointed out the question of the three-dimensional structure of the network in gels and of the type of interactions between crystals. They discussed the changes of the morphology of the crystals under different rates of cooling. In addition, other research showed that needle-shaped crystals were formed during slow cooling under static conditions, while platelet-like crystals with amorphous solid appeared under flow conditions (Dirand et al., 1998). Srivastava et al. (1993) found that the solid–liquid and solid–solid thermally induced phase transitions occurring in petroleum waxes play a great role in deciding the appearance of waxes and the deposition of waxes as temperature decrease causes sedimentation course. Pederson et al. (1991) established a thermodynamic model for the prediction of gel formation. However, these investigations mainly focused on the morphological and thermal behavior of the crystals under cooling conditions. This study firstly analyzed the composition of floccules. Then it found the cause of floccules formation in KH150BS at low temperature from the floccules formation mechanism and processing of KH150BS. Finally, the influence of floccules on properties of oil was investigated.

2. Experiment

2.1 Reagents and apparatus

Silica gel (60–100 mesh) was purchased from the Qingdao Chemical and Engineering Plant (Qingdao, China). The n-$C_{12}H_{26}$ to n-$C_{50}H_{102}$ n-Paraffins were all chromatographic grade, and other reagents used in this experiment were all analytical grade, including the petroleum ether, isooctane, carbon tetrachloride, urea, benzene, diethyl ether, butanone, toluene, and isopropyl alcohol. The petroleum ether was purified using percolation through the silica gel.

The analysis of the carbon number distribution of the floccules was performed on a gas chromatography (GC) 2010 (Shimadzu Company, Japan); the hydrocarbon type between the floccules and filtrated oils was analyzed by a QP5050 gas chromatography – mass spectrometry (GC-MS; Shimadzu Company, Japan); molecular weights were obtained using a K7000 Vapor Pressure Osmometer (Knauer Company, Germany); all degrees of branching were collected using a Nexus 670 infrared spectrometer (Nicolet Company, USA). For the other properties of original and filtrated oils, the apparatus are as follows: cloud point, BSQ-4D8K (Xi'an Jinghua Petrochemical Instruments Company, China); Pour point, CPP97-2 (CPP97-2 Company, France); freezing point, JH011208 (Xi'an Jinghua Petrochemical Instruments Company, China); Viscosity, HVM472 (Herzog Company, Germany); Aniline point, JSH2802 (Hunan Petrochemical Instruments Company, China) Oxidation stability, SYD-0193 (Wuhan Gelaimo Company, China) Antifoam, P643 (Normalab Analis Company, France).

The hydrotreated base oil sample (KH150BS) was provided by Karamay Refinery (Xinjiang Province, P. R. China).

2.2 Experimental procedures

2.2.1 Separation of the floccules from KH150BS

To separate the floccules from the KH150BS under low viscosity conditions conveniently, a method of solvent dilution and filtration was employed at low temperature. The detailed procedures were as follows: (1) The sample was diluted with an equal mass of a solvent made of a 1:1 mixture of butanone and toluene. (2) One container of the above solution was set in each of four cryostats whose temperature was previously set at -5°C, -10°C, -15°Cand -20°C. (3) The samples were separated into filtrated residue substances and filtrates with a vacuum pump, a grit filter (G4, aperture 5µm – 6µm) and a filter flask. (4) The solvents in the filtrates were again obtained with a rotary evaporator at negative pressure and a small amount of solvent in the filtrated residues was vaporized on a water bath. Finally, the floccules and filtrated oils were obtained by a series of above procedures.

2.2.2 Separation of n-paraffins from the floccules

The urea was dried for 2 hours at 105°C and then ground. One gram of floccules was dissolved in 30 grams of petroleum ether; then a certain amount of urea and isopropyl alcohol were added. The suspension was stirred in a round-bottomed flask for certain time at the certain temperature. After filtering with suction and washing the filter three times with 30 ml of petroleum ether, the mass was decomposed in hot water. The liberated hydrocarbon was dissolved up in 20 ml of diethyl ether, and the diethyl ether layer was separated. The evaporation of the diethyl ether and the residual petroleum ether left n-paraffins.

3. Results and discussion

3.1 Separation of floccules from KH150BS

The cloud point is the temperature at which haze was observed in the oil, and it can reflect the range of temperature when the floccules appeared. As is shown in Table 1, the pour point of KH150BS was 16°C; however, its cloud point was 12°C, and it decreased to 10°C after the floccules were separated from KH150BS at -5°C. As the experimental temperature decreased, the cloud point, pour point, and freezing point of filtrated oils obviously exhibited a decrease; the viscosity index of filtrated oils firstly decreased, but subsequently it increased to 84. The results indicate that the floccules were effectively separated from KH150BS by employing a method of solvent dilution and filtration at low temperature, and the floccules had high freezing points.

Items		KH150 BS	I	II	III	IV	Test method
Temperature/° C			-5	-10	-15	-20	
Floccules/wt.%			1.67	2.71	4.18	5.65	
Viscosity $/ mm^2 \cdot s^{-1}$	40° C	595.1	580.4	532.2	531.5	507.4	GB/T265-1988
	100° C	32.70	31.82	30.01	30.57	29.65	GB/T265-1988
Viscosity index		84	82	82	84	84	GB/T1995-1998
Cloud point/° C		+12	-10	-13	-14	-16	GB/T3535-1983
Pour point/° C		-16	-18	-21	-24	-30	GB/T510-1983
Freezing point/° C		-17	-20	-23	-25	-31	GB/T6986-1986

I, II, III and IV is experimental procedure at -5°C,-10°C,-15°C and -20°C respectively.

Table 1. The properties of KH150BS and filtrated oils.

Kane et al. (2003) found that gel formation appeared for very low amounts of crystals (around 1-2%) when the oil was cooled under quiescent conditions. In such conditions, a colloidal network, which embodies the oil, is formed, and the whole dispersion becomes a gel. As shown in Table1, the lowest content of the floccules separated from KH150BS was 1.67%, and the content of the floccules exhibited an increase with a decrease in the experimental temperature. This experimental result revealed the morphology of crystals formed in base oil during slow cooling under quiescent conditions. The common features of the morphology are the nuclei of crystals, which have the shape of discs and are a single molecule thick. The lateral extension of the nuclei depends on the crystallization conditions. Under quiescent conditions, the formation of extended, continuous lamellas is allowed, making a colloidal network, which covers the oil itself.

3.2 Analysis of the composition of the floccules

3.2.1 Average molecular weight

At lower temperature, the crystal growth depends largely on the molecular weight distribution, concentration of paraffinic component and the composition of the oil (Webber, 2001). The average molecular weights of the floccules, KH150BS and filtrated oils were firstly measured and the results were given in Table 2. From Table 2, it can be seen that the average molecular weight of the floccules was obviously higher than KH150BS and the corresponding filtrated oil, besides, the average molecular weight between the floccules and filtrated oil exhibited a decrease with a decrease of the experimental temperature. It indicate that its solubility decreased with an increase in average molecular weight of the floccules, so the floccules separated at higher experimental temperature were more difficult to dissolve in KH150BS. Webber (2001) pointed out that the hydrocarbon which had large molecular weight easily shaped crystals and deposited in the oil. Hence, it can be primarily concluded that the floccules were some hydrocarbon mixture which had larger average molecular weight and lower solubility compared with KH150BS.

Item	KH150BS	L_1	L_2	L_3	L_4	F_1	F_2	F_3	F_4	Test method
Molecular weight	590	582	569	562	532	819	726	631	617	SH/T0583-1994

F_1, F_2, F_3 and F_4 is the floccules separated from KH150BS at -5°C,-10°C,-15°C and -20°C respectively; L_1, L_2, L_3 and L_4 is the filtrated oil at -5°C,-10°C,-15°C and -20°C respectively.

Table 2. Average molecular weights of KH150BS, the floccules and filtrated oils.

3.2.2 Carbon number distribution of the floccules

3.2.2.1 Separation of n-paraffins from the floccules

The content of n-paraffins in the floccules was so low that the carbon number distribution of the floccules couldn't be directly obtained using GC. Bengen (1940) firstly found that urea had the interesting properties of forming solid complexes with straight-chain organic compounds. So the urea was selected for the separation of linear aliphatic compounds from the analogous branched and cyclic compounds by the complexes formation between urea and linear aliphatic compounds. Zimmerschied (1950) reported that formation of urea complexes was reasonably rapid at room temperature, where their stability was adequate for high recovery and convenient handling. Recovery of the components by heating or by stirring with water was practically quantitative. Hence, in this work, n-paraffins in the floccules were separated by above method and the detailed procedures were described in section 2.2.2. In addition, this work investigated the influence of each factor on the separation and the results are as follows:

a.   The impact of the urea

From Figure 1 (a), it can be seen the amount of n-paraffins separated from floccules increase with amount of urea increasing. When $m_{urea} : m_{KH150BS}$ is more than 0.4, the amount of n-paraffins has no obvious change with increase of urea. The excess urea can make complexation equilibrium move to the synthesis direction, thus the $m_{urea} : m_{KH150BS}$ is determined to be 0.6 in this study.

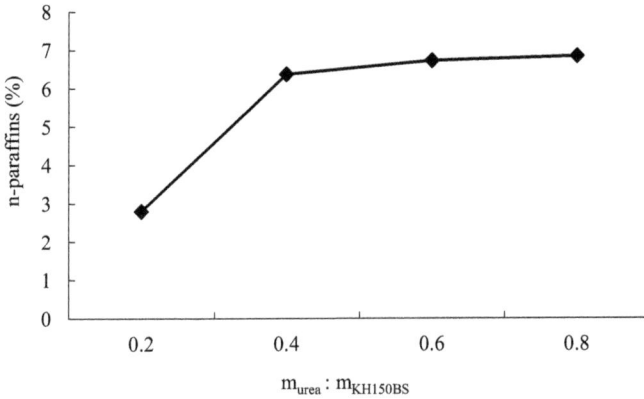

Fig. 1. (a) The impact of urea to separation of n-paraffins.

b.   The impact of activator

The activator isopropyl alcohol will affect the reaction effect. If isopropyl alcohol is less, the viscosity of reaction system will be too large, which will lead urea and sample cannot be fully engaged. Figure 1 (b) shows $m_{isopropyl\ alcohol} : m_{urea} = 0.2$, which can meet the response need. So this experiment determine $m_{isopropyl\ alcohol} : m_{urea}$ is 0.2.

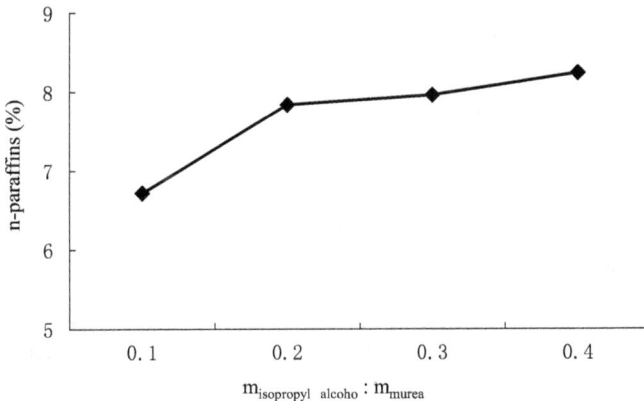

Fig. 1. (b) The impact of activator to separation of n-paraffins.

c.   The impact of reaction time

Figure 1 (c) is the relationship between the reaction time and content of n-paraffins and it shows when the reaction is 2 hours, the urea and n-paraffins has been completely complex. The content of n-paraffins has no obvious change with increasing of reaction hour after 2 hours. Therefore, it is determined the complex reaction time is 2 hours.

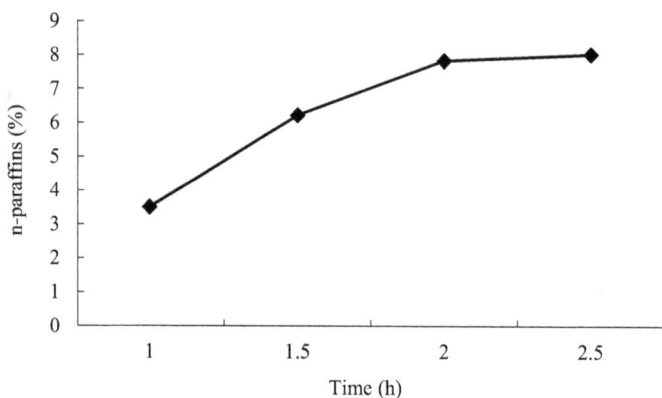

Fig. 1. (c) The impact of reaction time to separation of n-paraffins.

d.   The impact of temperature

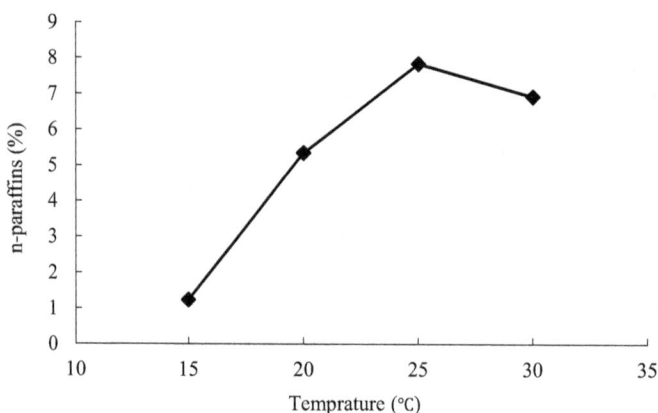

Fig. 1. (d) The impact of temperature to separation of n-paraffins.

The higher temperature can increase dissolving of each substance, decrease viscosity and improve the contact condition. However, the complex reaction is an exothermic reaction and lower temperature is conducive to this reaction. In addition, the higher temperature increases the stability of complexes and the content of n-alkanes will be reduced. From Figure 1 (d), it is determined the temperature was 25°C.

3.2.2.2 Analysis of carbon number distribution of the floccules by GC

The n-paraffins separated from the floccules at -20°C were dissolved in isooctane (0.01 g/mL), then the carbon number distribution of paraffinic molecules in the floccules was obtained by GC and the result was shown in Figure 2. Srivastava et al. (1997) found that the amount of deposition was connected with the phase-transition parameters of the wax, which in turn depended on its composition and the carbon number distribution. Wang and Dong (1995) studied the influence of carbon number distribution on the pour point of several Chinese crude oils, and the results demonstrated that if the molecular weight of hydrocarbon is heavier, then the pour point of hydrocarbon is higher. In these particular cases, there was not a defined maximum in the carbon number distribution curve, and bimodal behavior was not observed in simulated distillation chromatograms. Holder and Winkler (1965) inferred that the pour point depression for an n-paraffins binary blend begins when the concentration of the low molecular weight compound is higher than 25wt%. However, a very small amount of the heavier paraffin was required to activate the light one. Del Carmen Garcia (2000) reported that the abundance of large n-paraffins (n C24+) in paraffinic crude oils increased their tendency to the wax crystallization, which could be demonstrated by a linear correlation between the concentration of this hydrocarbon family and the crude oil cloud point. The carbon number of paraffinic molecules present in deposits was known to be higher than 15 atoms and to reach values of more than 80 carbon atoms, as reported by several authors (O'Donell, 1951; Woo et al., 1984). However, as is shown in Figure 2, the greatest atomic number of the carbon atoms of the floccules separated from KH150BS was 47. This is because KH150BS is produced using hydro processing technologies entirely. In this process, some n-paraffins with longer chains and iso-paraffins with a lower degree of branching are hydro cracked and turned into paraffins with shorter chains and a higher degree of branching. In addition, the hydrocarbon compounds with longer alkyl chains easily form crystals at ambient temperature, and they cannot form crystal line adducts with urea (Bengen, 1940). As is shown in Figure 2, the n-paraffins in the floccules had longer chains, the relative abundance of n C24+ paraffins was higher, and they easily formed crystals.

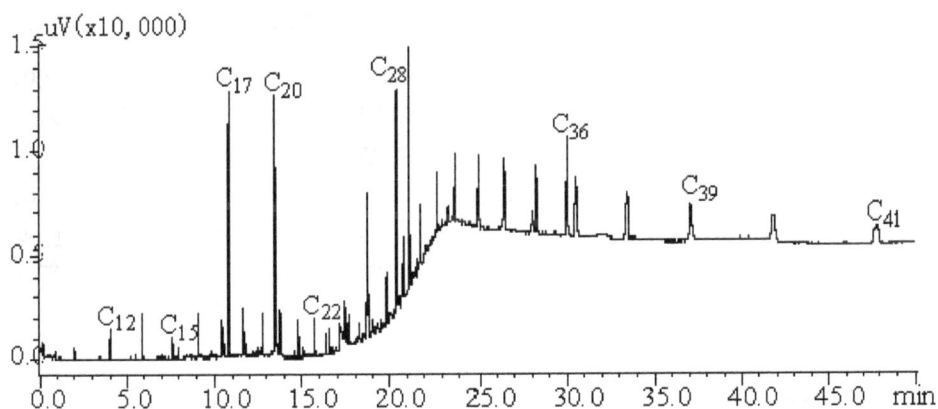

Fig. 2. Carbon number distribution of n-paraffins in the Floccules.

3.2.3 Hydrocarbon type of the floccules, KH150BS and filtrated oils

The floccules, KH150BS and filtrated oils underwent a silica separation to yield saturates, aromatics and resins fractions through petroleum ether, benzene and ethanol washing off, respectively. Joao (1997) pointed out that the aromatic compounds didn't crystallize with paraffins nor changed their crystallization behavior. The concentration of aromatic compounds in KH150BS was approximately 5%, in order to study the hydrocarbon type of the floccules, a method similar to ASTM D2786 was employed using GC-MS to analyze the saturates in the floccules, KH150BS and filtrated oils. The data of hydrocarbon type were given in Table 3.

Table 3 shows that there were some hydrocarbon compounds including a large amount of naphthenes except for some paraffins in the floccules. The content of paraffins in the floccules was higher than which in KH150BS and filtrated oils, in particular, the content of paraffins in the floccules separated at -5°C was 18.6%, which was higher than which in KH150BS by 6.85%. However, the content of naphthenes in the floccules was lower than which in KH150BS and filtrated oils. In addition, with an increase in the content of the floccules, the concentration of paraffins in the filtrated oils slightly exhibited a decreased, for the floccules, the content of paraffins and mononaphthenes exhibited a decrease concomitant with an obvious increase in total naphthenes. This result is consistent with the report of the literature (Turner et al., 1955), namely, for the crystal formed in the 500°C+ fractionation, it is composed of some naphthenes with longer side branch and lower ring numbers except for a number of paraffins with longer chain and lower degree of branching. Maria (2000) found that there was no change in the cloud point value when the concentration of cyclo+isoparaffins was lower than 40wt%, but cloud point of crude oil was increased when their concentration was greater than 50wt%. This increment in the cloud point was believed to arise as a consequence of the higher average molecular weight introduced by these components. From Table 3, it can be seen that the lowest concentration of naphthenes in the floccules was 77.36% and it could heighten cloud point of the oil.

Items	Concentration / wt.%								
	F_1	F2	F_3	F_4	KH150BS	L_1	L_2	L_3	L_4
Paraffins	18.6	16.39	15.61	15.31	11.75	10.79	10.25	10.07	9.7
mono-ring	15.75	14.01	13.87	13.79	12.39	11.1	13.11	12.9	11.04
di-ring	15.76	20.05	20.32	20.71	21.89	22.31	22.46	22.9	22.7
tri-ring	15.35	18.17	18.30	18.14	19.25	18.39	18.77	19.46	18.49
tetra-ring	16.49	17.02	16.78	16.56	18.04	19.43	17.32	16.83	18.69
pent-ring	9.05	9.35	9.62	9.88	10.39	11.3	11.49	10.83	11.88
naphthenes	77.36	78.57	78.89	79.05	81.91	82.48	83.14	82.97	82.8
aromatics	4.04	5.04	5.50	5.64	5.36	5.76	5.63	6	6.58
resins	0	0	0	0	0.98	0.97	0.98	0.96	0.90

F_1, F_2, F_3 and F_4 is the floccules separated from KH150BS at -5°C,-10°C,-15°C and -20°C respectively;
L_1, L_2, L_3 and L_4 is the filtrated oil at -5°C,-10°C,-15°C and -20°C respectively.

Table 3. Hydrocarbon type of the floccules, KH150BS and filtrated oils.

3.2.4 Degree of branching of saturates in the floccules

Li (1994) obtained some linear equations correlating the characteristic IR parameters to the η_{CH_2}/η_{CH_3} (1/degree of branching) using the standard n-paraffins in a proper solvent. Subsequently, he established a method for determining η_{CH_2}/η_{CH_3} of saturates in the petroleum wax by FTIR spectrometry. In this work, the above method was used to obtain η_{CH_3}/η_{CH_2} of saturates in the floccules. The results were shown in *Table 4*.

$$\eta_{CH_2}/\eta_{CH_3} = 3.07 \times A_{1460}/A_{1380} - 3.72 \tag{1}$$

Where A_{1460} and A_{1380} is absorbency of sample in the 1460 cm^{-1} and 1380 cm^{-1}.

From Table 4, it is observed that the degree of branching of saturates in the floccules separated at -5°C was lowest, and it increased with an increase in the content of the floccules. The results indicate that the degree of branching of hydrocarbon compound was lower, they were easier to increase wax crystallization. Ultimately, they formed deposition in the oil at low temperature.

Items	F_1	F_2	F_3	F_4
η_{CH_2}/η_{CH_3}	7.23	4.26	2.90	2.91
η_{CH_3}/η_{CH_2}	0.1383	0.2347	0.3448	0.3436

F_1, F_2, F_3 and F_4 is the floccules separated from KH150BS at -5°C,-10°C,-15°C and -20°C respectively;

Table 4. η_{CH_3}/η_{CH_2} of saturates in the floccules.

3.2.5 Carbon type between KH150BS and filtrated oils

It was investigated that the viscosity index, pour point, oxidation stability, and other related properties of base oil depended on the composition and chemical nature of the aromatic, paraffinic, and naphthenic carbon contents (Yates et al., 1992). In this study, carbon types between KH150BS and filtrated oils were obtained using some base data of physical properties, and the method was similar to SH/T0729-2004. The results are shown in Table 5.

Items	KH150BS	L1	L2	L3	L4
		1.4886	1.4886	1.4886	1.4886
n_D^{20}	0.8884	0.8885	0.882	0.8882	0.888
M	590	582	569	562	532
S/µg·g^{-1}	70	73	70	69	73
$C_A\%$	3.9148	3.9317	4.274	4.3528	4.848
$C_N\%$	31.3017	31.601	31.405	31.5451	31.8893
$C_P\%$	64.7835	64.4673	64.321	64.1021	63.2627
R_A	0.2859	0.2834	0.3005	0.3022	0.3181
R_T	3.2481	3.2308	3.1634	3.1408	3.0285
R_N	2.9622	2.9474	2.8629	2.8386	2.7204

L_1, L_2, L_3 and L_4 is the filtrated oil at -5°C,-10°C,-15°C and -20°C respectively.

Table 5. Carbon types of KH150BS and filtrated oils.

Table 5 shows that the C_P content of KH150BS was 64.78%, which was higher than that of filtrated oils. However, the C_N content was lower than that of filtrated oils. For the filtrated oils, the C_P content exhibited a decrease with a decrease in experimental temperature. The results confirm that the floccules are composed of paraffins with lower degrees of branching and naphthenes having longer side branches and lower ring numbers.

d_4^{20} -Density(20°C)/g $\cdot$ cm-3; n_D^{20} -Refractive index 20°C; M-Average molecular weight; C_A-Content of carbon atoms in aromatic rings; C_N-Content of carbon atoms in naphthenes ring; C_P-Content of carbon atoms in paraffins; R_A-Average number of aromatic rings in every molecular; R_T-Average number of total rings in every molecular; R_N-Average number of naphthene rings in every molecular.

3.3 Causes of floccules formation in hydro-treated lubricating base oil

3.3.1 Floccules formation mechanism

In order to find the cause of floccules formation, it is important to study the floccules formation mechanism. Moussan (2004) thought at high temperatures, the crude oils containing linear paraffins behave like Newtonian liquids, the paraffins being in the molten state. Below the wax appearance temperature (WAT), which correspond to incipient crystallisation of the paraffins during cooling in static conditions, oils turning to gels. Sighn and Dirand (1999) pointed out the question of the 3D structure of the network in gels and of the type of interactions between crystals. They inferred that needle shaped crystals were formed during slow cooling under static conditions, while under flow conditions platelet like crystals with amorphous solid appear. Chanda (1998) thought the rheological behavior of a crude oil is highly influenced by its chemical composition, temperature and the current, as well as previous thermal history. High waxy crudes exhibit a non-Newtonian character, often with a yield stress at and below their pour point temperature. At a sufficiently high temperature the crude oil, although chemically very complex, is a simple Newtonian liquid. If the waxy crude oil is allowed to cool, wax will crystallize, agglomerate and entrap the oil into its structure. This phenomenon often happens if the ambient temperature of the place is below the pour point of the crude oil. In addition, two types of wax are commonly encountered in crude oils. The first is the macro crystal line wax composed of mainly straight-chain paraffins (n-alkanes) with varying chain length (about C_{20}–C_{50}). The second is the micro crystal line or amorphous waxes containing high portion of iso-parafins (cyclo alkanes) and naphthenes with a molecular weight ranges from C_{30} to C_{60}. The presence of these solid particles causes a change in the flow behavior of crude oil from Newtonian to non-Newtonian (Rønningsen, 1991).

According to the analysis of section 2, the floccules were some hydrocarbon compounds with large average molecular weight, high paraffinic carbon content, and high freezing point. The floccules were composed of a large amount of naphthenes and some paraffins. The naphthenes had longer side branches and lower ring numbers, and the paraffins had longer chains and lower degrees of branching. During slow cooling under static conditions, these hydrocarbon compounds formed the nuclei of crystals, which had the shape of discs and were one molecule thick. Subsequently, the nuclei of crystals extended and became continuous lamellas. They formed acolloidal network and embodied the oil. Ultimately, they dispersed throughout the oil and formed the floccules.

3.3.2 The Impact of processing of KH150BS to floccules formation

KH150BS produced by the lubricating oil high-pressure hydrogenation plant which includes hydrogenation process, hydro-dewaxing and hydrogenation refining. Hydrogenation process is characterized that the hydrogenation components, aromatics, especially polycyclic aromatic hydrocarbons are saturated by hydrogenation in the hydroprocessing catalyst, the polycyclic naphthenes are open-loop and generate a single cycloalkane ring or single ring aromatics with long side chains to. The aim of this process is to increase the viscosity index of lubricating oil. The hydro-dewaxing is a new technology for lubricating oil production. The wax in oil are removed or transformed by shape-selective catalysts by hydrocracking or hydroisomerization. Its aim is to decrease the pour point of lubricating oil. The hydrogenation refining process remove the sulfur, oxygen, nitrogen and other impurities, which is to improve the stability and color of oil.

By analyzing the process of KH150BS, it can be seen that the hydro-dewaxing process has the significant relationship between floccules formation. The hydrodewaxing process, the key factor is to choose an ideal catalyst. This catalyst should have good selectivity, which is able to make paraffins with high melting point (n-paraffin hydrocarbons and isoparaffins with less side-chain) cracking into low molecular weight alkanes, or change paraffins with high melting point to the paraffin hydrocarbon with low melting point by isomerization. Finally, the freezing point of lubricating oil is decreased and the ideal lubricant component is not destroyed in order to ensure a high oil yield. The catalyst used in KH150BS hydrodewaxing stage used is RDW-1 catalyst developed by Research Institute of Petroleum in China. It is a dual function of ZSM-5 zeolite loaded Ni catalyst. The Ni catalytic is used to hydrogenation and dehydrogenation, but the role of ZSM-5 zeolite is to provide a acid site and the shape-selective function.

The selectivity of zeolite is that zeolite separates the molecules of different sizes and shape by its effective diameter. In the mixed raw materials, only the molecules whose diameter are smaller than the aperture of zeolite can enter the zeolite pore channels within the active site and participate in the reaction as a reactant, and the molecules whose diameter are larger than the aperture of zeolite will be excluded from the zeolite pore and cannot participate the reaction.

The floccules are some hydrocarbons with large molecular weight and low degree of branching. This may be due to pore of catalyst become more and more narrow in the end of hydrogenation catalysts, some naphthenic isoparaffin and long-chain alkyl with large molecular weight cannot enter the pore of cracking catalyst and participate in the reaction, which cause the oil was observed hazy and emerged floccules at low temperatures.

3.4 The Influence of floccules on properties of lubricating oils

In order to investigate the influence of floccules on properties of oils, the properties of KH150BS and filtrated oils were shown in Table 6. It shows that, after floccules was removed using the low-temperature leaching method, the cloud point, pour point and freezing point of KH150BS obviously decreased. For other properties, compared with KH150BS, the viscosity and viscosity index of filtrated oils decreased, which is due to removal of floccules with long-chain. The aniline point of the filtrated oils slightly decreased compared with KH150BS. The aniline point is determined by hydrocarbons type and chemical composition of oils. In general, for different hydrocarbons with the same number of carbon atoms, aromatic is lowest and alkanes is highest. From Table 5, it can be seen that percentage content of aromatic carbon atoms in the filtrated oils are higher than which in KH150BS, which led the aniline point of the filtrated oils decreased. As the floccules are mainly some hydrocarbons with the longer carbon chain, the lower the degree of branching and they are easily be oxidized. Thus, the oxidation stabilities of the filtrated oils are higher than KH150BS. After removing floccules, the antifoams of filtrated oils have no obvious change except for at 24°C. The antifoams of filtrated oils after removing floccules at -5°C and -15°C became 15 ml and 25 ml at 24°C. This is because the viscosity of KH150BS is higher than filtrated oils. The viscosity is higher, the bubble floating is slower and it is more difficult to break. Thus, the antifoams of filtrated oils are better than KH150BS.

Items		KH150BS	L_1	L_2	L_3	L_4	Test method
Cloud point/° C		+12	-10	-13	-14	-16	GB/T3535-1983
Pour point/° C		-16	-18	-21	-24	-30	GB/T510-1983
Freezing point/° C		-17	-20	-23	-25	-31	GB/T6986-1986
Viscosity	40° C	595.1	580.4	532.2	531.5	507.4	GB/T265-1988
$mm^2 \cdot s^{-1}$	100° C	32.70	31.82	30.01	30.57	29.65	GB/T265-1988
Viscosity index		84	82	82	82	84	GB/T1995-1998
Aniline point/° C		132.0	132.0	131.7	131.6	131.3	ASTM D661
Oxidation stability(rotary bomb oxidation test,150° C)/min		41.1	42.2	41.8	42.6	43.7	SH/T0193
Antifoam /ml/ml	24° C	90/0	90/10		80/20		
	93° C	25/0	25/0		30/0		GB/T12579
	24° C	40/0	15/0		25/0		

Table 6. The properties of KH150BS and filtrated oils.

4. Conclusions

In this study, firstly, the floccules were effectively separated from KH150BS base oil and the composition and structure of the floccules were investigated in detail. Then the formation cause of floccules was analyzed from floccules formation mechanism and processing of KH150BS base oil respectively. Finally, the properties of KH150BS and the lubricating oils after removing floccules were investigated. The results indicate that the floccules were some hydrocarbon compounds with large average molecular weight, high paraffinic carbon content, and high freezing point. They were composed of a large amount of naphthenes and some paraffins. The naphthenes had longer side branches and lower ring numbers, and the paraffins had longer chains and lower degrees of branching. This may be due to the pore of catalyst become more and more in the end of hydrogenation catalysts, some naphthenic isoparaffin and long-chain alkyl with large molecular weight cannot enter the pore of cracking catalyst and participate in the reaction, which cause the oil was observed hazy and emerged floccules at low temperatures. During slow cooling under static conditions, these hydrocarbon compounds formed the nuclei of crystals, which had the shape of discs and were one molecule thick. Subsequently, the nuclei of crystals extended and became continuous lamellas. They formed a colloidal network and embodied the oil. Ultimately, they dispersed throughout the oil and formed the floccules. In addition, compared with KH150BS, the properties of the lubricating oils after removing floccules basically became better.

5. References

Bengen, Ger. Patent application O.Z. 12438 (March 18, 1940). Technical Oil Mission Reel 6, frames 263-70 (in German), and Reel 143, pp. 135-9 (in English).

Chanda, D., Sarmah, A., Borthakur, A. (1998). Combined effect of asphaltenes and flow improvers on the rheological behaviour of Indian waxy crude oil. *Fuel.* 77 (11): 1163-1167.

Del Carmen Garcia, M. (2000). Crude oil wax crystallization: The effect of heavy n-paraffins and flocculated asphaltenes. *Energy & Fuels.* 14: 1043–1048.

Dirand, M., Chevallier, V., Provost, E., Bouroukba, M., and Petitjean, D. (1998). Multicomponent paraffin waxes and petroleum solid deposits : structural and thermodynamic state. *Fuel.* 77:1253-60.

Galiano-Roth, N., and Page, M. (1994). Effect of hydroprocessing on lubricant base stock composition and product performance. *Lubrication Engineering.* 50(8):659 –664.

Holder, G. A., and Winkler, J. (1965). Wax crystallization from distillate fuels. Part III. Effect of wax composition on response to pour depressant and further development of the mechanism of pour depression. *J. Inst. Pet.* 51(499): 243– 252.

Joao, A. P. Coutinho and Ve´ronique, R. M. (1997). Experimental Measurements and Thermodynamic Modeling of Paraffinic Wax Formation in Under cooled Solutions. *Ind. Eng. Chem. Res.* 36:4977-4983.

Kane, M., Djabourov, M., Volle, J-L., Lechaire, J-P., and Frebourg, G. (2003). Morphology of paraffin crystals in waxy crude oils cooled in quiescent conditions and under flow. *Fuel.* 82: 127-135.

Kane, M., Djabourov, M., Volle, J-L. (2004). Rheology and structure of waxy crude oils in quiescent and under shearing conditions. *Fuel.* 83: 1591-1605.

Morphology of paraffin crystals in waxy crude oils cooled in quiescent conditions and under flow. *Fuel.* 82:127-135.

Li, X. R., and Jiang, F. R. (1994). Determination the Ratio of Methylene to Methyl Group of the Saturate Fraction in Petroleum Wax by FTIR Spectrometry. *Petroleum Refinery and Chemical Industry.* 25:54-57.

Maria del Carmen Garcia. (2000). Crude Oil Wax Crystallization. The Effect of Heavy *n*-Paraffins and Flocculated Asphaltenes. *Energy & Fuels.* 14:1043-1048.

O'Donell, G. (1951). Separating Asphalt into Its Chemical Constituents. *Anal Chem.* 23 (6):894.

Pederson, K. S., Skovborg, P., and Ronningsen, H. P. (1991). Wax Precipitation from North Sea Crude Oils. 4. Thermodynamic Modeling. *Energy and Fuels.* 5:924.

Rønningsen, H. P., Bjørndal, B., Hansen, A. B., Pedersen, W. B. (1991). Wax precipitation from North Sea crude oils. 1. Crystallization and dissolution temperature, and Newtonian and non-Newtonian flow properties. *Energy and Fuels.* 5: 895–908.

Singh, P., Fogler, H.S., Nagarajan, N. (1999). Prediction of the wax content of the incipient wax-oil gel in a pipeline: An application of the controlled-stress rheometer. *J Rheol.* 43:1437–59.

Srivastava, S. P., Handoo, J., Agrawal, K. M., and Joshi, G.C. (1993). Phase-transiti- on studies in n-alkanes and petroleum-related waxes — A review. *Journal of the Physics and Chemistry of Solids.* 54:639.

Srivastava, S. P., Saxena, A. K., Tandon, R. S., and Shekher, V. (1997). Measurement and prediction of solubility of petroleum waxes in organic solvents. *Fuel.* 76(7):625- 630.

Turner, W. R., Brown, D. S., and Harrison D. V. (1955). Properties of Paraffin Waxes Composition by Mass Spectroometer Analysis. *Ind Eng Chem.* 47(6):1219.

Wang, B., and Dong, L.D. (1995). Paraffin characteristics of waxy crude oils in China and the methods of paraffin removal and inhibition. Paper SPE 29954, SPE International Meeting on Petroleum Engineering, Beijing, China. November 14–17, 33–48.

Webber, R. M. (2001). Yield Properties of Wax Crystal Structures Formed in Lubricant Mineral Oils. *Ind. Eng. Chem. Res.* 40:195-203.

Woo, F. T., Garbis, S. J., and Gray, T. C. (1984). Paper SPE 13126, presented at the 59th Annual Fall Technical Conference and Exhibition, Houston,TX.

Yates, N. C., Klovsky, T. E., and Bales, J. P. (1992, August). In Symposium on Processing, Characterisation and Applications of Base Oils, Division of Petroleum Chemistry, American Chemical Society, Washington, D C.

Zimmerschied, W. J., Dinerstein, R. A., Weitkamp, A. W., and Marschner, R. F. (1950). Crystalline Adducts of Urea with Linear Aliphatic Compounds. *Ind Eng Chem.* 42:1300-1306.

11

New Trends in Hydroprocessing Spent Catalysts Utilization

Hoda S. Ahmed and Mohammed F. Menoufy

Egyptian Petroleum Research Institute

Egypt

1. Introduction

The distillate of crude oil is an essential step in the petroleum refining practice. The yield and properties produced distillates depend on the properties of crude oil, distillation conditions and the type of distillation column. Primary distillates are subjected to an additional treatment to meet the environmental requirements and the performance of produced fuels. The schematics of a typical refinery operation processing a conventional crude shown in Fig. 1 lists four catalytic processes, i.e. reforming, hydrocracking, hydrotreating, catalytic cracking and alkylation. The residue from atmospheric distillation may be subjected to additional distillation under a vacuum to obtain valuable lubricant fractions which also require catalytic hydrotreatment. None- conventional refineries can process heavy oils and distillation residues. In this case, the catalytic hydro cracking of the heavy feed is usually the first step, followed by hydrotreating of the synthetic distillates. For the purpose of this review, the hydroprocessing will refer to both hydrocracking and hydrotreating. Light hydrocarbon which is byproducts of several refineries can converted to high octane fractions by catalytic alkylation and polymerization.

1.1 Refining catalysts [2]

Several operations employing a catalyst may be part of the petroleum refinery. The management of catalyst inventory represents an important part of the overall refinery cost. As shown in Fig. 2, the development of refining is closely connected with the growth of the use catalysts. In the past, refining catalysts accounted for more than half of the total worldwide catalyst consumption. Today because of the importance of environmental catalysis, refining catalysts account for about one third of the total catalyst consumption. Future advances in development of more active and stable catalysts may further decrease the overall consumption of refinery catalyst.

The solid catalysts are usually of none noble metal catalysts and noble types. None-noble metal catalysts include base metals and zeolites. Nobel metal catalyst includes a variety of precious metals from the platinum group. In many cases, catalytically active metals are combined with a solid support such as alumina, silica-alumina, zeolites, carbon, etc.

1.2 Catalyst demand worldwide [3-5]

The oil refining industry operation is analyzed in order to estimate the future catalyst market trends. The refining catalyst market corresponding to the main catalytic processes is estimated taking into account the following information: (i) the average refining capacity increases for the main catalytic processes since 1999, (ii) the additional refining capacity due to future plans of construction and expansion of refineries and process units that will be added by 2005, and (iii) the past refining catalyst market behavior. From this information, it has determined for the main catalytic processes a global average factor, expressed as processed oil barrels per dollar of catalyst. According to the catalyst Group Research Co, the global refining catalyst market has grown from $2.32 billion in 2001 to about $2.65 billion in 2005 (3.6% annual growth). Hydrotreating, fluid catalytic cracking, hydrocracking and isomerization represent about 74% of the total catalyst market and will grow by about $34, $32, $11 and $2.5 million per year, respectively. However, naphtha reforming catalyst market will not grow during 2001-2005. Higher catalyst spending growth is expected for the North America region ($27.5 million per year)

Worldwide refiners need to produce cleaner fuels and operate to meet environmental regulations. Therefore, refineries should convert more heavy feedstock's to satisfy regional fuel market needs, and minimize their profit margins which guides the developments in the refining catalyst market. Refiners require more processing capacity and higher efficiency to satisfy regional fuel market needs., therefore, more catalyst consumption is needed and in the same be more resistant to deactivate under sever operating conditions. Table 1 gives the refining catalyst market in tons and dollars by application for 1999 and 2005.

Process	1999			2005		
	10^3 tons	%	G$	10^3 tons	%	G$
Cracking	495	77	0.7	560	73.6	0.83
Hydrotreatments	100	15.5	0.72	135	17.7	0.96
Hydrocracking	7	1.1	0.10	9	1.2	0.12
Reforming	6	0.9	0.12	7	0.9	0.15
Others*	~35	5.5	0.56	~50**	6.6	0.64
Total solids	~640-650	100	2.2	~760	100	2.7
Alkylation	3100**	-	0.85	3700*	-	1

*Catalyst for H2 production, polymerization ,isomerization etherification ,claus ,lubes, etc
**Approximate values.

Table 1. World catalyst market in refining.

The refining catalyst market is the most competitive segment of the global catalyst. This market was increased at about 1.9% / year during 2001-2007. Two different catalyst market estimates 1 and 2 (Table 2) were developed according to the increasing refinery capacities or to the added capacities due to revamps, or expansion plans. Table 2 shows the future catalyst markets according to these estimates. The data indicate that the current catalytic processes, hydrotreating, FCC, naphtha reforming, hydrocracking and isomerization, represent about 77% of the total refining catalyst market. The future catalyst market will increase through 2007- 2008 to about $ 2.58 billion (2.8%/ year). Most growth was occurred in North America, Asia Pacific, and the Middle East. If the catalytic prices remain unchanged, then the future worldwide catalytic processes consumption will be about:

Hydrotreating	117,778 tons
FCC	357,612 tons
Naphtha reforming	7,273 tons
Hydrocracking	5,538 tons

	Estimate.1 2007 catalyst market Million $	Estimate..2 2007 catalyst market Million $
Hydrotreating	897	954
FCC	716	751
Naphtha reforming	139	144
Hydrocracking	132	144
Isomerization	58	64
Total	1,941	2,057

Table 2. Future catalyst market.

1.3 Catalysts deactivation [6-10]

In every catalytic operation, the activity of the catalyst gradually decreases. This decrease can be offset by changing some operational parameters. However, at a certain point, catalyst replacement is inevitable. The spent catalysts can be regenerated and returned to the operation. The regeneration of spent hydroprocessing, fluid catalytic cracking (FCC) and reforming catalysts has been performed commercially for several decades. These regeneration processes have been extensively reviewed by Furimsky and Massoth .Hughes and Fug.

Large quantities of catalysts are used in the refining industry for operation and upgrading of various petroleum streams and residues. The catalysts deactivate with time and the spent catalysts are usually discarded as solid wastes. The quantity of spent catalysts discharged from different processing units depends largely on the amount of fresh catalysts used, their life and the deposits formed on them during use in the reactors. In most refineries, a major portion of the spent catalyst wastes come from the hydroprocessing units. This is because the catalysts used in these processes deactivate rapidly by coke and metal (V and Ni) deposits, and have a short life due to fouling of the active catalytic centers by these deposits. Furthermore, technology for regeneration and reactivation of the catalysts deactivated by metal fouling is not available to the refiners.

The volume of spent hydroprocessing catalysts discarded as solid wastes has increased significantly in recent years due to the following reasons:

- rapid growth in the distillates hydro processing capacity to meet the increasing demand for ultra low-sulfur transportation fuels.,
- a steady increase in the processing of heavier feed stocks containing higher sulfur, nitrogen and metal (V&Ni) contents, and

- rapid deactivation and unavailability of a reactivation process for reside hydroprocessing catalysts.

In Kuwait's refineries, over 250 000 barrels of heavy residues higher in sulfur, and metals are upgraded and converted to high quality products by catalytic hydroprocessing, bringing substantial economic returns to the country. These operations generate a substantial amount of deactivated spent catalysts as solid waste every year. Currently, about 6000 tons of spent catalysts are discarded as solid wastes from Kuwait's refineries annually. This will increase further and exceed 10 000 tons/y when a fourth refinery is built to process heavy crudes and residues.

In Egyptian refiners, the solution is different, due to schematic refining processing such as, hydrotrating, isomerization and reforming process. The only full refining system was found is Medor refining which contains, beside the hydrotreting units, a conversion unit (hydrocracker) using less contaminants heavy residues. Therefore, the current discarded spent catalysts have fewer amounts, i.e. nearly 400-700 tons / y, but can be increased due to the future addition of new refining instillations.

1.4 Environmental considerations [10-12]

Environmental laws concerning spent catalyst disposal have becomes increasingly more severe in recent years. Spent hydroprocessing catalysts have been classified as hazardous wastes by the Environmental Protection Agency (EPA) in the USA. The most important hazardous characteristic of spent hydroprocessing catalysts is their toxic nature. Metals such as V, Ni, Mo and Co present in the catalyst can leached by water after disposal and pollute the environment. Besides the formation of leachates, the spent hydroprocessing catalysts, when in contact with water, can liberate toxic gases. The formation of dangerous HCN gas from the coke deposited on hydroprocessing catalysts that contains substantial amount of nitrogen has been reported. The solid spent refinery catalysts will refer to as non-regenrable catalysts and the hazardous nature of the spent catalysts is attracting the attention of environmental authorities in many countries and the refiners are experiencing pressures from environmental authorities for safe handling of spent catalysts.

In the USA, the disposal and treatment of spent refinery catalysts is governed by the Resource Conservation and Recovery Act (RCRA), which holds not only the approved dumb-site owner liable. but also the owner of the buried waste. This environmental responsibility continues for the life of the dumb-site. The current RCRA regulations require landfill to be built with double liners as well as with leachate collection and groundwater monitoring facilities. Thus, the landfill option is becoming expensive today. In Addition, it carries with it a continuing environmental liability. Treatment prior to land filling may be necessary in some cases, further increasing the cost.

1.4.1 Hazardous characteristics of spent hydroprocessing catalyst [13-16]

The hazardous nature of hydroprocessinig catalysts depends on operating conditions. However, the procedure applied during the catalyst withdrawal from the reactor at the end of the operation can be even more important. If a proper procedure can be applied, the hazardous cane be significantly minimized. For example, if a hydroprocessing catalyst cane be treated with either an inert gas or steam, and/ or CO_2 in the absence of H_2 and feed, and at a near operating temperature, the amount of the carried over liquids can be substantially decreased. The amount of entrapped volatile gases, which may include even H_2,can be decreased as well. Without a proper pretreatment prior to the catalyst withdrawal the concentration of flammable vapors above the solid material may reach dangerous levels. In some cases, e.g. when special precautions were not taken during the catalyst withdrawal, it may be appropriate classify the hazardous characteristics of spent hydroprocessing catalysts as that of the corrosive and flammable liquids. One information source indicates catalyst unloading under vacuum. It is stated that this method removes the catalyst without disturbing the operation, however, the type of catalyst and / or operation is not specified.

It a appears that there is no safe catalyst withdrawal procedure which could be generally accepted by all refiners.

Refiners usually apply their own procedure. The need for a commonly accepted and/or approved procedure may develop in the future. In this regards several patents describing the catalyst unloading techniques should be noted. These techniques can significantly reduce or even eliminate the self-heating character of spent catalyst. Otherwise, if spontaneous combustion begins, the inorganic sulphides and organic sulphur which are part of the spent catalysts may also contribute to the uncontrolled burn off. In such case, they will produce large quantities of SO_2. However, sulfides alone require temperatures exceeding 200°C for spontaneous combustion to occur. Part of nitrogen in coke will be converted to NOx during the spent catalysts burn off. Though the evolution of HCN and NH_3 is also possible. Fig. 3&4 show the formation of HCN and NH_3 during oxidation of spent CoMo, Ni Mo catalysts in 4% O_2.

2. New trends in utalization [17-19]

Several alternative methods such as disposal in land fills, reclamation of metals, regeneration/ rejuvenation and reuse, and utilization as raw materials to produce other useful products are available to the refiners to deal with the spent catalyst problem. The choice between these options depends on technical feasibility and economic consideration.

Nowadays, in order to alleviate the shortage of domestic resources and improve the environmental condition, many countries in the world pay much attention to the comprehensive utilization of the secondary resources. In Japan, recycling of the waste catalysts has been done since 1950s; the turnover of waste catalysts was already up to 500 million dollars in 1996.

Thus, the landfill option is becoming expensive today. In addition, it carries with it a continuing environmental liability. The potential future liability of landfills is estimated at about $200/ton. Treatment prior to land filling may be necessary in some cases increasing the cost. In the face of lower- sulfur regulations and regulations deterring catalysts from entering landfills, refiners have directed spent- catalysts traffic from disposal options to regeneration and reclamation plants.

2.1 Recovery of metals [20-29]

In recent years, increasing emphasis has been placed on the development of processes for recycling and recovering of the waste catalyst metals, as much as possible. In literature there are many applied researches for spent metals recovery, particularly for catalyst that contain high concentrations of valuable metals (Mo, Ni, V and Al_2O_3) However, fluctuations in the market prices of the recovered metals and their purity, together with the high costs of shipping significantly influence the economics of the metal reclamation process making it less attractive for spent catalysts that contain low metals concentrations.

As the environmental pressure increases, and as the cost of catalyst storage and disposal continues to rise, the utilization of spent refinery catalysts for metal recovery is becoming a viable pollution. Therefore, the refiners are ready to supply spent hydroprocessing catalysts free of charge in order to reduce their costs for storage and disposal. If the market value of the recovered materials is high enough, then it will offset, the processing cost yielding, a net profit to the reclaimer companies.

2.2 Reactivation/rejuvenation and reuse

Many literature review revealed that reactivation of spent catalysts technology did not reach to well developments. Spent catalysts lose their activities, and deactivated by pore blockage and fouling of the active surface with deposition of coke and metal contaminants. Therefore, many efforts were subjected to replace the conventional regeneration procedures in order to reactivate and rejuvenate the spent catalysts. The new procedures are conducted to remove contaminant metals selectively by chemical treatments without significantly affecting the chemical and physical characteristics of the original catalyst.

In some experimental works in EPRI (1), we were succeeded in rejuvenating spent catalyst Mo Ni/Al2O3 after re-refining of waste lube oil. Our data revealed that the treated spent catalyst can be restored to nearly the fresh HDS activity levels by application of oxalic acid leaching technique in addition to H2O2 as an oxidizing agent. The most effective leaching agent was 4% oxidized oxalic acid, and the extent of metals recovered was dependent on acid concentrations or the specific reuse of the spent catalyst. The rejuvenation process promotes the formation of a hydroprocessing catalyst due to improvements in surface area and average pore diameter (87% and 63% of fresh catalyst characteristics, respectively) as a result of metals recovery. These improvements caused recovery in the HDS activity of the treated catalyst in the range of 81–96% compared to the activity of the fresh, 95-98%, within reactor temperatures 340ıC-380ıC figure 1. Therefore, it is clearly possible to reuse the rejuvenated catalyst, especially in the refining processes, as a top-layer guard-bed or mixed with the same fresh catalyst (2-3).

Other improvements were also obtained in catalyst activities by leaching, especially prior to decoking, either using 2% or 4% leaching concentration. The results indicate that removing coke from the leached catalysts increases its cyclohexene conversions as well as its hydrogenation and isomerization activities figure 2. These results suggested that coke deposition has less effect than metals in deactivating hydrotreating catalysts. The treatments

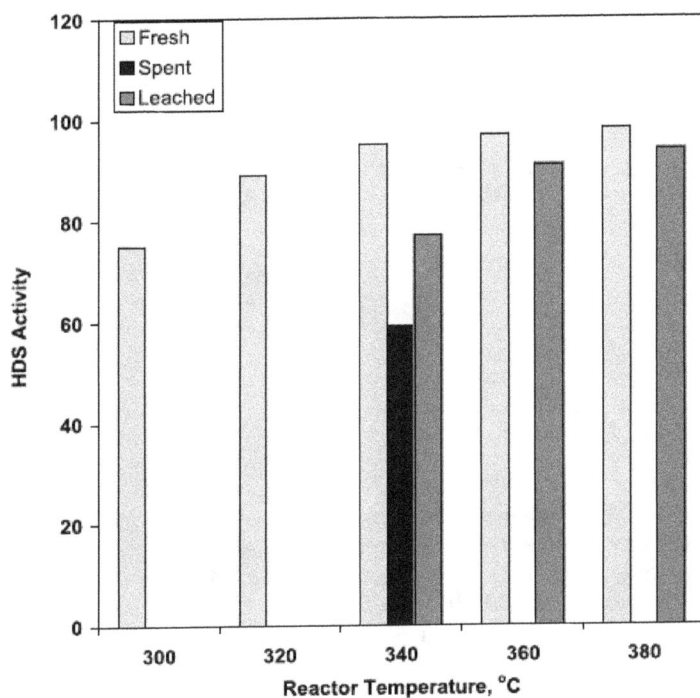

Fig. 1. Relative hydrodesulfurization (HDS) activity of fresh spent and leached catalysts under variable reactor temperature.

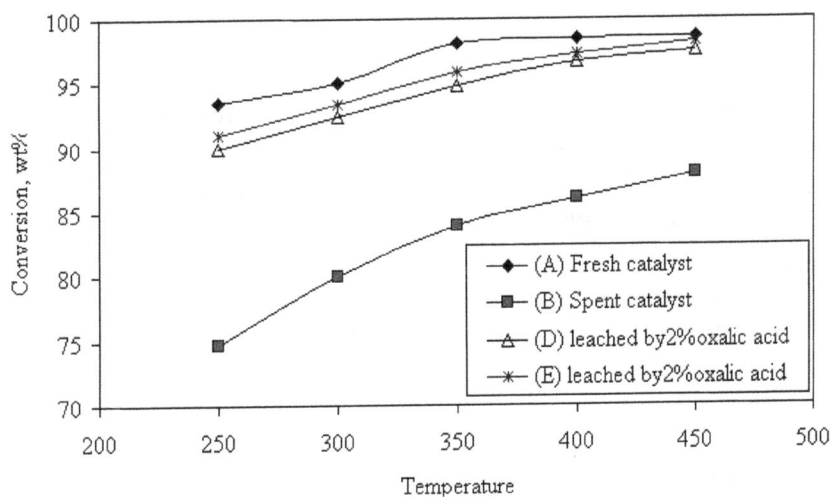

Fig. 2. Effect of reaction temperature on cyclohexene conversion when using catalysts: fresh (A) spent (B) decoked before (D) and after (E) leached by 2%.

developed in the present work appear to be very promising for reactivation (rejuvenation) of hydrotreating spent catalysts in order to recycle the solid spent catalysts and thereby reducing their environmental problems of disposal. (4).

2.3 Utilization of spent catalysts as raw materials

Utilization of spent catalysts as raw materials in the production of other valuable products is an attractive option from environmental and economical points of view. Spent fluid catalytic cracking (FCC) catalysts have been successfully utilized in cement production. In the U.S., cement kilns process about 60,000 tons/year of spent FCC catalysts.

3. Concluion

The volume of spent hydroprocessing catalysts in the Arab countries discarded, as solid wastes, has increased significantly due to:

1. Increasing in processing moderate and / or heavier feedstocks.
2. Rapid growth in the distillates hydroprocessing capacity.
3. Increase the environmental regulations for fuel emissions.
4. In Kuwait alone, around 7000-10000 t / y spent catalysts are generated every year more than the other Gulf Countries. In Egypt, the volume of spent catalysts is nearly around 400-700 t / y, depends on the type and volume of the refined feedstock. Therefore, the choice of the Kuwait refiners was focused now on two processes, one for rejuvenation of spent residue hydro treating catalysts, and the other for producing non-leachable synthetic aggregate materials or loaded as front end- reactors (HDM, guard beds).

Handling and utilization has been the subjected of some investigations in the Kuwait (KISR) and Egypt (EPRI) laboratories. In most of the previous studies in both laboratories, rejuvenation of spent hydroprocessing catalysts for reuse was addressed. Up to 70-80% of the spent catalyst was reclaimed, in the applied processes of both laboratories, with HDS activity as high as 94-95 % of fresh catalyst. Therefore, from the economic point of view, the rejuvenation and reuse of the catalyst is feasible with an internal rate of return.

In **KISR**, the researchers have been succeeded in preparing hydride- metallization (HDM) catalysts, by the addition of boehmite.

The resulted catalyst are used in the front-end of reactors used in petroleum residue hydrotreating processes to remove the foulant metals (V and Ni).The catalyst posses high activity for promoting HDM reactions together with some activity for HDS reactions.

4. References

[1] Inui, K .Res.Inst France Petrol (*IFP*).(49),5,1571 (1994).
[2] Martino, G. Bull. Soc. Chem.Fr.131, 444 (1994).
[3] Silvy,R.I. Refining catalyst business shows signs of strong recovery in 2004-07. Oil & Gas J. 102 (16), 58-65, April 26 (2004).

[4] Scott,A. Refining Catalysts. Chemical Week, 165 (12), 27-29, Mar. 26 (2003).

[5] Marcilly,C. Oil& Gas Science Technology-Rev. *IFP*, 56(5), 494-514, (2001).

[6] Furiamsk, E and.Massoth, F.E .Catal.Today.17 (19993).

[7] Hughes, R.Deactivation of Catalysts.Academic Press. Newyork.(19984).

[8] Worldwide Catalyst Report: Refining catalyst demand. Oil & Gas J, October, 9, 84-66, (2000).

[9] Mena,M., Antony, S.Applied Catalysis B:Environmental , 71, 199-206 (2007).

[10] Mena,M., Antony, S and Ezra,K.J of Environmental Managament ,86,665-681, (2008).

[11] Rapaport,D. Hydrocarbon process ,79,11. (2000).

[12] United Stated Environmental Protection Agency (USEPA). Hazardous waste management system fedral register, 86, (202), 59935 (2003).

[13] Lassner, J.A., Lasher, L.B., Koppel, R,L and Hamillon, I..N. Chem. Eng. Progs. Augest, 95 (1994).

[14] Kulions, T. US Pat. 4992071 (1990).

[15] Kawaskami. Jpn, Pat. 44755 (1977).

[16] Anonymous. Oil .Gas J, 56, 12 October (1992).

[17] Chang,T.J. Oil&Gas ,79-84 October (1998).

[18] Trimm, D.L. The regeneration or disposal of deactivated heterogeneous catalyst Applied Catalysis A, 212, 153–160, (2001).

[19] Huan-qun, LIU. Recovering of spent catalyst in the foreign country. J. Chinese Resource Comprehensive Utilization, (in Chinese) (12) , 35-37, (2000)

[20] Marafi, M., Stanislaus, A . Studies on rejuvenation of spent hydroprocessing catalysts by leaching of metal foulants. Journal of Molecular Catalysis A: Chemical 202, 117–125, (2003b).

[21] Berrebi, G., Dufresne, P., Jacquier, Y. Recycling spent hydro-processing catalysts: Eurecat Technology. Resources, Conservation and Recycling, 10, 1–9 (1994).

[22] Case, A., Garretson, G., Wiewiorowski, E. Ten years of catalyst recycling: a step to the future. In: Presented at the Third International Symposium on Recycling of Metals. Point Clear. Alabama, USA November 12–15,(1995).

[23] Lianos, Z.R., Deering, W.G.,. GCMC's integrated process for recovery of metals from spent catalysts. In: Presented at the Air and Waste Management Association's 90th. Annual Meeting Toronto. Canada, June 8–13,(1997).

[24] Kar, B.B., Datta, P., Misra, V.N. Spent catalyst: secondary source of molybdenum recovery. Hydrometallurgy ,72, 87–92,(2004)

[25] Marafi, M., Furimsky, E. Selection of organic agents for reclamation of metals from spent hydroprocessing catalysts. Erdol Erdgas Kohle ,121, 93–96,(2005).

[26] Chen, Y., Feng, Q., Shao, Y., Zhang, G., Ou, L., Lu, Y. Research\ on the recycling of valuable metals in spent Al2O3-based catalyst. Material Engineering ,19, 94–97,(2006).

[27] Chang, T. ABI/Inform Global. Oil and Gas J October 19,6,43,(1998).

[28] Menoufy,M.F and Ahmed,H.H. Tretment and reuse of spent hydrotreating catalyst.Energy Source. PartA. 30:000-000,(2008)

[29] Fang,Q., Chen,Y., Shao,Y.H., Zhang, G.F., Ou, L.M .,Lu,Y.P. J Csut,13,(2),(2006)

[30] E. Z. Hegazy., *M.Sc. Thesis*, Tanta University, Tanta (2003)

[31] M. F. Menoufy, H. S. Ahmed, in Proc. of the OAPEC Seminar on Energy Conservation and Environmental Protection in Petroleum Industries, Cairo (2004).

[32] M. F. Menoufy, H. S. Ahmed, .Energy Sources, Part A, 30:1213–1222, (2008)

[33] H S. Ahmed, M.F. Menoufy., *Chem. Eng Technol.*, 32, No. 6, 873–880, (2009)

12

Analysis of the Seismic Risk of Major-Hazard Industrial Plants and Applicability of Innovative Seismic Protection Systems

Fabrizio Paolacci[1], Renato Giannini[1] and Maurizio De Angelis[2]

[1]*University Roma Tre, Department of Structures, Rome,*
[2]*University of Rome "La Sapienza", Dep. of Structural and Geotechnical Eng., Rome,*
Italy

1. Introduction

Industrial plants are complex systems and it is such a complexity, due to numerous connections, equipment and components, together with the complexity of their operations that makes them particularly vulnerable (local vulnerability) to earthquakes (see Figure 1a).

Fig. 1. (a) Areal view of a refinery, (b) Damage in a process furnace (Izmit, 1999).

Activities carried out in process plants can also be arranged in series, which means that process activities are realized with specific sequence and boundary conditions. Consequently, the "failure" of a single element can get out of order the entire system. This is of fundamental importance for the seismic vulnerability of a plant (general vulnerability).

Seismic action can cause serious accidents to industrial plants as shown in several occasions. The actual worldwide situation of major-hazard plants against earthquakes should be considered as critical. For instance, in Italy about 30% of industrial plants with major-accident hazards are located in areas with a high seismic risk. In addition, in case of a seismic event, the earthquake can induce the simultaneous damage of different apparatus, whose effects can be amplified because of the failure of safety systems or the simultaneous generation of multiple accidental chains.

A representative example is certainly the Izmit Earthquake in Turkey (Erdik and Durukal, 2000), which induced severe damages to Tupras refinery (area of the plant, farm tanks, and landing place). An example of domino effect caused by a structural collapse was the breakdown of a concrete chimney that caused a big release of dangerous substances and damages to surrounding equipment (see Figure 1b).

In a plant, an earthquake can cause many dead as consequence of components collapses, similarly to what happens to buildings; moreover, the consequences deriving from a seismic event, such as economic losses for interruption of the production, environmental damages due to releases of dangerous substances, damages to persons due to explosions, fires and release of toxic substances, have also to be taken into account. Therefore, the usual safety requirements applied to civil buildings for ultimate and serviceability limit states, together with the consequences of exceptional actions, are generally unsuitable for structures belonging to industrial plants. As a matter of fact, a critical damage for a process safety that can cause even a modest release of inflammable substances, such as a flange opening or a welding breaking, can result unessential under structural point of view, but, at the same time, might cause considerable accidental chains. Consequently, for process industry it is necessary to associate the indirect consequences caused by possible accidents (i.e. a seismic event) to the direct structural damages.

During the last years, in order to increase safety against earthquakes, passive control techniques (PCT) have been developed, which are based on the concept of reducing the seismic action instead of increasing the strength (Housner et al., 1997). These techniques that for civil constructions are nowadays considered a consolidated alternative design tool, can also be used for seismic protection of industrial structures.

Unfortunately a very limited number of applications to industrial plants components have been realized. For this reason, in the present chapter the applicability of such a technique is investigated, aiming at providing general applicability criteria. An example of base isolation of a steel storage tank is also presented, whose effectiveness is investigated by a wide numerical and experimental activity.

2. Structural classification and seismic behavior of oil refinery components

A schematic representation of a refinery is shown in Figure 1(Moulein & Makkee, 1987). The raw material arrives in the refinery by different ways (by train, ship, pipelines, etc..), depending on the location, and then it is stored in big tanks.

The main operation in a refinery is the distillation of raw material in columns, named "topping", that divide the crude oil in a certain number of fractions (5-7) distinguished in 1) light fraction (gasoline and light components), 2) intermediated fraction (Kerosene and light oils) and 3) heavy fraction (residual oils). All these fractions are processed further in other refining units to obtain specific products. For example the heavy fraction is subjected to a specific "thermal cracking" or "Visbreaking" treatment to obtain a certain quantity of more refined material.

The equipment deputed to these operations are indicated in Figure 2. In particular; (A) tanks for the storage of raw and refined materials, (B) process equipment for the chemical treatment of crude oil and waste material, (C) piping systems for the transferring of the

Analysis of the Seismic Risk of Major-Hazard Industrial Plants and Applicability of Innovative
Seismic Protection Systems

235

material and (D) flares, used to eliminate waste gas. For a detailed description of the operation of each component that is beyond the aim of this chapter, the reader can refer to specialized publications (Mayers, 2004).

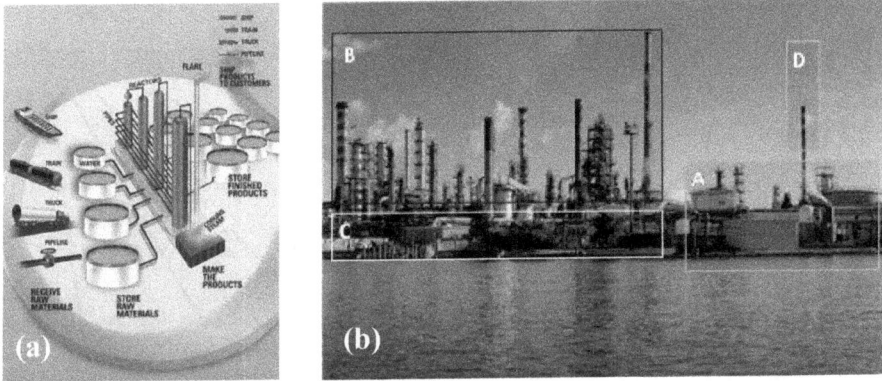

Fig. 2. (a) Schematic representation of an oil refinery, (b) Main components of a refinery.

During a transformation processes many dangerous substances are treated. Consequently, a refinery is equipped with numerous safety systems, some of which imposed by the codes and other adopted by designers. It is worth to highlight that during a seismic event they could fail, becoming useless. Therefore the seismic protection of a refinery must be mainly based on the reduction of the seismic risk of single components.

Earthquake	Data	M
Kern Kounty (California, USA)	21 July 1952	7.5
Anchorage (Alaska)	27 March 1964	8.6
Nigata (Japan)	16 June 1964	7.6
Valparaiso (Chile)	3 March 1985	7.8
San Fernando (California, USA)	9 Febbruary 1971	6.5
Loma Prieta (California, USA)	17 October 1989	6.9
Costa Rica	22 April 1991	7.6
Kocaeli (Izmit, Turkey)	17 August 1999	7.6
Bhuj (Gujarat, India)	26 January 2001	7.7
Tokachi-Oki (Japan)	26 September 2003	8.3
Honshu (Japan)	11 March 2011	9.0

Table 1. Important industrial plants struck by destructive earthquakes.

The experience derived from observing damages caused to industrial plants by past earthquakes can be very useful to identify the most exposed components to the seismic risk and the evaluation of the consequences. Despite the difficulties of obtaining and organizing data, detailed information on the behavior of the refineries in a certain number of earthquakes are available, in particular those listed in Table 1 (Kawasumi, 1968; Nilsen &

Kiremidijian, 1986; Scholl & Thiel, 1986; ATC-25, 1991; Showalter and Myers 1992; Beavers et al., 1993; Erdik & Durukal, 2000; Steinberg et al., 2000; Johnson et al., 2000; Suzuki, 2002; Kilic & Soren, 2003; Sezen & Whittaker, 2004; Hatayama 2008; Suzuki, 2006)

Based on this information, in the following, the main apparatus of process industrial plants are grouped into a restricted number of structural classes and the main observed damages caused by earthquakes are analyzed in detail.

2.1 Slim vessels

Cylindrical vessels with a high ratio height/diameter (between 5 and 30, and even higher) belong to this category. Among them, on the basis of the operation and the system of constraints to which they are subjected, it is possible to identify:

- Vertical cylindrical vessels which are directly anchored to foundations and free along the height. This category includes the distillation columns and many other reactors. The distribution of the mass along the height is usually rather uniform and it may be considered as continuous, even though some internal discontinuities could be present.
- Vertical cylindrical vessels which present additional constraints, besides at their basis, also along the height. This group includes very thin columns such as stacks and flares. Their mass is entirely due to the structure, because they contain just atmospheric pressure gas.
- Horizontal cylindrical vessels, supported by two or more saddles connected to a foundation platform. In this category many pressurized storage tanks and shell-and-tube heat exchangers are included.

As far as vertical slim vessels are concerned, the most common damages in case of seismic event are the failure at the foundation, due to the excessive stress, and the loss of contained fluids because of failure of connected flanges, due to excessive relative displacements. For example, during the *Loma Prieta* earthquake a refinery was seriously damaged, where the most important effect was on the anchor bolts of about 20 vertical vessels, on a total of 50 vessels. During the *Valparaiso* earthquake in a ventilation-stack with diameter 18″ the anchor bolts were subjected to a similar damage, with the yielding of bolts.

The Izmit earthquake induced serious damages in the Tupras refinery. An entire unit was destroyed by the collapse of a chimney of a furnace 105 m high (Figure 3a). The top of the chimney collapsed on the furnace, whereas the bottom part collapsed on a pipe system. Even if the furnace was designed according to ACI-307, it has been subjected to an important damage.

2.2 Above-ground squat equipment

These apparatus have similar dimensions in the principal directions and are characterized by heavy masses; they can be grouped in two main categories:

- Large cylindrical steel storage tanks with a height/diameter ratio between 0.2 and 2. The roof can be welded to the shell (fixed conic roof) or floating over the contained liquid. The operating volume varies from some tens to 200000 m³. The bottom plate is circular and placed directly on a granular backfill. When they are full filled, part of the mass oscillates under the seismic excitation producing high forces on the tank wall, whereas the remaining part induce a sloshing wave that might produce overtopping phenomena.

- Large process equipment like filters and decanters, or dynamic apparatus like pumps and compressors. They have large masses and are placed directly on high inertial foundations because of the dynamic action produced during the operating conditions.

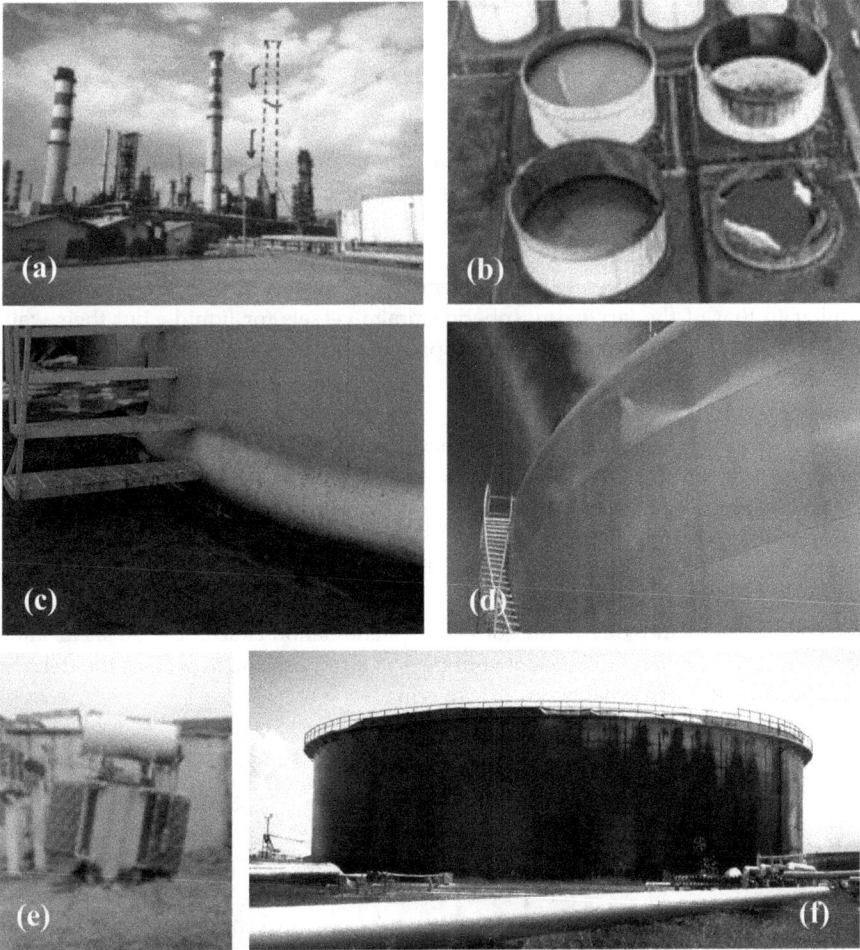

Fig. 3. (a) Collapse of a chimney at Tupras Refinery, (Itzmit, 1999), (b) Damages to tanks with floating roof (Tokachi-oki, 2003), (c) Elephant foot buckling (Anchorage, Alaska, 1964), (d) Elastic diamond buckling (San Fernando, California, 1971), (e) Collapse of an electric transformer, (f) Overtopping phenomenon in a big tank (Itzmit, 1999).

The typical damages associated to the first category are related to buckling phenomena of the wall (elephant foot buckling, sloshing buckling) or to failure of the wall-bottom plate joint. In Figure 3c and 3d examples of elephant buckling and elastic buckling are shown, respectively. But other possible damages are also possible, especially due to excessive sloshing motion, even in presence of floating roof, which can cause liquid overtopping and fire due to the crash between roof and wall. As a matter of fact, during the Itzmit and Tokachi-oki earthquakes, most of the tanks were destroyed for excessive sloshing motion (Figure 3b and 3f). Moreover, damages due to the uplift phenomenon of above ground tanks or ground settlements have been also observed during several seismic events, which have been capable to produce serious localized damage between wall and bottom plate.

2.3 Squat equipment supported by columns

In this category it can be included:

- Spherical storage vessels, essentially used for pressure liquefied gases. They are generally elevated with respect to ground by using steel columns placed along the circumference and welded to the shell at the equatorial level and normally linked each-other by diagonal braces (see Figure 4c).
- Vertical large storage vessels for cryogenic liquefied gas (LNG); their configuration is similar to that of the large atmospheric storage vessels for liquids, but their walls are realized by a double shell, in the inner-space of which an efficient thermal insulation is located; their bottom is anchored to a concrete plate, supported by short reinforced concrete columns (Figure 4a).
- Process furnaces and steam boilers. These equipments have the function to heat or vaporize large amounts of liquid products, according to the chemical process demand. Generally process furnaces are large structures, with few standardized shapes, mainly of cathedral type and vertical cylinder (Figure 4b). These furnaces are kept elevated from the ground by means of short reinforced concrete columns, according to the location of burners that requires pipes and space for maintenance. For these apparatus, the collapse is mainly due to the soft-story phenomenon caused by the shear failure of the short columns. The collapse of chimneys is also possible, as well as the detachment of pipes and of the internal wayward covering.

Fig. 4. (a) Damaged liquid oxygen and nitrogen tanks (Habas plant, Itzmit, 1999), (b) Cylindrical shape Furnaces (c) Spherical storage vessels.

During the Kern Kounty earthquake (California, 1952), at the Paloma Cycling Plant, two of the five spheres of butane collapsed, causing the cut of all the connecting pipes and creating a massive release of the content. The cloud formed quickly found a source of ignition in electrical transformers places a few hundred meters away. The result was a violent VCE (Vapor Cloud Explosion) followed by a fire of considerable proportions, see Figure 5a.

During the Itzmit earthquake some LNG tanks sustained by reinforced concrete columns were subject to important damages followed by the collapse of the basement, due to the insufficient shear strength of the columns (Figure 5b) (Sezen et al., 2007)

Fig. 5. (a) Butane sphere fire (K. Kounty, 1952), (b) Collapse of LNG tanks (Itzmit, 1999).

2.4 Piping systems

Piping systems connect all equipment involved in the process, transferring fluids within the plant (Figure 6a); as mentioned before, in a large refinery, hundreds kilometers of pipes of different size are installed; they are mainly realized with steel, but in some cases also with ceramics, glass, concrete, etc., if a specific performance against corrosion is required.

Fig. 6. (a) A piping system in a refinery, (b) Breakage of a piping flanged connection.

Metallic pipes themselves are not particularly vulnerable to seismic actions, but they can suffer the effects of differential displacements, which could not be compatible with the pipe deformations. Moreover, a collapse of the support structure can cause a catastrophic collapse of pipes, as shown in Figure 6b where the breakage of a piping flanged connection is shown.

2.5 Support structures

Piping systems, heaters, pumps, fans and other equipment, require a support structure. The geometrical configuration depends on different factors; consequently, they present different structures with irregular distribution of stiffness and strength. In addition, the structure is often modified according to the production requirements that can be necessary during the life-cycle of the plant. The support structures are mainly realized with steel frames, often stiffened by diagonal bracings. Nevertheless, structures realized with a different material are not rare, especially reinforced concrete and steel-concrete composite structures.

In the several reports dedicated to seismic effects in industrial plants, description of damages suffered by support structures are rare. During the Loma Prieta earthquake service frames of a reactor were subjected to limited damages. On the contrary, the support frame of a group of fans was subjected to severe damage as shown in Figure 7a. The structure was initially designed to support a piping system, but later was used for supporting the fans. Consequently, during the seismic event some of the beams collapsed for elastic instability.

Finally, collapse of the entire supporting structure due to the failure of the surrounding structures cannot be excluded. For example during the Itzmit earthquake in 1999 the chimney of Fig. 3a fallen down on a piping system causing important damages (Figure 7b)

Fig. 7. (a) Damage to the supporting structure of fans (Loma Prieta, 1989), (b) Collapse of a piping system (Itzmit, 1999).

3. Applicability of passive control systems for the seismic protection of oil refinery components

3.1 Passive control techniques (PCT) for the seismic protection of industrial components

Innovative seismic control systems belongs to the world of the vibration control techniques of structures, which includes passive, semi-active, active and hybrid systems (Housner et al., 1997; Spencer, 2003; Christopoulos and Filiatrault, 2007). The experiences acquired during experimental activities and worldwide applications have indicated the passive control techniques as the most suitable solutions for the seismic protection of structures. These systems modify the stiffness and/or the dissipative properties of the structure, favoring the reduction of the dynamic response to seismic actions. They can be classified on the basis of

Analysis of the Seismic Risk of Major-Hazard Industrial Plants and Applicability of Innovative
Seismic Protection Systems

241

the physical phenomenon adopted, and specifically: a) Seismic isolation, b) Energy dissipation, c) Tuned mass damper (TMD).

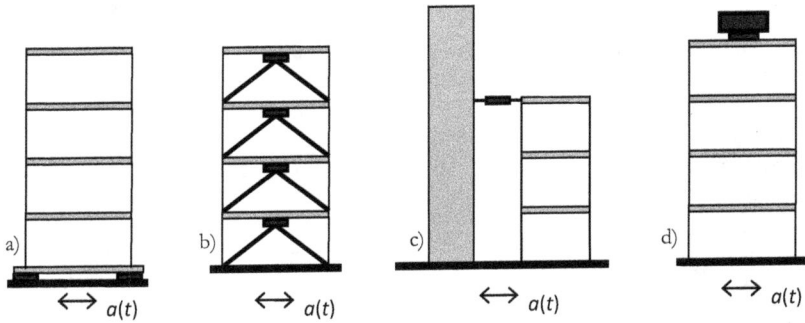

Fig. 8. PCS : a) base isolation, b) dissipative bracings, c) dissipative coupling, d) Tuned mass.

Base isolation systems produce a certain level of uncoupling between structure and ground. At this end, devices named "isolators" are usually placed at the foundation. Consequently, the structure is divided in two distinct parts: substructure and superstructure (Figure 8a); these devices are characterized by a high deformability in the horizontal direction and, at the same time, they are capable to transmit the vertical forces without appreciable deformations.

Isolators modify the dynamic characteristics of the structure, obtaining a high reduction of inertia forces. Unfortunately, there is a price to be paid: an appreciable increasing of displacements between substructure and superstructure. This inconvenient can be limited by increasing the energy dissipation capability of the structure using external dissipation devices or by a specific dissipative mechanism included in the isolators.

The isolation systems are usually subdivided in two categories: a) conventional devices, characterized by an elastic behavior, which realizes the elongation of the fundamental period of the structure, b) dissipative isolators, which in addition to the period elongation effect, increase the dissipation capability of the devices.

Dissipation control systems realize an artificial increasing of the structural damping with a consequent reduction of structural forces and displacements. Energy dissipation devices can be placed within the same structure through dissipative bracings (Figure 8b (Ciampi et al, 1995; Renzi et al., 2007), or between adjacent structures (dissipative coupling, Figure 8c) with different dynamic characteristics (Basili et al., 2007a, 2007b; Fraraccio et al., 2006).

Finally, using mass damping systems, namely TMD (Tuned Mass Dampers), the control of the seismic vibrations is carried out by adding auxiliary masses connected to the structure (conventional TMD). The additional damping is due to the transfer of energy form the structure to the auxiliary mass; this energy is then dissipated through dissipative devices placed between the structure and the auxiliary mass (Hoang et al., 2008). A non conventional TMD can also be defined: masses already present on the structure are converted into tuned masses, retaining structural or architectural functions beyond the mere control function (De Angelis et al, 2001).

The above techniques cannot be indifferently applied to each type of equipment of a plant, because their effectiveness depends on the characteristics of the structure to be protected. As a matter of fact, for a cathedral-type furnace (Figure 9a) the seismic isolation can be profitably used by inserting isolators between the superstructure and the base columns.

A typical situation to which the techniques of Figure b, c and d can be easily applied is shown in Figure 9b. In fact, dissipative bracings can highly reduce the seismic response of the steel frame. Alternatively, the equipment placed at the several floors of the frame can be used as TMDs connecting them to the frame using dampers. Moreover, being distillation column and frame adjacent structures, they can be coupled using dissipative devices (Paolacci et al., 2009; Paolacci et al., 2010)

Until now, passive control techniques have been used for a very limited number of industrial applications; for example, in Europe the isolation technique has been adopted only in a few cases: the seismic protection of Petrochemical LNG terminal of Revythousa, Greece (Tajirian, 1998) and of ammoniac tanks, at Visp, in Switzerland by means of elastomeric isolators (Marioni, 1998), Figure 20. Friction Pendulm devices have also been used for the seismic isolation of an elevated steel storage tank of the petrochemical plant of Priolo Gargallo in Sicily (Italy) (Santangelo et al., 2007).

It is worth to observe that damages like leakage of a flanged joint or breakage of a welding, usually considered critical for the safety of the transformation process, are instead treated as unessential under structural point of view, even if might generate important accidental chains. Therefore, for the process industry, unlike civil structures, it is necessary to associate the direct damage inflicted by the earthquake, with the indirect consequences of possible incidental chains.

The experience provided by the observation of seismic damages suffered by industrial plants allows recognizing the most vulnerable components. Assuming that one of these components is damaged and considering this as an initiator event, using a classic analysis of "event tree", the risk of the various consequences could be assessed.

The first step is then to identify the most likely initiator events of accidental chains that can lead to serious consequences. Table 2 summarizes the typical observed damage and accidents due to earthquakes of some major industrial equipment together with the suggested passive techniques for their seismic protection. For slim vessels, the most likely damage in case of earthquake is the yielding of anchorage bars at the foundation level and the leakage of fluid due to failure of flanged joints caused by excessive displacements. In this case, the most appropriate passive control technique seems to be the dissipative coupling between vessel and adjacent structures (Ciampi et al., 2006). As far as above ground broad tanks is concerned, typical observed damages (failure of wall-bottom plate welding, elephant foot buckling, elastic buckling of wall, settlements of the ground under the tank) can be eliminated using base isolation technique (De Angelis et al., 2010). Other accidents and possible damages due to sloshing motion in presence of floating roof (overtopping, fire due to impact between the roof and mantle), are neither amplified nor reduced by the isolation. A possible solution to reduce the effects of the impact between floating roof and tank wall can be represented by spacers placed between roof and wall or by inserting a TMD system into the roof (for eg. a Tuned Mass Damper Column, (Sakai and Inoue, 2008).

Fig. 9. a) Cathedral furnace where the seismic isolation can be easily applied, b) Distillation
column and adjacent steel frame in which is possible to realize: dissipative bracings in the steel
frame, dissipative coupling between column and frame, and Tuned Mass Damper, linking the
equipment placed at several levels of the frame with the same frame using proper isolators.

Fig. 10. a) Base isolation of a LNG tank at Revythousa Island (Grecee), b) Base isolation of an
elevated tank ammoniac tanks at Visp, Switzerland, c) Dissipative bracings solution for
elevated tanks, d) dissipative coupling solution for elevated tanks.

For squat equipment placed on short columns the base isolation technique appears to be
highly effective. The unique drawback is represented by the increasing of relative
displacement between the apparatus and the surrounding ones. For this reason proper
counter-measures have to be adopted to preserve the integrity of the connected pipes, for
example flexible joints.

The use of dissipative bracings can be effective (Figure 10c) (Drosos et al., 2005). These
devices reduce stresses and displacements, but can be invasive and limit the operation of the
equipment. In some cases, like spherical tanks, the dissipative coupling can also be used in
conjunction with auxiliary reaction structures (Figure 10d) (Addessi et al., 2001).

For process furnaces, beyond the base isolation technique, a sophisticated solution may
consist in designing the chimney as a TMD system, adding an auxiliary mass at the top
(Balendra et al., 1995).

Structural typology	Critical equipment	Typical seismic observed damages	Other possible dameges	Passive control techniques
Slim vessels	Columns Reactors Chimney Torch	• Leakage of fluid in flanged joints • Yielding of anchor bars	Overturning	Dissipative coupling
Above-ground squat equipment	Big broad tanks with fixed and floating roof	Failure of wall-bottom plate welding Elephant foot buckling Diamond buckling of tank wall Settlements of ground Impact of floating roof to tank wall.	Uplifting Overtopping Torch fire	Base isolation dissipative spacers between roof and wall, TMD
Squat equipment placed on short columns	Spherical tanks	Collapse of structure due to shear failure of columns		Dissipative bracings Base isolation Dissipative coupling
	Process Furnaces	Collapse of structure due to shear failure of columns Collapse of the chimney Detachment of internal pipes Detachment of the internal refractory material	Leakage from pipes; Increase of temperature of Furnace wall	Base isolation Dissipative bracings TMD
	Cryogenic tanks	Collapse of structure due to shear failure of columns		Base isolation
Piping systems and support structure	Steel or R.C. frames	Collapse for excessive stresses	Damages to supported equippment (pipes, tanks,..)	Dissipative bracings Dissipative coupling TMD (using the same supported equipment)

Table 2. Seismic damages of industrial process components and passive control systems.

Finally supporting frames can be effectively protected by several control techniques. For example, to reduce forces and displacements, dissipative bracings and dissipative coupling techniques can be profitably used, as for the slim vessels. To avoid damage in the supported equipment (compressors, pumps, tanks, etc..), the TMD technique can also be adopted. In this case, it is possible to reduce the stress level in the frame and contemporarily to protect the equipment used as TMD. In this case, special dampers like wire-ropes could be profitably used (Paolacci and Giannini 2008).

The innovative technologies for the vibration control are also effective to preserve integrity and operational continuity of equipment.

An effective way to protect seismic vulnerable equipment consists in implementing an isolation system between the internal apparatus and the supporting structure. There are two configurations proposed in literature: the apparatus to be isolated may correspond either to an individual raised floor (Hamidi and El Naggar, 2007), on which a group of several equipment is anchored (isolated raised-floor systems or floor isolation systems, (Alhan and Gavin, 2005), or to a single equipment itself (equipment isolation systems), especially when having a large mass (De Angelis et al., 2011). As a result, the absolute accelerations transmitted to equipment are considerably reduced and the damages due to excessive inertial forces are prevented.

Passive linear and nonlinear isolation systems have been proved to be effective and practical to protect acceleration sensitive equipment from earthquake hazard [Reggio and De Angelis, 2011].

4. Example: Application of two isolation systems for the seismic protection of above-ground steel storage tanks

4.1 Introduction

In this section, the effectiveness of the base isolation on steel storage tanks is investigated through numerical models and then checked by shaking table tests on a reduced scale (1:14) physical model of a real steel tank (diameter 55m, height 15.6 m), typically used in petrochemical plants. In the experimental campaign the floating roof has also been taken into account.

In practice, in order to evaluate the effect of the sloshing, many seismic design codes adopt the formula relative to free surface conditions, (European Committee for Standardization, 2003; American Petroleum Institute 2007). This is justified by theoretical studies (Sakai and Nashimura, 1984), where is shown that the effect of the floating roof has a negligible influence on the frequency and amplitude of the first vibration mode and it is confirmed by the experimental study presented in the following.

The tests have been carried out using the six d.o.f. 4 x 4 m shaking table installed in the laboratory of ENEA (Italian National Agency for New Technologies, Energy and the Environment) Research Centre "La Casaccia" at Rome, Italy. The tests have been performed on the physical model both in fixed and isolated base configurations; in particular two alternative base isolation systems have been used: high damping rubber bearings devices (HDRB) and PTFE-steel sliding isolation devices with c-shaped elasto-plastic dampers (SIEPD).

In the following, after a brief presentation of the dynamic behavior of tanks, with and without base isolation systems, the main results of the experimental tests are shown and discussed. Finally, a comparison between experimental and numerical results has been illustrated. A detailed description of the experimental campaign can be found in (De Angelis et al. 2010).

4.2 Dynamics of liquid storage tanks

4.2.1 Fixed base tanks

The dynamics of cylindrical tanks subjected to a base motion has been extensively studied by several authors. Starting from the earliest work of Housner (1963), the hydrodynamic pressure induced by the liquid on the tank wall due to the base motion has been determined, taking into account the deformability of the tank wall; see for example (Fisher, 1979; Haroun & Housner, 1981; Velestos & Tang 1987).

In brief, the liquid mass can be imagined subdivided in two parts: an impulsive component, which follows the base motion and the deformability of the tank wall, and a convective component, whose oscillations cause superficial wave of different frequency with a very low percentage of mass ($\approx$4%) relative to the higher modes; moreover, while in the slender tanks the most part of the liquid moves rigidly with the tank, in the broad tanks most of mass oscillates in the convective modes.

Under the hypothesis of rigid tank, the impulsive and convective part of hydrodynamic pressure can be easily evaluated. On the contrary, the part, which depends on the deformability of the tank wall, can be determined solving a fluid-structure interaction problem, whose solution depends on the geometrical and mechanical characteristics of the tank: radius R, liquid level H, thickness s, liquid density ρ and elastic modulus of steel E. The problem can be uncoupled in infinite vibration modes, but only few of them have a significant mass. Thus, the impulsive mass is distributed among the first vibration modes of the wall.

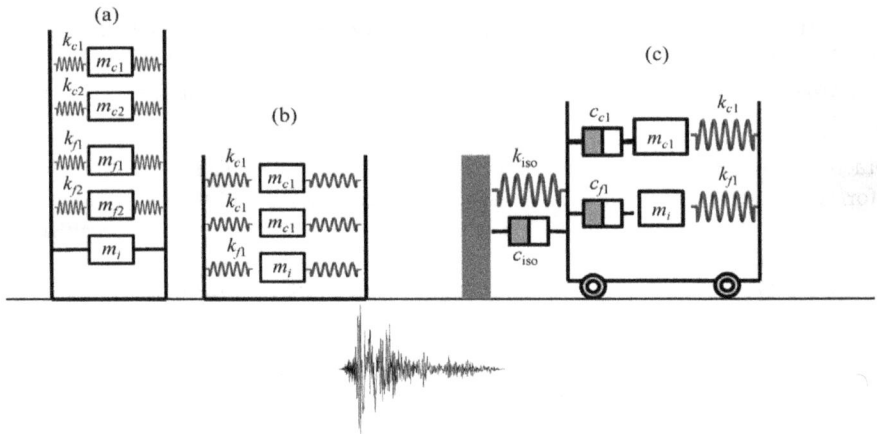

Fig. 11. (a) Equivalent spring-mass model: (a) general, (b) broad tanks (c) Dynamic model of a base isolated broad tank.

On the basis of the above observations it can be drawn that the study of the hydrodynamic pressure in tanks subjected to a seismic base motion can be easily performed using the simple model shown in Figure 11, in which the liquid mass is lumped and subdivided in three components: rigid, impulsive and convective masses named m_i , m_{ik} (mass of k-th mode of the wall vibrations), m_{ck} (mass of k-th convective mode). The impulsive and

Analysis of the Seismic Risk of Major-Hazard Industrial Plants and Applicability of Innovative
Seismic Protection Systems

247

convective masses are connected to the tank wall by springs of stiffness k_{ik} and k_{ck}. The total pressure is given by adding the effects of the mass mi subjected to the base motion acceleration, of the masses m_{ik} subjected to the acceleration of the wall relative to the bottom of the thank, and of the masses m_{ck} subjected to the absolute acceleration.

In case of broad tanks the model of Figure 11(a) can be updated by the simplest model shown in Figure 11(b). In fact, the contribution of the higher order vibration modes is negligible and the entire impulsive mass is practically equal to the mass of the first vibration mode; moreover, because the distributions of the impulsive pressure, with and without wall deformability, are almost coincident, the effects of the impulsive action are simply taken into account by the response in terms of absolute acceleration of a simple oscillator of mass m_i and stiffness k_i. Neglecting the higher convective modes effect, the model becomes a simple two degrees of freedom model. The frequencies of the convective and impulsive modes are generally very different (tenths of a second against tens of seconds). This justifies the usual choice of neglecting the interaction between these two phenomena.

4.2.2 Base isolated tanks

As shown in section 2, the idea of seismic protection of tanks through base isolation technique is not really new, but the important amount of numerical investigation present in literature has proved its high effectiveness. Unfortunately, a limited number of experimental activity has been performed so far (Bergamo et al., 2007; Summers et al., 2004; Calugaru and Mahin, 2009).

On the basis of the observations of the previous section a dynamic model of a base isolated tank can be easily built. For example a simple model of base isolated broad thanks is shown in Figure 11(c).

The vibration period of the impulsive component of pressure generally falls in the maximum amplification field of the response spectrum, whereas the convective period T_c is usually very high and thus associated with a low amplification factor. This implies a high effectiveness of the base isolation system, which can reduce highly the base shear due to the impulsive pressure component. Neglecting the influence of the wall deformation, the period of the isolated structure is approximately given by:

$$T_{iso} \approx 2\pi \sqrt{\frac{m_i + M_s + M_b}{k_{iso}}} \tag{1}$$

in which m_i is the impulsive part of the liquid mass, M_s and M_b are respectively the wall and base tank masses, and k_{iso} is the elastic stiffness of the isolators.

For broad tanks m_i is a relatively small part of the total mass and this allows a significant reduction of the devices dimensions.

Moreover, in the case of big tanks, for which $T_c >> T_{iso}$, the first period of the convective motion is practically unmodified. Consequently, the base isolation system does not show any important mitigation effects on the sloshing pressures. This is not relevant, since the convective pressure is very small because of the long period of the fluid oscillations.

The negative effect of the sloshing is related only to the superficial motion, because either the height of the wave can exceed the upper limit, causing overtopping phenomenon, or the floating roof motion could cause a breaking of the gaskets and the leakage of dangerous vapours of inflammable substances. Unfortunately, the base isolation does not modify this phenomenon. Moreover, the base isolation can cause high displacements between tank and ground, which may induce dangerous damage to the pipes-tank connections.

4.3 Shaking table tests

4.3.1 Description of the physical model

The full scale structure is a big steel liquid storage tank typically installed in petrochemical plants. The dimensional characteristics of the tank are the following: radius R=27.5m, height Hs=15.60 m, liquid level H=13.7 m, liquid density ρ=900 kg/m3, and wall thickness variable from 17 to 33 mm. The scale model has radius R=2 m, height Hs=1.45 m and wall thickness s=1 mm (Figure 12a). Thus, the scale ratio is about 13.7.

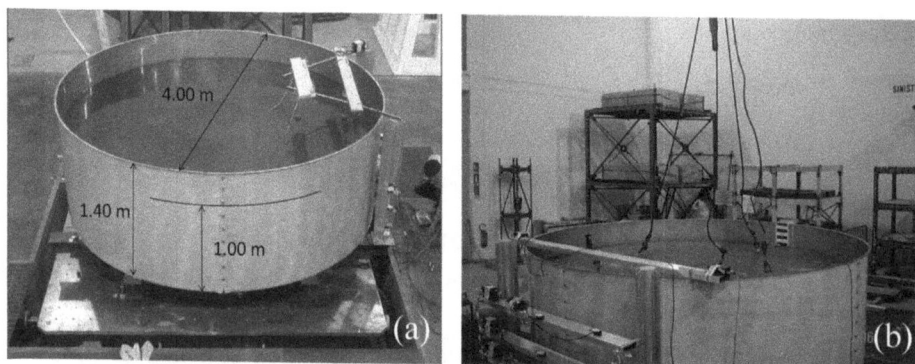

Fig. 12. (a) Tank without floating-roof, (b) Floating roof installed on the mock-up.

Because the period of the sloshing mode is a function of the square root of the dimensions (European Committee for Standardization, 2003), the time scale of the convective motion is 3.7. In order to obtain the same scale ratio for the impulsive frequencies it would be necessary to reduce the thickness of 140 times with respect to the real one. This is obviously impossible and only one time scale can be respected. In particular, the convective motion scale has been adopted during the test (De Angelis et al., 2010).

The floating roof of the full scale tank is realized through a truss structure, which sustains steel plates. In order to maintain the mass ratio, the floating roof of the mock-up has been realized with a wood structure (Figure 12b). The gasket, which in the full scale tank is composed by a more complex mechanism, has been here realized with a rubber tube applied along the circumference of the roof.

4.3.2 Design of base isolation systems

For the seismic protection of the tank two isolation systems has been used: high damping rubber bearings (HDRB) and PTFE-steel isolation devices with metallic c-shaped dampers

(MD). Both the base isolation systems have been designed by choosing a properly isolation period (T_{iso}) and then evaluating the stiffness k_{iso} using the equation (1). For the high damping rubber bearings the damping ratio ξ has been assumed equal to 10% whereas the yielding force of the metallic dampers has been designed using the method proposed by (Ciampi et al., 2003).

The prototypes of the isolator devices, properly realized for the experimental activity by the Company Alga Spa (Milan), are shown in Figure 13a and 13b. The isolators have been characterized by cyclic imposed displacement tests carried out in the experimental laboratory of the University of Roma Tre. The main characteristics of the devices are summarized in Table 3.

In order to check the effectiveness of the control systems, for each isolator typology four devices have been used, which have been placed at the cross of the bottom metallic beams.

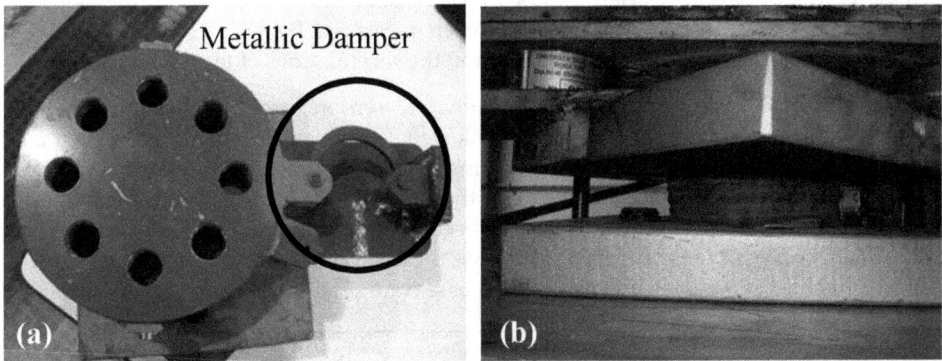

Fig. 13. (a) Sliding bearing with dissipative damper, (b) High Damping Rubber Bearing.

	HDRB	MD
Isolation period in real scale (sec)	2.8	1.6
Initial stiffness(kN/m)	900	1617
Yielding strength (kN)	3.00	7.880
hardening ratio	0.40	0.05
Fiction coefficient	--	2-3%

Table 3. Isolation devices characteristics.

4.3.3 Test set-up

A series of dynamic tests have been carried out on the tank using the shaking table installed at the Research Center of ENEA "La Casaccia" (Rome). The mock-up has been tested in four different configurations: fixed base tank without floating roof (case A), fixed base tank with floating roof (case B), isolated tank with HDRB and floating roof (case C), isolated base tank with SIEPD devices and floating roof (case D).

Six different base motion histories have been used in each configuration (four natural and two synthetic accelerograms, generated by Simqke (Vanmarcke and Gasparini, 1976), according to the European code spectra (soil C), and scaled to different intensity levels). For all accelerograms the time scale has been changed according to the indications of section 3.1. The natural accelerograms have been selected from the Pacific Earthquake Engineering Research center database, between time histories recorded for soil C and generated by seismic events with magnitude lower than 8 and epicenter distance lower than 50 km. Another more selective criterion, adopted here, consists of a selection of signals filtered using a high-pass filter with cut-off frequency greater than 0.1 Hz. As already seen, this is almost the frequency of the sloshing motion of the full scale tank. The records used during the tests are reported in Table 4. The tank has been also tested using white-noise and harmonic signals with variable frequency (sine-sweep) for the identification of its dynamic characteristics.

The response signals were measured using numerous sensors: pressure transducers, strain-gauges and laser transducers placed on the tank wall. Laser transducers have also been used for the sloshing motion of the liquid or floating roof, whereas the motion of the table has been monitored by several accelerometers. In the isolated base configurations the motion of the structure with respect to the base has been measured by wire LVDT sensors. The arrangement of the sensors has changed between the several series of tests.

The response signals were measured using numerous sensors: pressure transducers, strain-gauges and laser transducers placed on the tank wall. Laser transducers have also been used for the sloshing motion of the liquid or floating roof, whereas the motion of the table has been monitored by several accelerometers. In the isolated base configurations the motion of the structure with respect to the base has been measured by wire LVDT sensors. The arrangement of the sensors has changed between the several series of tests. In Figure 14 a sketch of the main sensors used in the isolated base configuration is shown.

Accelerogram	M	Distance (km)	PGA (g)	Duration (s)
Irpinia, 11/23/80,Sturno, 270	6.5	32	0.313	40.00
Duzce 11/12/99, Duzce, 180	7.2	8.2	0.482	25.88
Kocaeli 08/17/99, Arcelik,090	7.4	17.0	0.149	28.00
Chi-Chi 09/20/99, TCU120, w	7.3	8.1	0.268	90

Table 4. Characteristics of the natural records used during the experimental test.

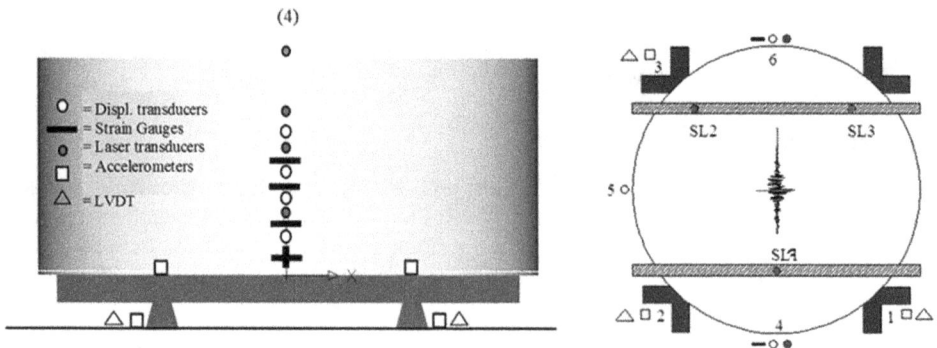

Fig. 14. Experimental Set- up.

4.3.4 Analysis of results

4.3.4.1 Identification of dynamic characteristics

Before evaluating the seismic effects, the main dynamic characteristics of the tank (frequencies and damping) have been identified. The frequencies of the systems have been determined by means of the transfer functions between the input (base acceleration) and output signals (e.g. liquid pressure). The damping ratio has been evaluated using the logarithmic decrement method applied to free vibrations time histories. For each vibration mode, the signal has been filtered using a pass-band filter around its frequency.

In Figure 15 the transfer functions between the table acceleration and one of the pressure sensors (near the bottom of the tank) are shown for the fixed base configurations (A and B), without (Figure 15a) and with floating roof (Figure 15b). For both cases, the frequency of the sloshing motion does not change and is almost coincident with the theoretical value (0.4 Hz), whereas a resonant frequency of around 18 Hz is also shown, which correspond to the main natural frequency of the motion due to the deformability of the wall.

The damping of the sloshing motion has been measured on the basis of the amplitude decrement of free vibrations. In case of free liquid surface the damping is very low and is not in practice measurable; using the floating roof a 1% of damping has been identified. This value, greater than 0.5%, usually adopted in numerical model, is probably due to the interaction between the liquid surface and the floating roof and the friction between rubber gasket and tank wall-

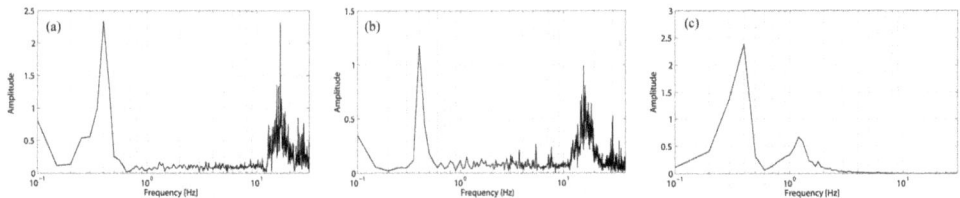

Fig. 15. Transfer functions: (a) pressure transducer, case A, White Noise, (b) pressure transducer, case B, White Noise, (c) LVDT, case C, ARC090.

The free vibrations of the floating roof for the isolated tank are shown in Figures 15c. The signals have been filtered in order to remove the frequencies greater than 1 Hz. Comparison with the results of non isolated case results, a considerable increasing of the sloshing damping, variable in the range 2.5-3.0%, for both isolated configurations has been detected. This increment was expected by numerical models, carried out with non-classical damping theory, but with lower values than the experimental ones.

4.3.4.2 Response of the base isolated tank

The evaluation of seismic effects on the tank wall has been made in terms of both maximum base shear induced by the dynamic pressure and relative impulsive and convective components.

These resultants have been evaluated interpolating the experimental values of the pressure with the relative theoretical functions and then integrating along the height. The separation of the sloshing and impulsive plus flexible components was obtained by filtering the signals around the corresponding frequencies.

The responses of the table, measured in the different experimental tests, because of some problems with the control system, were different, even if the same accelerograms were used. Because of these differences, comparison was difficult. To overcome this problem the comparison was done in terms of mean spectra accelerations Sa, (5% of damping), calculated in the period range 0.05 and 0.1 s (10 ÷ 20 Hz) for the impulsive component, and 2.2-2.8 sec (~ 0.35 ÷ 0.45 Hz) for the sloshing component. Impulsive and sloshing forces have been represented as functions of the relative spectra acceleration (Figures 16a and 16b).

In the same figures the interpolation curves of the experimental data are shown for the fixed base (case B) and isolated base cases (case C, case D). For the fixed base case the trend is linear, whereas the isolated base cases clearly show a nonlinear behavior of the phenomenon, especially for high values of spectral acceleration.

The total shear has been plotted as function of the same spectra ordinate as that used for the impulsive component, the latter being a significant part of the total one (Figure 16c). The figure clearly shows the high effectiveness of both the isolation systems, which induce considerable reductions of the total shear affecting the tank wall. However, whereas the effectiveness of HDRB isolators does not practically vary with S_a, the effectiveness of SIEPD devices is reduced for low values of spectral acceleration, and it is practically negligible for Sa<0.3g.

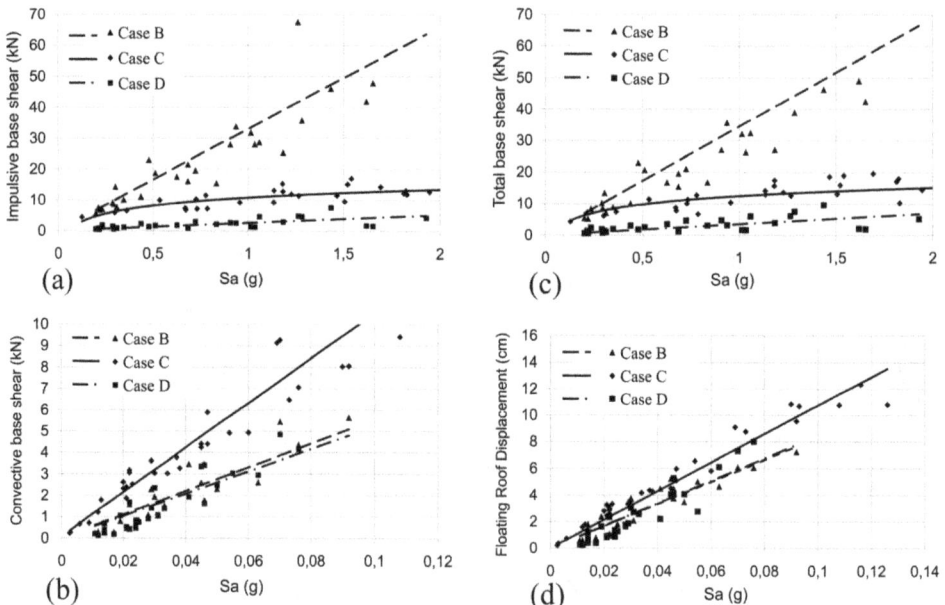

Fig. 16. (a) Impulsive base shear versus spectral acceleration of the impulsive motion, (b) convective base shear versus spectral acceleration of the convective motion, (c) total base shear versus spectral acceleration of the impulsive motion, (d) maximum displacements of the floating roof versus spectral acceleration of the convective motion.

Analysis of the Seismic Risk of Major-Hazard Industrial Plants and Applicability of Innovative
Seismic Protection Systems

253

This is due to the friction effects, often neglected for this kind of device, but particularly significant for the tanks, in which the gravity mass, represented by the entire liquid volume, is greater than the dynamic mass, which is represented by the impulsive mass only (in the present case only 30% of the total mass). Therefore, to overcome the friction appreciable levels of acceleration are needed. This problem is not present for elastomeric devices.

This behavior, if appropriately controlled, could represent an interesting advantage recognizable to SIEPD devices compared to HDRB devices. In fact, for low levels of seismic intensity, it could be convenient to use an isolation device with a reduced amount of slip, which instead is required for more intense events in order to achieve a full employment of the dissipation capability of the devices.

For the impulsive component (Figure 16a), the reduction is quite important, as already observed for the total base shear, especially for high values of Sa; on the contrary, the sloshing component (Figure 16b) remain practically unchanged for HDRB, whereas a slight increasing has been observed using SIEPD. This increasing has no influence on the stresses of the tank wall, at least for large broad tanks. The negative effects are related to the floating roof oscillations, which could generate, in some cases, the breaking of the gaskets and the leakage of inflammable materials. Unfortunately, the base isolation does not reduce the amplitude of the superficial waves, although the sensible increasing of damping reduces its duration and then probably the risk of failure of the gaskets. However, for case D only an increasing of the convective pressure has been measured, which has induced a rather moderate increasing of the amplitude displacement of the floating roof that does not seem to be not particularly worrying (Figure 16).

Finally, the maximum values of displacements of the tank base, recorded during the tests of case D and here not shown for brevity (De Angelis et al. 2010), are lower than the displacements of case C. This is another advantage for the SIEPD isolation system, since large displacements, which have to be absorbed by the pipes-tank connections, represent a problem, the solution of which is not always easy.

4.3.4.3 Comparison between numerical and experimental results

The numerical model defined in section 3.2 has also been used to carry out step-by-step non linear analyses using as base motion the accelerograms recorder on the shaking table during the tests. The isolator devices have been modeled by the classical Bouc-Wen model. The aim is to confirm the reliability of the analytical and numerical formulation of the problem, also when the floating roof is present, generally missing in the evaluation of the dynamic response of tanks.

As a matter of fact, hereafter a numerical-experimental comparison of the results of cases B, C and D, is presented and discussed. For brevity only the results of the Arcelik accelerogram, scaled to the nominal acceleration nearest to the design value (0.5g) are shown.

Figure 17 shows the comparison, in terms of base shear for case B. Firstly, a good agreement between the experimental and numerical response may be noted. A good predictive capacity of the model emerges when the floating roof is present as well.

The numerical-experimental agreement can be better highlighted by analyzing the single components of motion. Figure 17a shows, for case B, the base shear due to impulsive pressure. The agreement is quite good and sufficient to correctly describe the phenomenon.

For the convective component of motion, although the value of the fundamental frequency was very close to the theoretical value (0.4 Hz), the comparison between numerical and experimental results proved unsatisfactory, because of non-zero initial conditions due to a low damping of the liquid motion. The numerical-experimental agreement is improved in the cases C and D. In fact, in these cases the fluid-structure interaction become less important and the high damping of the liquid motion arrests oscillations more rapidly, allowing zero initial conditions.

For the isolated tank Figures 17b and 17c show the comparison between theoretical and experimental results. The good agreement proves the reliability of the model.

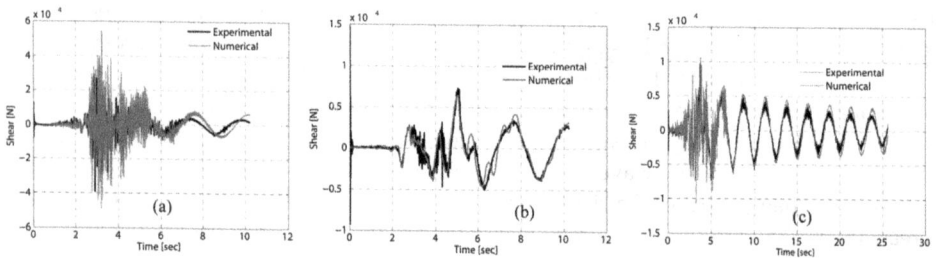

Fig. 17. Total base shear. Numerical-Experimental comparison, (a) case B, (b) case C, (c) case D, Arcelik 0.5g.

To summarize, the comparison between the shaking table test results and the response of simple numerical models described in section 2 shows the suitability of the latter to simulate the behavior of the fluid-structure system, both for fixed base and isolated base tank, also considering the floating roof as well. Actually, the slight discrepancies between numerical and physical models are probably due to some drawbacks in the experimental activity, caused by the high mass of the filled tank and the frequencies range investigated.

5. Conclusions

Recently, the attention paid on the protection of industrial facilities against natural phenomena is increasing, especially for the catastrophic consequences of strong events that induced severe damages to people and environment.

Among the natural phenomena capable to determine serious hazards to industrial plants, earthquakes should be taken into account especially because they are capable to generate multiple sources of releasing of dangerous substances and domino effects within the same plant, determining the complete destruction of the site. The analysis of past accidents induced by earthquakes has shown the high vulnerability of some typical industrial components and the severity of the consequences.

Accordingly, in this chapter, earthquakes effects on major-hazard industrial plants have been analyzed. In particular, typical equipment and components of a refinery were identified and gathered in a limited number of structural typologies; furthermore, their seismic vulnerability was analyzed, looking into historical events, i.e. concerning the damages caused by past earthquakes to several industrial components.

Subsequently, the applicability of some innovative techniques for the seismic protection of industrial equipment was studied and the most suitable passive control techniques were identified for each structural typology. As an example, the main results of a study on the effectiveness of yielding-based and HRDB isolators for the seismic protection of above-ground steel storage tanks were presented. In particular, after a brief introduction to the problem of designing a base isolation system for storage tanks, the outcome of a wide numerical and experimental investigation on a broad tank with floating roof were deeply discussed. Shaking table test results have shown the high efficiency of both the isolation systems and, at the same time, the reliability of lumped mass model for the prediction of the seismic response of isolated above-ground tanks.

6. References

Addessi D., Ciampi V. (2001), Ongoing studies for the application of innovative anti-seismic techniques to chemical plant components in Italy, 7th International Seminar on Seismic Isolation, Passive Energy Dissipation and Active Control of Vibrations of Structures, Assisi.

Alhan, C., Gavin, H.P. . Reliability of base isolation for the protection of critical equipment from earthquake hazards. Eng Struct 2005;27:1435-49.

ATC-25. (1991) Seismic vulnerability and impact of disruption of lifelines in the conterminous United States, (Applied Technology Council, Redwood City, CA).

API 650, (2007), Welded Tanks for Oil Storage, Edition: 11th Ed., American Petroleum Institute.

Balendra T, Wang C. M., Cheon N. F., (1995). Effectiveness of tuned liquid column dampers for vibration control of towers. Engineering Structures 17(9), 668-675.

Basili M; De Angelis M. (2007a). A reduced order model for optimal design of 2-MDOF adjacent structures connected by hysteretic dampers. Journal of Sound and Vibration, vol. 306; p. 297-317, ISSN: 0022-460X

Basili M; De Angelis M. (2007b). Optimal passive control of adjacent structures interconnected with non-linear hysteretic devices. Journal of Sound and Vibration, vol. 301; p. 106-125, ISSN: 0022-460X

Beavers, J.E., Hall, W.J. and Nyman, D.J. (1993) "Assessment of Earthquake Vulnerability of Critical Industrial Facilities in the Central and Eastern United States," In: Proceedings of the National Earthquake Conference, (Memphis, Tennessee), Vol.II, pp. 81-90.

Bergamo G.,.Castellano M.G, Crespo M., Forni M., Gatti F., Karabalis D., Martì J., Novak H., Poggianti A., Silbe H., Strauss A., Summers P., S.Triantafillou. (2005). Guidelines for the application of passive control techniques in petrochemical plants. Bologna. Rapporto finale delle attività INDEPTH.

Calugaru, Vladimir; Mahin, Stephen A., (2009), Experimental and analytical studies of fixed-base and seismically isolated liquid storage tanks, 3rd International Conference on

Advances in Experimental Structural Engineering, October 15-16, 2009, San Francisco.

Ciampi V., De Angelis M., Paolacci F., (1995). "Design of yelding or friction based dissipative bracings for seismic protection of buildings", Engineering Structures, Vol.17 n°5, 381-391.

Ciampi V., (1998), A methodology for the design of energy dissipation devices for the seismic protection of bridges, U.S.-Italy Workshop on Seismic Protective Systems for Bridges, New York, 27-28 Apr. 1998

Christopoulos C., Filiatrault A., (2006), Principles of Passive Supplemental Damping and Seismic Isolation, IUSSpress, Pavia, Italy

De Angelis M, Giannini R., Paolacci F., (2010), Experimental investigation on the seismic response of a steel liquid storage tank equipped with floating roof by shaking table tests, Earthquake Engineering & Structural Dynamics, 39: 377-396. DOI: 10.1002/eqe.945

De Angelis M., Perno S., Reggio A., (2011). Dynamic response and optimal design of structures with large mass ratio TMD. Earthquake Engineering And Structural Dynamics. Published online (wileyonlinelibrary.com). DOI: 10.1002/eqe.1117.

Drosos J.C., Tsinopoulos S.V., Karabalis D. L., "Seismic response of spherical liquid storage tanks with a dissipative bracing system," 5th GRACM International Congress on Computational Mechanics, Limassol, 29 June-1 July, 2005, 313-319.

Erdik M. and Durukal E. (2000) ``Damage to and Vulnerability of Industry in the 1999 Kocaeli, Turkey Earthquake''.

European Committee for Standardization (CEN). Eurocode 8: Design of structures for earthquake resistance – Part 4: Silos, tanks and pipelines, 2003.

Fischer D. Dynamic fluid effects in liquid-filled flexible cylindrical tanks. Earthquake Engineering and Structural Dynamics 1979; 7: 587-601.

Fraraccio G., De Angelis M., (2006). Adjacent structures controlled by steel elastoplastic devices (In Italian). Ingegneria Sismica, Anno XXIII, N.2, maggio-agosto 2006.

Hamidi, M., El Naggar, M.H. . On the performance of SCF in seismic isolation of the interior equipment. Earthq Eng Struct D 2007;36:1581–604.

Haroun MA, Hausner GW. Earthquake response of deformable liquid storage tanks. Journal of Applied Mechanics 1981; 48: 411-418.

Hatayama K., (2008). Lessons from the 2003 Tokachi-oki, Japan, earthquake for prediction of long-period strong ground motions and sloshing damage to oil storage tanks. J Seismol (2008) 12: 255–263.

Hoang N., Fujino Y., Warnitchai P., 2008. Optimal tuned mass damper for seismic application and practical design formulas. Engineering Structures 30, 707-715.

Housner G.W., Bergman L.A., Caughey T.K., Chassiakos A.G., Claus R.O., Masri S.F., Skelton R.E., Soong, T.T., Spencer B.F., Jao J.T.P. (1997) "Structural Control: Past, Present and Future", ASCE J. of Eng Mech, 123, 897-971.

Housner GW. The dynamic behavior of water tanks. Bulletin of the Seismological Society of America 1963; 53: 381-387.

Johnson, G.H., Aschheim, M. and Sezen, H. (2000) ``Chapter 14: Industrial Facilities'' In Kocaeli, Turkey Earthquake of August 17, 1999 Reconnaissance Report. Earthquake Spectra 16 (Supplement A), pp. 311-350.

Kawasumi, H. (ed.) (1968) ``Introduction'' In: General Report on the Niigata Earthquake of 1964. Tokyo Electrical Engineering College Press, Tokyo.

Kilic, S.A. and Sozen, M.A. (2003) ``Evaluation of Effect of August 17, 1999, Marmara Earthquake on Two Tall Reinforced Concrete Chimneys''. ACI Structural Journal 100 (3), pp.357--364.

Marioni, A, (1998), The use of high damping rubber bearings for the protection of the structures from the seismic risk, Jornadas Portuguesas de Engenharia de Estruturas, Lisboa, LNEC, 25 - 28 de Novembro de 1998

Meyers R.A., (2004), Handbook of petroleum refining processes, McGraw-Hill handbooks, ISBN 0071391096, 9780071391092

Moulein, J.A. and Makkee, M. (1987) "Chapter 4: Processes in the oil refinery," Lecture notes for Proceskunde, TU Delft, The Netherlands, pp.1-60.

Nielsen, R. and Kiremidjian, A.S. [1986] ``Damage to Oil Refineries from Major Earthquakes'' ASCE Jnl Struct Eng 112 (6), pp.1481--1491.

Paolacci F, Giannini R., De Angelis M., Ciucci M., (2009), Seismic vulnerability of major-hazard industrial plants and applicability of innovative seismic protection systems for its reduction, 11WCSI, November 17-21, Guangzhou, China

Paolacci F., Giannini R., De Angelis M., Ciucci M., (2010). Applicability of passive control systems for the seismic protection of major-hazard industrial plants, Sustainable Development Strategies for Constructions in Europe and China, 19-20, Rome, Italy

Paolacci F., Giannini R., (2008), "Study of the effectiveness of steel cable dampers for the seismic protection of electrical equipment", 14 World Congress on Earthquake Engineering, Beijing, China

Reggio, A., De Angelis, M. . Feasibility of a passive isolation system with nonlinear hysteretic behaviour for the seismic protection of critical equipment. In: Proceedings of the XIV Covegno ANIDIS Associazione Nazionale Italiana di Ingegneria Sismica. Bari, Italy; 2011.

Renzi E., Perno S., Pantanella S., Ciampi V., (2007). Design, test and analysis of a light-weight dissipative bracing system for seismic protection of structures. Earthquake Engineering And Structural Dynamics. vol. 36, pp.519-539 ISSN: 1096-9845

Sakai F., Inoue R, (2008) Some considerations on seismic design andcontrols of sloshing in floating-roofed oil tanks, The 14th World Conference on Earthquake Engineering October 12-17, 2008, Beijing, China

Sakai F., Nishimura M., Ogawa H. Sloshing behavior of floating-roof oil storage tanks. Computer & Structures 1984, 19:1-2, 183-192.

Santagelo A, Scibilia N., Stadarelli R., (2007), Seismic Isolation of a tanks at Priolo Gargallo (in Italian), Giornate AICAP 2007, 4-6 Ottobre, 2007.

Showalter, P.S, Myers, M.F., (1992) Natural Disasters as the Cause of Technological Emergencies: A Review of the Decade 1980–1989, Working Paper #78, Boulder, Natural Hazards Research and Applications Information Center, Univ. of Colorado, Colorado.

Scholl, R.E. and Thiel Jr., C.C. (eds.) (1986) ``The Chile Earthquake of March 3, 1985 - Industrial Facilities'' Earthquake Spectra 2 (2), pp.373-409.

Sezen, H. and Whittaker, A.S. (2004) ``Performance of industrial facilities during the 1999, Kocaeli, Turkey Earthquake'' In: Proceedings of the 13th World Conference on Earthquake Engineering (Vancouver, Canada; Paper No. 282).

Sezen, H., Livaoğlu, R., Doğangün, A., (2008), Dynamic analysis and seismic performance evaluation of above-ground liquid-containing tanks, Engineering Structure, 30: 794-803

Spencer B. F., Nagarajaiah S., (2003). State of art of structural control. Journal of Structural Engineering, 129, 845-855.

Steinberg, L.J., Cruz, A.M., Vardar-Sukan, F. and Ersoz, Y. (2000) ``Risk management practices at industrial facilities during the Turkey earthquake of August 17, 1999: Case Study Report''.

Summers P., Jacob P., Marti J., Bergamo G., Dorfmann L., Castellano G., Poggianti A., Karabalis D., Silbe H., and Triantafillou S., 2004. Development of new base isolation devices for application at refineries and petrochemical facilities. 13th World Conference on Earthquake Engineering, Vancouver, B.C., Canada, August 1-6, 2004 Paper No. 1036.

Suzuki, K., (2002) ``Report on damage to industrial facilities in the 1999 Kocaeli earthquake, Turkey'' Journal of Earthquake Engineering 6 (2), pp.275-296.

Suzuki, K., (2006), Earthquake Damage to Industrial Facilities and Development of Seismic and Vibration Control Technology – Based on Experience from the 1995 Kobe (Hanshin-Awaji) Earthquake – Journal of Disaster ResearchVol.1 No.2, 2006

Tajirian F. F., (1998), Base isolation design for civil components and civil structures, Proceedings of Structural Engineers World Congress, San Francisco, California, July

Vanmarcke EH, Gasparini DA. (1976). Simulated earthquake motions compatible with pre-scribed response spectra. Technical Report R76-4, Dept. of Civil Engineering, Massachusetts Inst. of Technology Cambridge, 1976-01, 99 pages (420/G32/1976).

Veletsos AS, Tang Y. Rocking response of liquid storage tanks. Journal of Engineering Mechanics 1987; 113: 1774-1792.

Part 4

Remediation and Safety Measures

13

Prediction of the Biodegradation and Toxicity of Naphthenic Acids

Yana Koleva and Yordanka Tasheva
University "Prof. Assen Zlatarov"- Burgas,
Bulgaria

1. Introduction

Crude oil is a complex mixture of hydrocarbons, basically composed of aliphatic, aromatic and asphaltene fractions along with nitrogen, sulfur and oxygen-containing compounds. The constituent hydrocarbon compounds are present in varied proportion resulting in great variability in crude oils from different sources (Speight, 1999). There are several reports indicating the recalcitrance and potential health hazards of the different constituents of crude oil (Kanaly&Harayana, 2000). These compounds have been reported to be carcinogenic, mutagenic and have immunomodulatory effects on humans, animals and plant life (van Gestel et al., 2001; Miller& Miller, 1981). The sites contaminated with hydrocarbons are ecologically important locations as one may encounter microbial flora of diverse nature, which may be potential candidates for important industrial processes (Jain et al., 2005).

The microorganisms possess the greatest enzymatic diversity found on earth and metabolize millions of organic compounds to capture chemical energy for growth. This metabolism, called catabolism or biodegradation, is the principal driving force in the degradative half of the earth's carbon cycle (Dagley, 1987). Microorganisms are increasingly used in engineered systems to biodegrade hazardous, xenobiotic compounds, an application commonly known as bioremediation (Alexander, 1994).

There have been numerous efforts to predict both the biodegradability or the pathway(s) of biodegradation for a given compound under a given set of conditions, typically either aerobic or anaerobic (Boethling et al., 1989; Parsons& Govers, 1990; Howard et al., 1991; Klopmann et al., 1995; Damborsky, 1996; Punch et al., 1996). Most of the efforts have been rule based, drawing general conclusions about what structures would or would not be readily biodegraded. Some only address whether a compound will be biodegraded and, if so, will biodegradation proceed slowly or quickly. The expert system projects, META (Klopmann et al., 1995) and BESS (Punch et al., 1996), also seek to determine at least one plausible biodegradation pathway.

Biodegradation pathway prediction requires the use of biochemical knowledge sometimes called metabolic logic. This requires knowledge of:

- organic functional groups to match a new chemical structure to one whose metabolism is already known;

- intermediary metabolism pathways to deduce how a new biodegradation can funnel a metabolite into a common pathway most efficiently;
- microbial enzymatic reactions to match a given reaction with a known enzyme;
- organic chemistry reactions to deduce what new reactions are chemically plausible to decompose a compound when precedents are not available.

Scientists studying biodegradation acquire this knowledge and these skills through many years of study and experimentation. This requires a means of organizing biodegradation reactions in some systematic fashion (Wackett& Ellis, 1999).

The University of Minnesota Biocatalysis/ Biodegradation Database (UM-BBD) began in 1995 and now contains information on almost 1200 compounds, over 800 enzymes, almost 1300 reactions and almost 500 microorganism entries. Besides these data, it includes a Biochemical Periodic Table (UM-BPT) and a rule-based Pathway Prediction System (UM-PPS) that predicts plausible pathways for microbial degradation of organic compounds (Gao et al., 2010). The inherent biodegradability of these individual components is a reflection of their chemical structure, but is also strongly influenced by the physical state and toxicity of the compounds. Therefore, the physical state is that strongly influences their biodegradation (Bartha& Atlas, 1977).

Naphthenic acid are most significant environmental contaminants. They are comprised of a large collection of saturated aliphatic and alicyclic carboxylic acids found in hydrocarbon deposits (petroleum, oil sands bitumen, and crude oils). Moreover, they are toxic components in refinery wastewaters and in oil sands extraction waters. In addition, there are many industrial uses for naphthenic acids, so there is a potential for their release to the environment from a variety of activities. Studies have shown that naphthenic acids are susceptible to biodegradation, which decreases their concentration and reduces toxicity.

They are described by the general chemical formula $C_nH_{2n+Z}O_2$, where n indicates the carbon number and Z is zero or a negative, even integer that specifies the hydrogen deficiency resulting from ring formation. Naphthenic acids have dissociation constants, which is typical of most carboxylic acids. Naphthenic acids are non-volatile, chemically stable, and act as surfactants (Seifert, 1975).

The presence of naphthenic acids in the environment is seldom studied and little is known about their fate. The investigations that used actual naphthenic acids focused on the biodegradation of these compounds as a group, because current analytical methods do not allow the study of individual compounds in the complex mixture. The aim of this study was to predict the biodegradation of the individual naphthenic acids and the possible toxicity of the parent structure and their metabolites. The software used for prediction of the microbial metabolism (biodegradation) of the naphthenic acids is the OECD (Q)SAR Application Toolbox (OECD (Q)SAR Project). The Toolbox is a software application intended to be used by governments, chemical industry and other stakeholders in filling gaps in (eco) toxicity data needed for assessing the hazards of chemicals. Degradation pathways used by microorganism to obtain carbon and energy from 200 chemicals are stored in a special file format that allows easy computer access to catabolic information. Most of pathways are related to aerobic conditions. Single pathway catabolism is simulated using the abiotic and enzyme-mediated reactions via the hierarchically ordered principal molecular transformations extracted from documented metabolic pathway database. The hierarchy of the transformations is used to control the propagation of the catabolic maps of the chemicals. The simulation starts with the search for match between the parent molecule and the source fragment associated with the transformation having the highest hierarchy.

2. Biodegradation of naphthenic acids

Naphthenic acids are highly toxic, recalcitrant compounds that persist in the environment for many years, and it is important to develop efficient bioremediation strategies to decrease both their abundance and toxicity in the environment. However, the diversity of microbial communities involved in naphthenic acid-degradation, and the mechanisms by which naphthenic acids are biodegraded, are poorly understood. This lack of knowledge is mainly due to the difficulties in identifying and purifying individual carboxylic acid compounds from complex naphthenic acid mixtures found in the environment, for microbial biodegradation studies.

2.1 Microbial degradation of naphthenic acids

Due to the high degree of complexity of the natural naphthenic acid mixtures and a lack of sources of individual naphthenic acid compounds, surrogate naphthenic acids were used in early microbial degradation studies (Herman et al., 1993; Herman et al., 1994, Lai et al., 1996). Naphthenic acids are acutely toxic to a range of organisms (Clemente& Fedorak, 2005; Headley& McMartin, 2004). MacKinnon and Boerger (MacKinnon& Boerger, 1986) demonstrated that with chemical and microbiological treatment approaches, the toxicity of tailings water could be reduced, presumably by removal or biodegradation of Naphtehic acids, although this was not shown directly. Herman et al. (Herman et al., 1994) followed biodegradation of naphthenic acids extracted from Mildred Lake Settling Basin (Syncrude) in laboratory cultures and also observed detoxification, as determined by the Microtox method. Clemente et al. (Clemente, MacKinnon& Fedorak, 2004) used enrichments of naphthenic acid-degrading microorganisms to biodegrade commercially available naphthenic acids (Kodak Salts and Merichem). Microtox analyses of culture supernatants revealed a reduction in toxicity after less than 4 weeks of incubation (Clemente, MacKinnon& Fedorak, 2004).

2.2 Microbial degradation prediction of naphthenic acids

The software used for prediction of the microbial metabolism (biodegradation) of petroleum thiophene is the OECD (Q)SAR Application Toolbox. The Toolbox is a software application intended to be used by governments, chemical industry and other stakeholders in filling gaps in (eco) toxicity data needed for assessing the hazards of chemicals. The Toolbox incorporates information and tools from various sources into a logical workflow (OECD (Q)SAR Project).

Degradation pathways used by microorganism to obtain carbon and energy from 200 chemicals are stored in a special file format that allows easy computer access to catabolic information. The collection includes the catabolism of C1-compounds, aliphatic hydrocarbons, alicyclic rings, furans, halogenated hydrocarbons, aromatic hydrocarbons and haloaromatics, amines, sulfonates, nitrates, nitro-derivatives, nitriles, and compounds containing more than one functional group. Most of pathways are related to aerobic conditions. Different sources including monographs, scientific articles and public web sites such as the UM-BBD (Ellis, Roe & Wackett, 2006) were used to compile the database.

The original CATABOL simulator of microbial metabolism is implemented in the OECD (Q)SAR Application Toolbox (Jaworska et al., 2002; Dimitrov et al., 2002; Dimitrov et al., 2004). Single pathway catabolism is simulated using the abiotic and enzyme-mediated reactions via the hierarchically ordered principal molecular transformations extracted from documented metabolic pathway database. The hierarchy of the transformations is used to control the propagation of the catabolic maps of the chemicals. The simulation starts with the search for match between the parent molecule and the source fragment associated with the transformation having the highest hierarchy. If the match is not found search is performed with the next transformation, etc. When the match is identified, the transformation products are generated. The procedure is repeated for the newly formed products. Predictability (probability that the metabolite is observed, given that the metabolite is predicted) evaluated on the bases of documented catabolism for 200 chemicals stored in the database of "Observed microbial catabolism" is 83%.

In this work will be researched the possible metabolites (observed and predicted) for some naphthenic acids. For this aim we will use the OECD (Q)SAR Application Toolbox system. Predictions are based on biotransformation rules that, in turn, are derived from reactions found in the UM-BBD and the scientific literature. The UM-PPS most accurately predicts compounds that are similar to compounds with known biodegradation mechanisms, for microbes under aerobic conditions and when the compounds are the sole source of energy, carbon, nitrogen or other essential elements for these microbes. Results in the OECD (Q)SAR Application Toolbox system are presented in Table 1.

3. Toxicity of naphthenic acids

Napthenic acids likely behave as surfactants as they consist of a hydrophilic head and a hydrophobic tail giving them unique solubility properties (Ivanković & Hrenović, 2010). These compounds are commonly found in detergents or cleaning products used in mining, oil, food and textile industries (Sandbacka, Christianson & Isomaa, 2000). Untreated industrial effluents often contain surfactants or surfactant-like compounds in concentrations sufficient to elicit acute toxicity in aquatic organisms (Ankley& Burkhard, 1992). Investigation into the influence that molecular structure exerts on the toxicity of naphthenic acid revealed that the observed acute toxicities for naphthenic acid-like surrogates to aquatic organisms rise with increasing Molecular weight and decreased with greater carboxylic acid content (Frank et al., 2009). These results suggested that the acute toxicity of naphthenic acid was influenced by hydrophobicity, thereby supporting narcosis as the probable mode of action (Kőnemann, 1981). Chemicals acting by a nonpolar narcotic mode of action are biologically unreactive, and their toxicity acts as a function of their concentration at the site of action, typically the cellular membrane (Cronin & Schultz, 1997).

Persistent Organic Pollutants (POPs) and Persistent, Bioaccumulative and Toxic (PBT) substances are carbon-based chemicals that resist degradation in the environment and accumulate in the tissues of living organisms, where they can produce undesirable effects on human health or the environment at certain exposure levels (Pavan & Worth, 2006).

№	CAS number	Name of compound	Observed Microbial metabolism	Predicted Microbial metabolism
1	142-62-1	Hexanoic acid	No metabolite	8 metabolites
2	334-48-5	Decanoic acid	No metabolite	16 metabolites
3	98-89-5	Cyclohexane carboxylic acid	4 metabolites	13 metabolites
4	5962-88-9	Cyclohexane pentanoic acid	No metabolite	21 metabolites
5	110-15-6	Succinic acid	No metabolite	1 metabolite
6	124-04-9	Adipic acid	No metabolite	6 metabolites
7	1076-97-7	1,4-Cyclohexane dicarboxylic acid	No metabolite	13 metabolites

Table 1. Observed and predicted microbial metabolism of some selected naphthenic acids.

3.1 Toxic prediction of naphthenic acids

Naphthenic acids have been reported to be acutely toxic to various organisms. However, the critical mechanism of toxicity remains largely unknown. Naphthenic acids are persistent in aquatic environments and are acutely toxic to aquatic bacteria, invertebrates, fish, and plants (Clemente & Fedorak, 2005]. Narcosis has been suggested to be the probable mode of acute toxicity by naphthenic acids, particularly for lower molecular weight of naphthenic acids. Higher molecular weight of naphthenic acids are less acutely toxic than lower molecular weight of naphthenic acids. Toxicity of naphthenic acids is inversely proportional to carboxylic acid content within naphthenic acid structures of higher molecular weight of naphthenic acids (Frank et al., 2009). Toxicity of naphthenic acids is also related to the amount of naphthenic acid that can be accumulated into the organisms as well as their inherent toxic potency.

In general, the complex and changing nature of mixtures of naphthenic acids make it difficult to predict toxicity. By determining the critical mechanism of toxicity of naphthenic acids, it might be possible to develop more effective predictive relationships to account for the toxic effects observed in living organisms exposed to naphthenic acids.

3.1.1 Use methods for toxic prediction of naphthenic acids

The PBT Profiler is a screening-level tool that provides estimates of the persistence, bioaccumulation, and chronic fish toxicity potential of chemical compounds. It is designed to be used when data are not available. In order to help interested parties make informed decision on a chemical's PBT characteristics, the PBT profiler automatically identifies chemicals that may persistent in the environment and bioaccumulate in the food chain. These chemicals are identified using thresholds published by the EPA (PBT Profiler).

The PBT Profiler combines the persistence criteria for water, soil, and sediment and highlights chemicals with an estimated half-life ≥ 2 months and < 6 months as persistent and those with an estimated half-life ≥ 6 months as very persistent. The half-life in air is not used in the PBT Profiler's Persistence summary (chemicals with an estimated half-life > 2 days are considered as persistent). The PBT Profiler uses 30 days in a month for its comparisons.

The PBT Profiler combines the bioaccumulation criteria and highlights chemicals with a BCF ≥ 1000 and < 5000 as bioaccumulative and those with a BCF ≥ 5000 as very bioaccumulative.

To highlight a chemical that may be chronically toxic to fish, the PBT profiler uses the following criteria: Fish ChV (Chronic Value) > 10 mg/l (low concern), Fish ChV = 0.1 - 10 mg/l (moderate concern) and Fish ChV < 0.1 mg/l (high concern).

Toxicity values of some naphthenic acids to Tetrahymena pyriformis were obtained from the literature (Schultz, 1997) and reported in Table 3. Population growth impairment was assessed after 40h with the common ciliate *Tetrahymena pyriformis*. The experimental data for rat (oral LD50 values) were collected from the literature (ChemIDplus).

Data for the logarithm of the 1-octanol-water partition coefficient (log P) were obtained from the KOWWIN software (US EPA, KOWWIN). Where possible measured log P values were verified and used in preference to calculated values.

In this study several models were used for non-polar compounds to aquatic and terrestrial species to determine the acute toxicity of selected naphthenic acids (Tables 3).

Baseline model (saturated alcohols and ketones) of *Tetrahymena pyriformis* (Ellison et al., 2008):

$$\log(1/IGC50) = 0.78*\log P - 2.01 \tag{1}$$

$$n = 87 \quad R2 = 0.96 \quad s = 0.20 \quad F = 2131$$

Baseline model (saturated alcohols and ketones) of Rat (oral) (Lipnick, 1991):

$$\log(1/LD50) = 0.805*\log P - 0.971*\log(0.0807*10\log P+1) + 0.984 \tag{2}$$

$$n = 54 \quad R2 = 0.824 \quad s = 0.208 \quad F = 35.3$$

The property - excess toxicity - was used to define the toxicity of chemicals (reactive or nonrective) (Lipnick, 1991). The extent of excess toxicity was determined as the toxic ratio (TR), which was calculated by the following equations 3-4 (Lipnick, 1991, Nendza & Müller, 2007):

$$TR = \log(1/C)\exp - \log(1/C)calc \tag{3}$$

or

$$TR = (predicted \; baseline \; toxicity) \, / \, (observed \; toxicity) \tag{4}$$

3.1.2 Results and discussion

The results of estimation of naphthenic acids for persistence, bioaccumulation and toxicity are presented in Table 2. The components of naphthenic acids are commonly classified by their structures and the number of carbon atoms in the molecule.

Naphthenic acids are a family of carboxylic acid surfactants, primarily consisting of cyclic terpenoids used in source and geochemical characterisation of petroleum reserves (Brient, Wessner & Doyle, 1995). The compound group is composed predominately of alkyl-substituted cycloaliphatic carboxylic acids with smaller amounts of acyclic aliphatic (paraffinic or fatty) acids. Aromatic olefinic, hydroxyl and dibasic acids are also present as minor components of naphthenic acids. The cycloaliphatic acids include single rings and fused multiple rings.

The PBT profiler uses a well-defined set of procedures to predict the persistence, bioaccumulation, and toxicity of chemical compounds when experimental data are not available. The persistence, bioaccumulation, and fish chronic toxicity values estimated by the PBT profiler are automatically compared to criteria published by the EPA.

Analysis of data in Table 2 reveals that as all naphthenic acids are not persistence, bioacculumative and toxic (Fish ChV), but some compounds are exception about their chronic toxic. The compounds with moderate toxic (0.1-10 mg/l) are decanoic and cyclohexanepentanoic acids.

№	Name of compound	Persistence		Bio-accumu-lation	Toxicity
		Media (water, soil, sediment, air) Half-life (days)	Percent in Each Medium	BCF	Fish ChV (mg/l)
1	Hexanoic acid	8.7; 17; 78; 2.9	30%; 65%; 0%; 5%	3.2	84
2	Decanoic acid	8.7; 17; 78; 1.4	25%; 69%; 3%; 3%	3.2	2.7
3	Cyclohexanecarbo xylic acid	8.7; 17; 78; 1.6	31%; 65%; 0%; 4%	3.2	51
4	Cyclohexanepen tanoic acid	15; 30; 140; 1	22%; 70%; 6%; 2%	3.2	1.6
5	Succinic acid	8.7; 17; 78; 5.8	34%; 66%; 0%; 0%	3.2	21,000
6	Adipic acid	8.7; 17; 78; 2.9	34%; 66%; 0%; 0%	3.2	3,800
7	1,4-Cyclohexanedicar boxylic acid	8.7; 17; 78; 1.5	31%; 69%; 0%; 0%	3.2	1,100

Table 2. PBT Profiler estimate of the naphthenic acids.

All organic chemicals have the potential to cause narcosis. Their ability to do so is mainly governed by their concentration and their ability to cause more serious toxic effects, which would mask any narcotic effect the chemical may cause (van Wezel & Opperhuizen, 1995). The toxicity is not observed to be related to hydrophobicity and is in excess of baseline toxicity for the more compounds (Fig. 1 and Table 3).

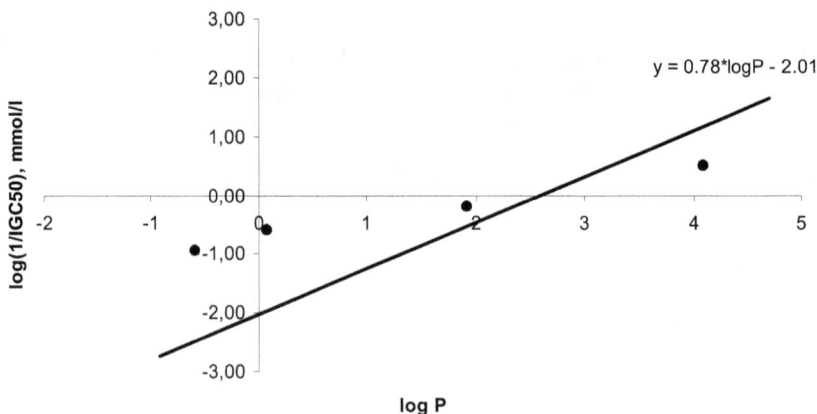

Fig. 1. Plot of toxicity to Tetrahymena pyriformis vs log P for naphthenic acids showing baseline toxicity.

CAS number	Name of compound	EcoSAR classification	log P	Exp. T. pyriformis log(1/IGC$_{50}$), mmol/l	Pred. T. pyriformis log(1/IGC$_{50}$), mmol/l /TR	Exp. oral Rat LD$_{50}$ mmol/ kg	Pred. oral Rat LD$_{50}$ mmol/ kg / TR
142-62-1	Hexanoic acid	Neutral organic-acid	1.92[a]	-0.208	-0.512/ 0.30	17.65	46.57/ 2.64
334-48-5	Decanoic acid	Surfactant -anionic- acid	4.09[a]	0.506	1.180/ -0.67	58.05	23.23/ 0.40
98-89-5	Cyclohe xane carboxy lic acid	Neutral organic- acid	1.96[a]			25.47	46.38/ 1.82
5962-88-9	Cyclohe xane penta noic acid	Neutral organic- acid	4.32[b]				
110-15-6	Succinic acid	Neutral organic- acid	-0.59[a]	-0.94	-2.470/ 1.53	19.14	3.16/ 0.16
124-04-9	Adipic acid	Neutral organic- acid	0.08[a]	-0.606	-1.948/ 1.34	75.27	10.22/ 0.14
1076-97-7	1,4-Cyclohe xane dicarbo xylic acid	Neutral organic- acid	0.95[b]				

[a]Experimental value of log P; [b]Calculated value of log P.

Table 3. Experimental and predicted values of acute toxicity for some naphthenic acids.

On the basis of calculated and experimental values for acute toxicity, the toxicity ratio (TR) as the ratio of the calculated baseline toxicity over the experimentally determined value was calculated (Table 3). A TR-value less than one could indicate rapid hydrolysis and/or biotransformation of the parent compound by the organism to non-toxic metabolites (Aptula & Roberts, 2006).

3.2 Toxic prediction of their metabolites

The reactions should be considered as an approximation of the real catabolism by mixed cultures of bacteria. Each chemical transformation includes source fragment (transformation target) and its bidegradation products. Results of possible metabolism (observed and predicted) for some selected naphthenic acids, the EcoSAR classification of their metabolites and protein binding in the OECD (Q)SAR Application Toolbox system are presented in Table 4.

Hexanoic acid (142-62-1)					
Predicted Microbial metabolism					
Number of metabolite	1	2	3	4	5	6
EcoSAR classifi cation	Neutral organics-acid	Neutral organics-acid	Neutral organics-acid	Neutral organics-acid	Neutral organics-acid	Neutral organics-acid
Possible mechanism of action			Nucleophilic addition to ketone			

Predicted Microbial metabolism					
Number of metabolite	7	8			
EcoSAR classifi cation	Neutral organics-acid	Neutral organics-acid			
Possible mechanism of action	Nucleophilic addition to ketone				

Decanoic acid (334-48-5)					
Predicted Microbial metabolism					
Number of metabolite	1	2	3	4	5	6
EcoSAR classifi cation	Neutral organics-acid	Neutral organics-acid	Neutral organics-acid	Neutral organics-acid	Neutral organics-acid	Neutral organics-acid
Possible mechanism of action			Nucleophili c addition to ketone			

Predicted Microbial metabolism					
Number of metabolite	7	8	9	10	11	12

EcoSAR classifi cation	Neutral organics-acid	Neutral organics-acid	Neutral organics-acid	Neutral organics-acid	Neutral organics-acid	Neutral organics-acid
Possible mechanism of action	Nucleophilic addition to ketone				Nucleophilic addition to ketone	
Predicted Microbial metabolism						
Number of metabolite	13	14	15	16		
EcoSAR classifi cation	Neutral organics-acid	Neutral organics-acid	Neutral organics-acid	Neutral organics-acid		
Possible mechanism of action			Nucleophili c addition to ketone			

	Cyclohexane carboxylic acid (98-89-5)					
Observed Microbial metabolism						
Number of metabolite	1	2	3	4		
EcoSAR classifi cation	Neutral organics-acid	Neutral organics-acid	Neutral organics-acid	Neutral organics-acid		
Possible mechanism of action			Nucleophili c addition to ketone			
Predicted Microbial metabolism						
Number of metabolite	1	2	3	4	5	6
EcoSAR classifi cation	Neutral organics-acid	Neutral organics-acid	Neutral organics-acid	Neutral organics-acid	Neutral organics-acid	Neutral organics-acid
Possible mechanism of action			Nucleophili c addition to ketone			
Predicted Microbial metabolism						

Number of metabolite	7	8	9	10	11	12
EcoSAR classification	Neutral organics-acid	Neutral organics-acid	Neutral organics-acid	Neutral organics-acid	Neutral organics-acid	Neutral organics-acid
Possible mechanism of action	Nucleophilic addition to ketone				Nucleophilic addition to ketone	
Predicted Microbial metabolism	HO–C(=O)–CH₃					

Number of metabolite	13
EcoSAR classification	Neutral organics-acid
Possible mechanism of action	

Cyclohexane pentanoic acid (5962-88-9)

Predicted Microbial metabolism						

Number of metabolite	1	2	3	4	5	6
EcoSAR classification	Neutral organics-acid	Neutral organics-acid	Neutral organics-acid	Neutral organics-acid	Neutral organics-acid	Neutral organics-acid
Possible mechanism of action			Nucleophilic addition to ketone			

Predicted Microbial metabolism						

Number of metabolite	7	8	9	10	11	12
EcoSAR classification	Neutral organics-acid	Neutral organics-acid	Neutral organics-acid	Neutral organics-acid	Neutral organics-acid	Neutral organics-acid
Possible mechanism of action	Nucleophilic addition to ketone				Nucleophilic addition to ketone	

Predicted Microbial metabolism						

Number of metabolite	13	14	15	16	17	18
EcoSAR classifi cation	Neutral organics-acid	Neutral organics-acid	Neutral organics-acid	Neutral organics-acid	Neutral organics-acid	Neutral organics-acid
Possible mechanism of action			Nucleophili c addition to ketone			
Predicted Microbial metabolism	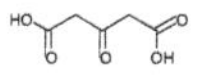	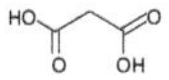	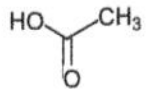			
Number of metabolite	19	20	21			
EcoSAR classifi cation	Neutral organics-acid	Neutral organics-acid	Neutral organics-acid			
Possible mechanism of action	Nucleophilic addition to ketone					

Succinic acid (110-97-7)

Predicted Microbial metabolism	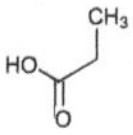					
Number of metabolite	1					
EcoSAR classifi cation	Neutral organics-acid					
Possible mechanism of action						

Adipic acid (124-04-9)

Predicted Microbial metabolism	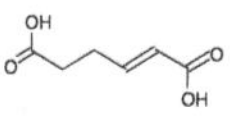	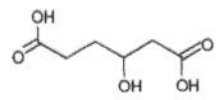	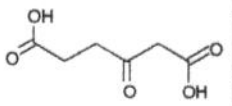	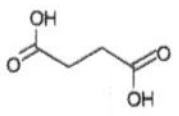	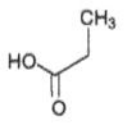	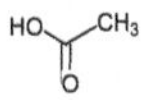
Number of metabolite	1	2	3	4	5	6
EcoSAR classifi cation	Neutral organics-acid	Neutral organics-acid	Neutral organics-acid	Neutral organics-acid	Neutral organics-acid	Neutral organics-acid
Possible mechanism of action			Nucleophilic addition to ketone			

1,4-Cyclohexane dicarboxylic acid (1076-97-7)					
Predicted Microbial metabolism						
Number of metabolite	1	2	3	4	5	6
EcoSAR classification	Neutral organics-acid	Neutral organics-acid	Neutral organics-acid	Neutral organics-acid	Neutral organics-acid	Neutral organics-acid
Possible mechanism of action			Nucleophilic addition to ketone			

Predicted Microbial metabolism						
Number of metabolite	7	8	9	10	11	12
EcoSAR classification	Neutral organics-acid	Neutral organics-acid	Neutral organics-acid	Neutral organics-acid	Neutral organics-acid	Neutral organics-acid
Possible mechanism of action	Nucleophilic addition to ketone				Nucleophilic addition to ketone	

Predicted Microbial metabolism					
Number of metabolite	13				
EcoSAR classification	Neutral organics-acid				
Possible mechanism of action					

Table 4. Predicted microbial metabolites, EcoSAR classification and possible protein binding (mechanism of action).

4. Conclusion

Naphthenic acids can enter the environment from both natural and anthropogenic processes. Naphthenic acids are highly toxic, recalcitrant compounds that persist in the environment for many years, and it is important to develop efficient bioremediation strategies to decrease both their abundance and toxicity in the environment. However, the diversity of microbial communities involved in naphthenic acid-degradation, and the mechanisms by which naphthenic acids are biodegraded, are poorly understood. This lack of knowledge is mainly due to the difficulties in identifying and purifying individual carboxylic acid compounds from complex naphthenic acid mixtures found in the environment, for microbial biodegradation studies. Further, the persistence and fate of naphthenic acids in the environment is not well documented due to a lack of adequate analytical methods to determine the concentration, composition and extent of these crude-oil based compounds in environmental samples. This lack of knowledge constitutes a critical gap in scientific understanding. Acute toxicity is one of endpoints used in environmental risk assessment to determine the safe use and disposal of organic chemicals.

Degradation by the action of microorganisms is one of the major processes that determines the fate of organic chemicals in the environment. Quantitative Structure-Activity Relationships (QSAR) methods can be applied to biodegradation. Such relationships, often referred to as Quantitative Structure-Biodegradability Relationships (QSBRs), relate the molecular structure of an organic chemical to its biodegradability and consequently aid in the prediction of environmental fate.

5. References

Alexander, M. 1994. *Biodegradation and Bioremediation*, pp. 159-176. Academic Press, San Diego, California.

Ankley, G.T. & Burkhard, L.P. 1992. Identification of Surfactants as Toxicants in a Primary Effluent. *Environmental Toxicology & Chemistry*, Vol. 11, pp. 1235-1248.

Aptula, A.O. & Roberts, D.W. 2006. Mechanistic applicability domains for nonanimal-based prediction of toxicological End Points: General principles and application to reactive toxicity. *Chemical Research in Toxicology*, Vol. 19, No 8, pp. 1097-1105.

Bartha, R. & Atlas, R.M. 1977. The microbiology of aquatic oil spills. *Adv Appl Microb*, Vol. 22, pp. 225-266.

Boethling, R.S., Gregg, B., Frederick, R., Gabel, N.W., Campbell, S.E. & Sabljic, A. 1989. Expert systems survey on biodegradation of xenobiotic chemicals, *Ecotoxicol Environ Safety*, Vol. 18, pp. 252-267.

Brient, J.A., Wessner, P.J. & Doyle, M.N. 1995. Naphthenic acids, In: *Encyclopedia of Chemical Technology*, Kroschwitz, J.I. (Ed.)., pp. 1017-1029, 4th ed., John Wiley & Sons, New York.

Clemente, J.S., MacKinnon, M.D., Fedorak, P.M. 2004. Aerobic biodegradation of two commercial naphthenic acids preparations. *Environmental Science & Technology*, Vol. 38, pp. 1009-1016.

Clemente, J.S., Fedorak, P.M. 2005. A review of the occurrence, analyses, toxicity, and biodegradation of naphthenic acids. *Chemosphere*, Vol. 60, pp. 585-600.

Criteria used by the PBT Profiler: http://www.pbtprofiler.net/criteria.asp

Cronin, M.T.D. & Schultz, T.W. 1997. Validation of *Vibrio fisheri* acute toxicity data: mechanism of action-based QSARs for non-polar narcotics and polar narcotic phenols. *Science of the Total Environment*, Vol. 204, No 1, pp. 75–88.

Dagley, S. 1987. Lessons from biodegradation, *Annual Review in Microbiology*, Vol. 41, pp. 1-23.

Damborsky, J. 1996. A mechanistic approach to deriving quantitative structure–activity relationship models for microbial degradation of organic compounds. *SAR &QSAR in Environmental Research*, Vol. 5, pp. 27-36.

Dimitrov, S., Breton, R., Mackdonald, D., Walker, J., Mekenyan, O. 2002. Quantitative prediction of biodegradability, metabolite distribution and toxicity of stable metabolites, *SAR & QSAR in Environmental Research*, Vol. 13, pp. 445-455.

Dimitrov, S., Kamenska, V., Walker, J.D., Windle, W., Purdy, R., Lewis, M., Mekenyan, O. 2004. Predicting the biodegradation products of perfluorinated chemicals using Catabol, *SAR & QSAR in Environmental Research*, Vol. 15, pp. 69-82.

Ellis, L.B.M., Roe, D., Wackett, L.P., 2006. The University of Minnesota Biocatalysis/Biodegradation Database: the first decade, *Nucleic Acids Research*, Vol. 34, pp. D517- D521.

Ellison, C.M., Cronin, M.T.D., Madden, J.C., Schultz, T.W. 2008. Definition of the structural domain of the baseline non-polar narcosis model for Tetrahymena pyriformis. *SAR and QSAR in Environmental Research*, Vol. 19, No 7, pp. 751-783.

Frank, R.A., Fischer, K., Kavanagh, R., Burnison, B.K., Arsenault, G., Headley, J.V., Peru, K.M., Van Der Kraak G. & Solomon, K.R. 2009. Effect of Carboxylic Acid Content on the Acute Toxicity of Oil Sands Naphthenic Acids. *Environmental Science & Technology*, Vol. 43, pp. 266–271.

Gao, J., Ellis, L.B.M. & Wackett, L.P. 2010. The University of Minnesota Biocatalysis/ Biodegradation Database: improving public access, *Nucleic Acids Research*, Vol. 38, D488-D491.

Headley, J.V., McMartin, D.W. 2004. A review of the occurrence and fate of naphthenic acids in aquatic environments. *Journal of Environmental Science & Health, Part A: Environmental Science & Engineering*, Vol. A39, pp. 1989-2010.

Herman, D.C., Fedorak, P.M., Costerton, J.W. 1993. Biodegradation of cycloalkane carboxylic acids in oil sand tailings. *Canadian Journal of Microbiology*, Vol.39, No 6, pp.576-580.

Herman, D.C., Fedorak, P.M., Mackinnon, M.D., Costerton, J.W. 1994. Biodegradation of naphthenic acids by microbial populations indigenous to oil sands tailings. *Canadian Journal of Microbiology*, Vol. 40, No 6, pp. 467-477.

Howard, P.H., Boethling, R.S., Stiteler, W., Meylan, W. & Beauman, J. 1991. Development of a predictive model for biodegradability based on BIODEG, the evaluated biodegradation data base. *The Science of the Total Environment*, Vol. 109/110, pp. 635-641.

Ivanković, T. & Hrenović, J. 2010. J. Surfactants in the Environment. *Archives Of Industrial Hygiene and Toxicology*, Vol. 61, pp. 95-110.

Jain, R.K., Kapur, M., Labana, S., Lal, B., Sarma, P.M., Bhattacharya, D. & Thakur, I.S. 2005. Microbial diversity: Application of microorganisms for the biodegradation of xenobiotics. *Current Science*, Vol. 89, No 1, pp. 101-112.

Jaworska, J., Dimitrov, S., Nikolova, N., Mekenyan, O. 2002. Probabilistic assessment of biodegradability based on metabolic pathways: CATABOL system. *SAR & QSAR in Environmental Research*, Vol. 13, pp. 307-323.

Kanaly, R. A. & Harayama, S. 2000. Biodegradation of high molecular weight polycyclic aromatic hydrocarbons by bacteria. *Journal of Bacteriology*, Vol.182, No8, pp. 2059–2067.

Klopmann, G., Zhang, Z., Balthasar, D.M. & Rosencranz, H.S. 1995. Computer-automated predictions of metabolic transformations of chemicals. *Environmental Toxicology & Chemistry*, Vol. 14, No3, pp. 395-403.

Kőnemann, H. 1981. Quantitative Structure-Activity Relationships in Fish Toxicity Studies. Part 1. Relationship for 50 Industrial Chemicals. *Toxicology*, Vol. 19, pp. 209-221.

Lai, J.W.S., Pinto, L.J., Kiehlmann, E., Bendell-Young, L.I., Moore, M.M. 1996. Factors that affect the degradation of naphthenic acids in oil sands wastewater by indigenous microbial communities. *Environmental Toxicology and Chemistry*, Vol. 15, pp. 1482-1491.

Lipnick, R.L. 1991. Outliers: their origin and use in the classification of molecular mechanisms of toxicity. *Science of the Total Environment*, Vol. 109, pp. 131-153.

MacKinnon, M., Boerger, H. 1986. Description of two treatment methods for detoxifying oil sands tailings pond water. *Water Pollution Research Journal of Canada.*, Vol. 21, pp. 496-512.

Miller, E.C. & Miller, J.A. 1981. Search for the ultimate chemicals carcinogens and their reaction with cellular macromolecules. *Cancer*, Vol. 47, No 10, pp. 2327-2345.

Nendza, M. & Müller, M. 2007. Discriminating toxicant classes by mode of action: 3. Substructure indicators. *SAR and QSAR in Environmental Research*, Vol. 18, No 1-2, pp. 155-168.

OECD Quantitative Structure-Activity Relationships [(Q)SARs] Project: http://www.oecd.org/document/23/0,3343,en_2649_34379_33957015_1_1_1_1,00.html

Pavan, M. & Worth, A. 2006. Review of QSAR Models for Ready Biodegradation, European Commission Directorate – General Joint Research Centre Institute for Health and Consumer Protection.

Parsons, J.R. & Govers, H.A.J. 1990. Quantitative structure–activity relationships for biodegradation, *Ecotoxicology & Environmental Safety*, Vol. 19, No 2, pp. 212-227.

Punch, B., Patton, A., Wight. K., Larson, B., Masscheleyn, P. & Forney, L. 1996. A biodegradability and simulation system (BESS) based on knowledge of biodegradability pathways, In: *Biodegradability Prediction*, Peijnenburg, W.J.G.M. & Damborsky, J. (Eds), pp. 65-73, Dordrecht: Kluwer Academic Publishers.

Sandbacka, M., Christianson, I., Isomaa, B. 2000. The acute toxicity of surfactants on fish cells, *Daphnia magna* and fish: a comparative study. *Toxicology in Vitro*, Vol. 14, pp. 61-68.

Seifert, W.K. 1975. Carboxylic acids in petroleum and sediments. *Fortschritte der Chemie Organischer Naturstoffe*, Vol. 32, pp. 1-49.

Schultz, T.W. 1997. TETRATOX: Tetrahymena pyriformis population growth impairment Endpoint-A surrogate for fish lethality. *Toxicological Methods*, Vol. 7, pp. 289-309.

Speight, J. G. 1999. *Chemistry and Technology of Petroleum*, Vol. 4, pp. 145-146. Marcel Dekker Inc, New York.

US EPA, KOWWIN; software available at http://www.epa.gov/oppt/exposure/pubs/episuite.htm

van Gestel, C. A. M., van der Waarde, J.J., Derksen, J.G.M., van der Hoek, E.E., Veul, M.F.X.W., Bouwens, S., Rusch, B., Kronenburg, R., Stokman, G.N.M. 2001. The use of acute and chronic bioassays to determine the ecological risk and bioremediation efficiency of oil-polluted soils. *Environmental Toxicology & Chemistry*, Vol. 20, No 7, pp. 1438–1449.

van Wezel, A.P. & Opperhuizen, A. 1995. Narcosis due to environmental pollutants in aquatic organisms: Residue-based toxicity, mechanisms, and membrane burdens. *Critical Reviews in Toxicology*, Vol. 25, pp. 255-279.

Wackett, L.P. & Ellis, L.B.M. 1999. Predicting biodegradation, *Environmental Microbiology*, Vol. 1, No 2, pp. 119-124.

Website for data of rat: http://chem.sis.nlm.nih.gov/chemidplus/

14

Research on Remediation of Petroleum Contaminated Soil by Plant-Inoculation Cold-Adapt Bacteria

Hong-Qi Wang*, Ying Xiong, Qian Wang,
Xu-Guang Hao and Yu-Jiao Sun
*College of Water Sciences Key Laboratory for Water
and Sediment Sciences of Ministry of Education
Beijing Normal University, Beijing,
China*

1. Introduction

With the rapid development of the petroleum industry, a large number of oil contaminants leaked into the soil in the petrochemical complex areas, gas stations, automobile factory and other places, which resulted in serious soil contamination. Micro-Biological degradation of oil pollution was the focus of the present study. Many hydrocarbon-contaminated environments were characterized by low or elevated temperatures, acidic or alkaline pH, high salt concentrations, or high pressure. Hydrocarbon-degrading microorganisms, adapted to grow and thrive in these environments, played an important role in the biological treatment of polluted extreme habitats.

The biodegradation of many components of petroleum hydrocarbons has been reported in a variety of terrestrial and marine cold ecosystems. Cold-adapted microorganisms are potentially interesting for use in environmental biotechnology applications since a large part of the biosphere has low temperatures during at least parts of the year. Many studies have shown that both oil-contaminated and uncontaminated soils in the Arctic, the Antarctic and the Alps contain microbes that can degrade different hydrocarbons deriving from oils[1].For application at cold climate sites, bioremediation approaches are appealing because they have potential to be more efficient and cost-effective than alternative, more energy intensive approaches. Several bioremediation approaches have been reported to be successful for petroleum hydrocarbon-contaminated soils at cold climate sites [2]. Two psychrotrophic bacterial strains isolated from Antarctic seawaters were investigated for their capability to degrade commercial diesel oil at 4 and 20°C over a period of two-months. The result suggested the possible exploitation of two bacterial strains in future biotechnological processes, directly as field-released micro-organisms both in cold and temperate contaminated marine environments [3].

The cold-adapted bacterial communities were also studied. Some research suggested that geographical origin of the samples, rather than petroleum contamination level, was more

* Corresponding Author

important in determining species diversity within these cold-adapted bacterial communities [4]. Predominant populations from different soils often included phylotypes with nearly identical partial 16S rRNA gene sequences (i.e., same genus) but never included phylotypes with identical ribosomal intergenic spacers (i.e., different species orsubspecies)[5]. Twenty-two polycyclic aromatic hydrocarbon (PAH)-degrading bacterial strains were isolated from Antarctic soils with naphthalene or phenanthrene as a sole carbon source and all these strains showed a high efficiency to degrade naphthalene at 4°C, and some additionally degraded phenanthrene. Phylogenetic analysis showed that all belonged to the genus Pseudomonas except one that was identified as the genus of Rahnella[6]. 32 cold-adapted, psychrophilic and cold-tolerant, yeast strains isolated from alpine habitats with regard to their taxonomy, growth temperature profile, and ability to degrade phenol and 18 phenol-related mono-aromatic compounds at 10°C [7].

Cold-adapted microorganisms inhabit in most parts of the surface of earth and the annual average temperatures of these regions are below 15°C, but these temperatures are far lower than the optimum growth temperatures of mesophilic microorganisms which have been widely used in bioremediation technology at present [8]. In most parts of China, it was cold in winter and cool in spring and autumn, when the microbial survival rate was low and the natural degradation capacity was poor. Cold-adapted microorganisms which had special genetic background and metabolic channels were widely distributed in the nature at low temperatures. The cold-adapted microorganisms could produce a lot of active substances, and show unique characters in many areas at low temperatures [9].

On the other hand, bioremediation was considered to be the most cost-effective and environmental friendly technology to treat oil-contaminated soil currently. And the combination of microbe and plant would be practical and effective.

Plants could be used to remediate organic or inorganic polluted soils. This technique had been applied widely in practice, named as phytoremediation. The plant roots improved the water content in the polluted soil. The increased microbial activities induced by the improved rhizosphere microecosystem could be used to enhance biodegradation of petrochemicals and bioremediation of polluted soil[10].

In this study, oil-contaminated soil in Tianjin and was used for domestication at different temperatures. Two different communities of bacteria depend on crude oil as sole carbon source were found and the main foundation of temperature on the microbial community screening was researched and revealed. The biodegradation characteristic was studied then. The effect of combined remediation by winter wheat and cold-adapted degrading bacterial was examined. In order to explore the effect of combined remediation by bacterial and plant at low temperature on petroleum hydrocarbons contaminated soil, the TPH removal and the catalase activity were examined at different incubation time in different treatments.

2. Materials and methods

2.1 Domestication and purification of petroleum – Degradation strains

Crude oil and petroleum contaminated soil samples were taken from Dagang oil field (1m deep) in Tianjin. Well drainage was also taken from the well in Dagang oil field. Magnesium sulfate (AR.), Di-hydrogen phosphate ammonia (AR.), Sodium nitrate (AR.), Potassium nitrate (AR.) and di-Potassium hydrogen phosphate (AR.) were obtained from the Beijing Chemical Factory. All the chemicals were used as received. The restriction endonuclease was purchased from Promega. The nucleic acid electrophoresis apparatus (Bio-Rad) and GDS gel imaging system (Gene company) was prepared.

Inorganic drainage liquid medium: NaCl 5 g/L, K_2HPO_4 1 g/L, $NH_4H_2PO_4$ 1 g/L, $(NH_4)_2SO_4$ 1 g/L, $MgSO_4$ $7H_2O$ 0.2 g/L, KNO_3 3 g/L. Well drainage was used as solvent and pH adjusted to 7.0. 1mL sterilized crude oil was joined to 100mL inorganic liquid medium to make crude oil inorganic liquid medium. 1g soil sample polluted by petroleum hydrocarbons was dissolved with saline water and cultured in a liquid medium supplemented with petroleum oil as the sole source of carbon and energy for 10 days on 10°C and 25°C respectively. Then a certain amount of the above culture solution was inoculated and cultured in a mineral liquid medium supplemented with petroleum oil as the sole source of carbon and energy for 10 days again. The step was repeated again. The culture solution was diluted (10^{-1} ~ 10^{-8}) separately. 50ul culture solution was coated to the solid culture medium and cultured for 5 days. The flats of proper concentration were selected in different temperatures. All the colonies were inoculated to flat plate to sieve again. Purified strains were cultured in test-tube and stored in refrigerators at 4°C.

2.2 Polymerase Chain Reaction (PCR) assays

The purified cultivated strains were used as template and the primers used for the PCR were 8F and 1492R [11]. (Table 1).

Primer	Position	Primer sequences
8F	8-27	5'-AGAGTTTGATCCTGGCTCAG-3'
1492R	1492-1511	5'-CGGTTACCTTGTTACGACTT-3'

Table 1. The sequence and position of primer.

PCR reaction system contained 2 x PCR mixture 25ul (0.1 U Taq Polymerase/ul, 500uM dNTP, 20mM Tris-HCl [pH8.3]), 100mM KCl, 3 mM $MgCl_2$, 2ul primer 8F (5uM), 2ul primer 1492R (5uM), 2ul template and 19ul double steamed water in a final volume of 50ul.

The thermal cycling was as follows: initial denaturation at 94 °C for 6min, followed by 30 cycles of 30s at 94°C, 40s at 52 °C, 1.5 min at 72 °C, and a final extension of 10 min at 72°C. The PCR products were analyzed by running 5ul aliquots of the reaction mixtures in 1.2% agarose gels.

2.3 Genotyping of individual isolates by ARDRA (Amplified Ribosomal DNA Restriction Analysis)

Strains were analyzed by ARDRA using two restriction endonuclease enzymes (RsaI and MspI). Restriction analysis of the PCR products was performed in a 10ul reaction containing: 1ul restriction enzyme buffer, 0.3ul restriction endonuclease and 5ul PCR product. Reactions were then incubated at 37°C for 3h and stopped at 60°C. Restriction fragments were separated on 3% agarose gels. Gels were photographed after staining with ethidium bromide with the imaging system Bioprint.

2.4 Biodegradation of petroleum by cold-adapted petroleum-degrading bacteria

Each triangle flask was filled with 200mg sterilized crude oil and 50mL inorganic salt liquid medium (4g/L), pH7.0. Strains were cultured in the beef extract peptone liquid medium for 48h at 25°C and then centrifuged at 4000 r / min for 10 min to prepare suspending solutions. 2mL suspension was inoculated in each triangle bottle. The sample with no suspension was blank sample. All the samples were cultured at 150 r / min at 10°C. Bacteria density was measured at different time. The strains which grew well were chosen and for the next experiments.

60g crude oil was delivered to 1000g clean soil and blended for many times and then the mix was sterilized to prepare petroleum contaminated soil (6%). Different test conditions were set respectively: soil with bacteria D17 added, soil with bacteria D24 added, soil with mixed bacteria (D17 + D24) added and blank contrasts. Petroleum hydrocarbons degradation rate was determined in different conditions at different time. The oil in soil was extracted by accelerated solvent extraction system (APLE, Beijing Titian Instruments Co., Ltd, China) and determined by weight method. The extraction condition by APLE was set as follow: extraction pressure -10 M Pa, heating time- 100 s, static time- 720 s, elution volume- 60%, purge time- 60 s, preheating temperature-170°C, heating temperature-170°C and followed by a cycle. The solvent was hexane/ acetone =1 : 1 (V/V).

2.5 Combination remediation of petroleum contaminated soil by degrading bacterialD17 and winter wheat

The experimental soils used for this study were collected from an agriculture field in Beijing, China. The soil was dispersed and mixed, then passed through a 5-mm mesh. The soil material was first thoroughly mixed with of oil by different mixing ration and then dried. And the oil content was measured 20 days later while the concentration was 1409.13mg/kg, 2060.26mg/kg and 3273.40mg/kg respectively. The basic soil properties were determined using standard methods recommended by the Chinese Society of Soil science. The basic properties of the soil were shown in table 2. The cold-adapted bacterial used in this study was D17. The experiment condition was showed as Table 3.

The soil of every pollution level was treated as below: first, combined remediation: planted winter wheat and added degrading bacteria. Second, microbial remediation: added the degrading bacteria; Third, the control: no planting and no bacteria. All treatments were shown in Table2.

The study was performed in the incubator which is 1 meter wide, 1.2 meter long and 1meter deep. The soils were put into the incubator and the depth of the soil was 60cm. Transplanted the winter wheat into the incubator with the row spacing of 17cm and added N, P, K slow-release fertilizer to the soil. 360ml bacteria liquid was added to the soil per square meter on February 5, and the OD value was 0.8366. Then 150 ml bacteria liquid was added to the soil per square meter again on April 7th, and the OD value was 0.8433. Simulated conventional irrigation farming conditions during the whole incubation time.

All treatments were incubated in the greenhouse for 126 days, from February 5th to June 10th. The varieties of soil enzymes and oil content with incubation time under different treatments were analyzed. The rhizosphere soil and non-rhizosphere soil of winter wheat

Oil content(mg/kg)	Total N	Total P	Total K	pH
1409.13	0.028%	0.046%	1.74%	8.13
2060.26	0.096%	0.058%	2.02%	8.26
3273.40	0.016%	0.048%	1.65%	8.29

Table 2. Some physical and chemical properties of tested soil.

The catalase was tested by the reduction of potassium permanganate. The results were expressed as an average of three parallel determinations of soil samples.

sample	Pollution level	Oil content(mg/kg)	Experiment treatment
A1	1	1409.13	Combined remediation
A2	2	2060.26	Combined remediation
A3	3	3273.40	Combined remediation
B1	1	1409.13	microbial remediation
B2	2	2060.26	microbial remediation
B3	3	3273.40	microbial remediation
C1	1	1409.13	control
C2	2	2060.26	control
C3	3	3273.40	control

Table 3. Experimental treatment.

were taken on February 5th, March 4th, April 28th, May 8th and June 10th respectively. Record February 5th the first day of the experiment, then the sampling time were the 1st day, the 53rd day, the 83rd day, the 93rd day and the 126th day respectively.

Petroleum Hydrocarbon in soil samples was analyzed by using an APLE-2000 system. Data treatments were performed using Excel 2003 and SPSS 17.0 for windows.

3. Results and discussion

3.1 Microbial community of oil-contaminated soil at different temperature conditions

Strains was analyzed by ARDRA and different genotyping was obtained in different temperature conditions. Each genotyping was conducted as an operational taxonomic unit (OTU). All the strains were analyzed with 16srDNA and the sequence subjected to BLAST GeneBank analysis (Table 4 and Table 5).

Considering from the composition point of view, 8 OTUs has been domesticated at 25 °C. Analyzed and retrieved in GeneBank Database, *Bacillus*, *Sinorhizobium*, *Rhizobium*, *Chryseobacterium* and *Bartonella* were homologous and the dominant bacteria were *Rhizobium* and *Bacillus*. Meanwhile, 11 OTUs has been domesticated at 10°C, which were *Arthrobacter*, *Micrococcus*, *Rhodobacter*, *Bacterium*, *Paracoccus*, *Rhizobium* and *Halomonas* and the dominant bacteria were Arthrobacter and Halomonas. Some studies had showed that parts of Rhizobium strains could survive by using organic sulfur instead of carbon [12] for energy. It was reported that Bacillus had the ability of degrading petroleum hydrocarbon. It was also

OTU	Species with most similar sequence (Accession number)	Similarity /%	Number of Strains
OTU 1-C1	Bacillus sp. KDNB4(EU835566)	99	5
OTU2- C3	Sinorhizobium sp. CCBAU 05296(EU571251)	98	1
OTU3 -C4	Rhizobium sp. SL-1(EU556969)	98	8
OTU4 -C6	Bacillus sp. cp-h46(EU584545)	99	1
OTU5- C10	Chryseobacterium sp. LDVH 42(AY468475)	99	1
OTU6- C13	Bacillus sp. cp-h45(EU584544)	99	1
OTU7- C14	Bacillus megaterium(EU931553)	99	2
OTU8- C15	Bartonella elizabethae(AB246807)	98	1

Table 4. The cultured bacteria and their most closely matched species in Gene Bank(25°C).

OTU	Species with most similar sequence (Accession number)	Similarity /%	Number of Strains
OTU1'-D2	*Arthrobacter bergerei*(AJ609633.2)	98	7
OTU2'-D8	*Micrococcus* sp. BSs20065(EU330348)	95	2
OTU3'- D9	*Rhodobacter changlensis*(AM399030.1)	97	1
OTU4'-D10	*Bacterium* Ips_3(DQ836709)	98	1
OTU5'- D11	*Micrococcus* sp. PB7-11B(EU394442)	97	3
OTU6'- D14	*Paracoccus* sp. WPCB175(FJ006918)	98	5
OTU7'- D17	*Paracoccus* sp. 428(EU841535)	98	1
OTU8'-D18	*Rhizobium* sp.28(DQ310471)	98	1
OTU9'-D19	*Halomonas* sp. Claire(AJ969933)	98	8
OTU10'-D22	*Paracoccus* sp. B-1018(DQ270725)	98	1
OTU11'-D24	*Halomonas* sp. ice-oil-302(DQ533958)	98	1

Table 5. The cultured bacteria and their most closely matched species in Gene Bank (10°C).

reported that lots of *Arthrobacter bergerei* could degrade oil effectively [13, 14]. Some studies suggested that degradation rates could reach 50% after 4d while others found a 90.8% degradation rates after 20d [13, 14]. *Arthrobacter bergerei* were likely to formed biological emulsifier when they used petroleum as carbon sources. The results showed that *Halomonas* was also feasible for high-sodium soil contaminated by petroleum and could be used for bioremediation in future.

It was obvious that the community composition was more diverse under the conditions of 10°C than of 25°C (Fig. 1). The microbial community diversity was significantly higher under the conditions of 10 °C. The dominant bacteria were different in different temperature conditions and Rhizobium was the only same genus. 40% of the total community composition was Rhizobium at 25°C and the proportion at 10°C was 3%.On the other hand, the total bacterial concentration at 10°C was 6.2×10^5 CFU/mL and which under the conditions of 25°C was 4×10^8 CFU/mL. The bacteria abundances was higher at 25°C.Low temperature condition would inhibit growth of microbes. As a result, the temperature had a strong impact on the composition of soil microbial communities.

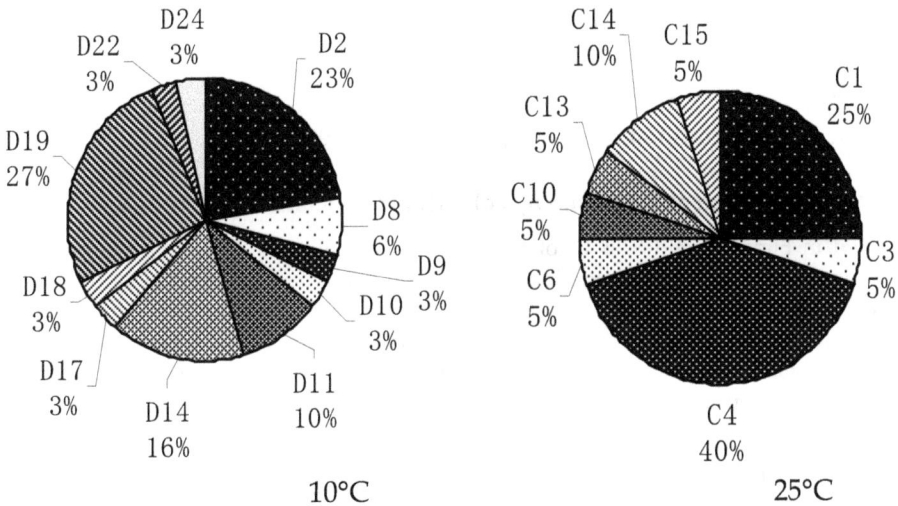

Fig. 1. Comparison of microbial community structure of 10°C and 25°C.

3.2 The bacteria growth situation in liquid medium at 10°C

31 strains were domesticated and purified at 10°C and 11 genotypes were obtained. The growth situation of the 11 kinds of bacteria cultured in a liquid medium supplemented with petroleum oil as the sole source of carbon and energy at 10°C was researched. Bacteria density was detected after being cultivated for 4d, 11d and 20d respectively (Table 6). After being cultivated for 20d, crude oil was emulsified by some of the bacteria while D17 and D24 could produce emulsification to a greater degree. Some studies showed that a lot of strains which used petroleum hydrocarbons as the sole source of carbon and energy could secrete various biological emulsifier and biosurfactant [15]. The increment of D2, D9, D17 and D24 was clear in liquid medium while which of D17 and D24 was obvious especially. Therefore, D17 and D24 were selected and their ability of degradation of petroleum was researched then.

Representative strains	Bacteria density after 4d (A_{600})	Bacteria density after 11d (A_{600})	Bacteria density after 20d (A_{600})
blank	0.0018	0.0152	0.0189
D2	0.0351	0.0838	0.5309
D8	0.0362	0.0402	0.0377
D9	0.0236	0.0473	0.4652
D10	0.0282	0.0143	0.3822
D11	0.0907	0.017	0.324
D14	0.0095	0.0335	0.3463
D17	0.8591	1.3843	1.6063
D18	0.0051	0.0715	0.6154
D19	0.132	0.0299	0.3453
D22	0.0125	0.0635	0.3249
D24	0.1064	1.1922	1.5301

Table 6. The bacteria growth situation in liquid medium at 10°C.

3.3 The biodegradation of petroleum by cold-adapted bacteria

As showed in Figure 2 and 3, both of the strains had ability of degrading petroleum at different temperature. The degradation rates were analyzed at 0d, 25d, 50d, 75d, 100d respectively. The degradation rate was 25.8% by D17, 25.3% by D24 and 21.9% by D17+D24 at 25°C while which was 11.3%, 10.2% and 10.3% respectively at 100d. Low temperature would affect the growth and enzyme activity of bacteria .However, it was suggested that the degrading of petroleum by cold-adapted bacteria was feasible at lower temperature.

In the fourth group of experiments, the effect of D17 and D24 on the biodegradation of crude oil was studied. Throughout the course of the experiment, D17 had a better effect not

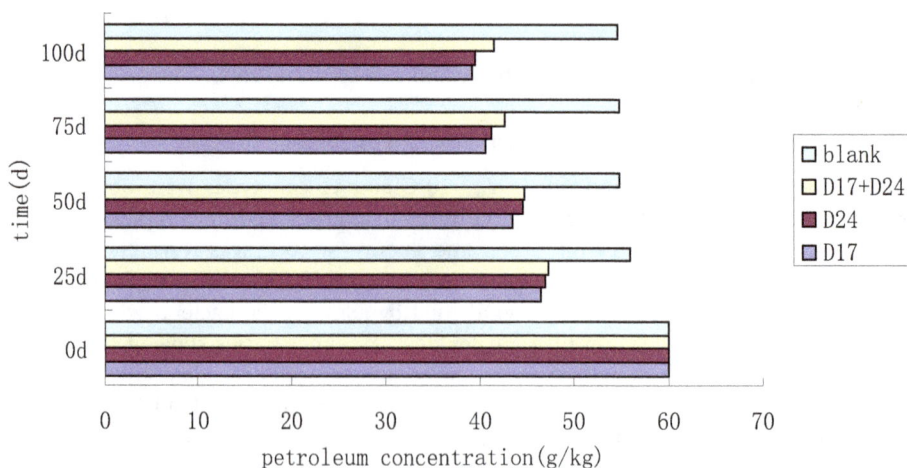

Fig. 2. The biodegradation of petroleum by cold-adapted bacteria at 25°C.

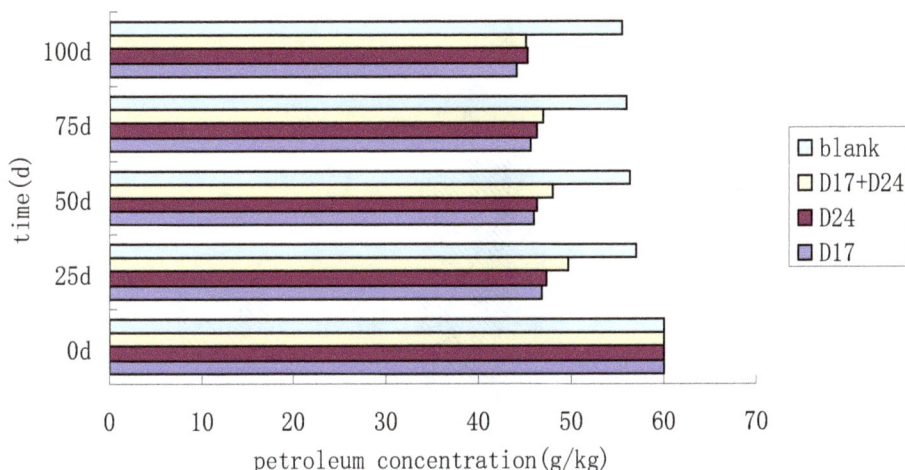

Fig. 3. The biodegradation of petroleum by cold-adapted bacteria at 10°C.

only at 25°C but also at 10°C. As showed in Figure 3, the trend of degradation rate by the strains D17, D24 and D17+D24 respectively was also similar with which showed in Figure 2. However, the general degradation rate by D17+D24 was lower than which by D17 and D24 respectively. It was concluded that the degradation may be inhibited when the strains existed at the same time.

3.4 The effect of petroleum pollution of soil on growth of the winter wheat

When petroleum pollutant leaked into the soil environment, it would damage the plant ecosystem structure and function of the microbial, soil, the plant growth, seed and fruit quality, etc [16]. The higher the petroleum concentration became ' the more rapidly the germination, growth and yield of crops declined[17].

The height of winter wheat was similar when transplanting. After transplanting, the winter wheat grew in the same environment. With the increase of culture time, the height and stem width of wheat varied. The stem width decreased in accordance with the increase of the TPH content. Figure 4 showed the average height of wheat when harvested in different contaminated concentration soil. The figure showed that with the increase of the initial concentration of oil, the height of wheat decreased. All theses indicated that the petroleum pollution had certain inhibition effect on the growth of the winter wheat.

3.5 Changes of soil catalase activities with time

The responses of soil enzymes activities to pollutant exposure could be used to evaluate the soil microbial properties. Enzyme activity was widely used to monitor soil pollution and remediation process [18]. Catalase in the soil can undermine the hydrogen peroxide which was toxic to the organisms. Catalase could be induced by environmental harmful factors, and the activity status reflects the stress situation of the environment to some extent [19].

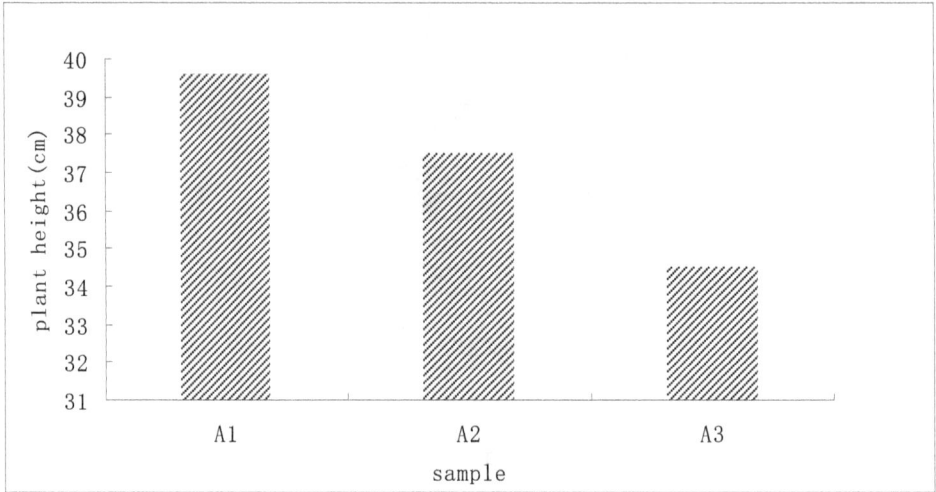

Fig. 4. Height of winter wheat when harvested.

Figure 5 showed the catalase activity of each soil sample at different incubation time in the combined remediation. The figure showed that the catalase activity increased after the combined remediation, however there was a decrease in catalase activity in April. The catalase activity was closely related to the aerobic microorganism quantity and the soil fertility. So the reason for this phenomenon might be related to the decline of the microbe quantity or the shortage of soil fertility. The decline of the microbe quantity might be caused by the competition of the inoculating microbes and the indigenous microorganisms.

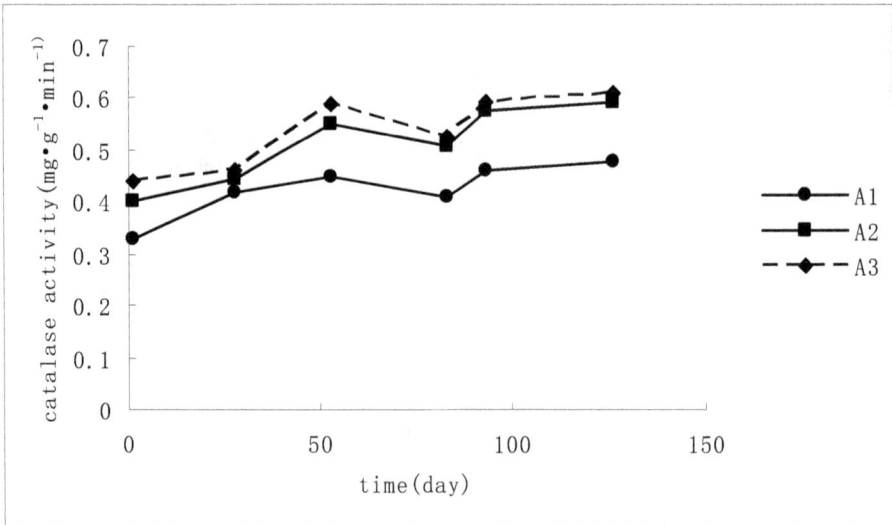

Fig. 5. Soil catalase activities in different soil samples of combined remediation at each incubation time.

3.6 Comparison of degration between combined remediation and microbial remediation

The degradation of petroleum hydrocarbon in soil was the combined effect of physical, chemical and biological processes. In general, the effect of natural process was limited [20]. In this study，the natural degradation was limited too.

There were indigenous microorganisms in the soil, the species and quantity of which were larger than the inoculating microbes. In these conditions, the inoculating microbes needed to adapt the complex environment and compete with the indigenous microorganisms [21].

The TPH removal after remediation in each treatment was shown in Figure 6. The figure showed that the TPH removal of combined remediation was higher than that of microbial remediation, which indicated that the winter wheat promoted the TPH degradation by microorganism.

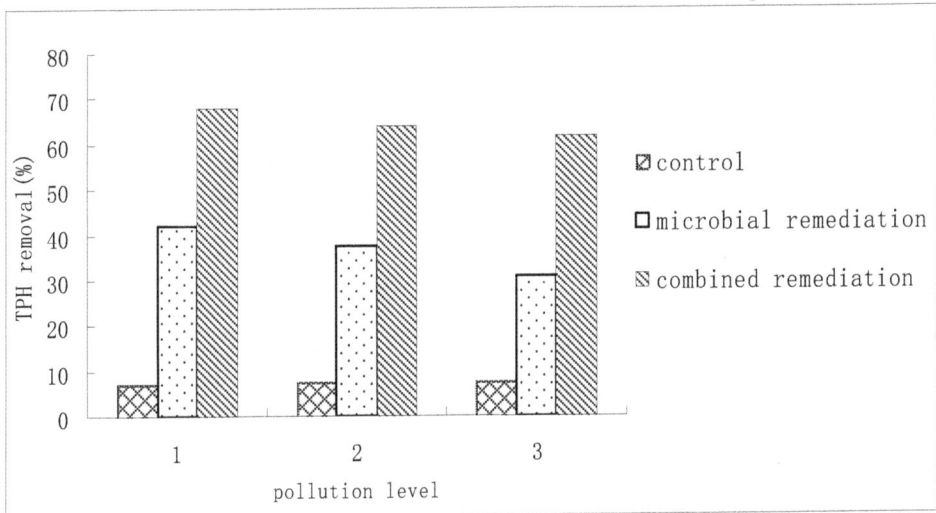

Fig. 6. TPH removal in different treatment after remediation.

3.7 Combined remediation of the petroleum polluted soil

Figure 7 showed the oil content of each soil sample at different incubation times. From the figure, it could be concluded that the degradation trend of petroleum is similar. And there was positive correlation between the catalase activity and the TPH removal. Figure 8 showed the removal rate of TPH by combined remediation at different time. In the initial month, the ground temperature was low when the enzyme activity and the degradation ability of microorganisms were at a low level. Thus the TPH removal rate was slow. With the increase of the ground temperature, the wheat began to turn green when the wheat root began to active. In the same time the TPH removal rate of the oil increased.

Some studies indicated that root quantity was the minimal in earlier winter stage, latter with the winter wheat grew, to flowering stage, reached the largest. Then the root quantity continuously reduced, and to the ripe stage, achieved roughly the same level with that in the over wintering stage. And there were two times that the root quantity changed fastest. One was the rapid rising period from the reviving stage to the flowering stage, the other was the rapid decline period from the flowering stage to the ripe stage[22].

Figure 8 showed a sharp decrease in the TPH removal rate in April when the winter wheat was in the jointing stage. According to the variation of root quantity and TPH removal rate, it could be concluded that the TPH removal rate increased in the reviving stage, reduced in the jointing stage, recovered in the flowering stage, and then reduced again in the ripe stage.

The reason for the decline in the jointing stage might be related to the reducing of microorganisms quantity caused by the added of inoculating microorganisms, or the shortage of soil fertility which would influence the degradation of TPH by the microorganisms. The specific reasons couldn't be determined, so a further study was needed to analyze the influencing factors on the TPH removal in the jointing stage of the winter wheat. After the florescence stage, the TPH removal rate was at a low level. It might be related to the reducing of root quantity. From the study, it could be inferred that the main period for the TPH removal was from the turning green stage to the flowering stage.

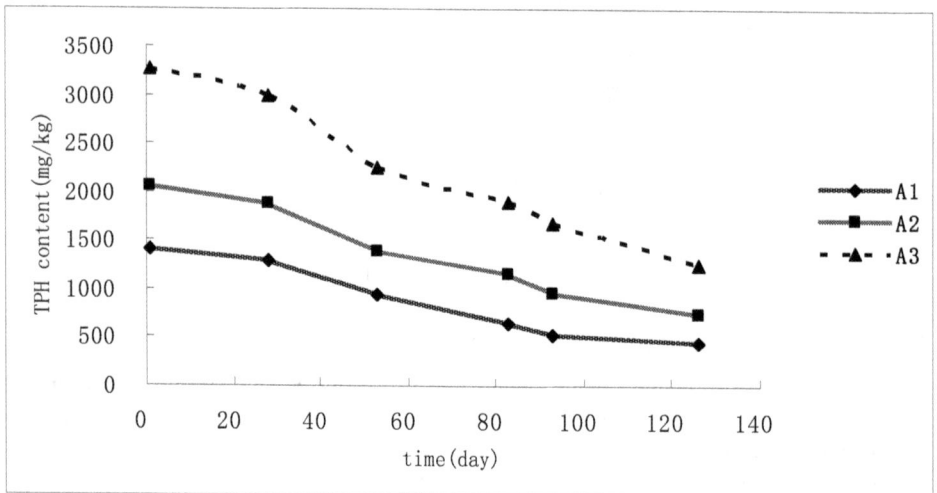

Fig. 7. Changes of TPH content in each treatment at different incubor time.

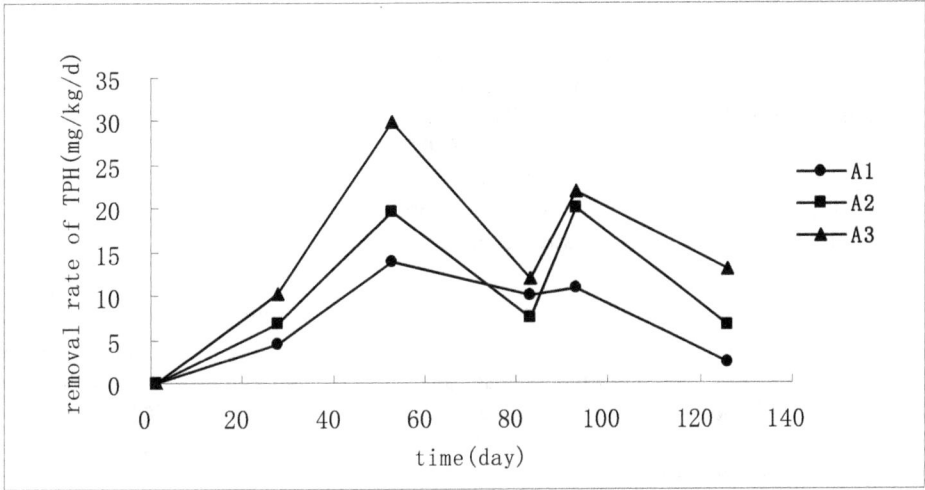

Fig. 8. Changes of TPH removal rate with time in the combined Remediation.

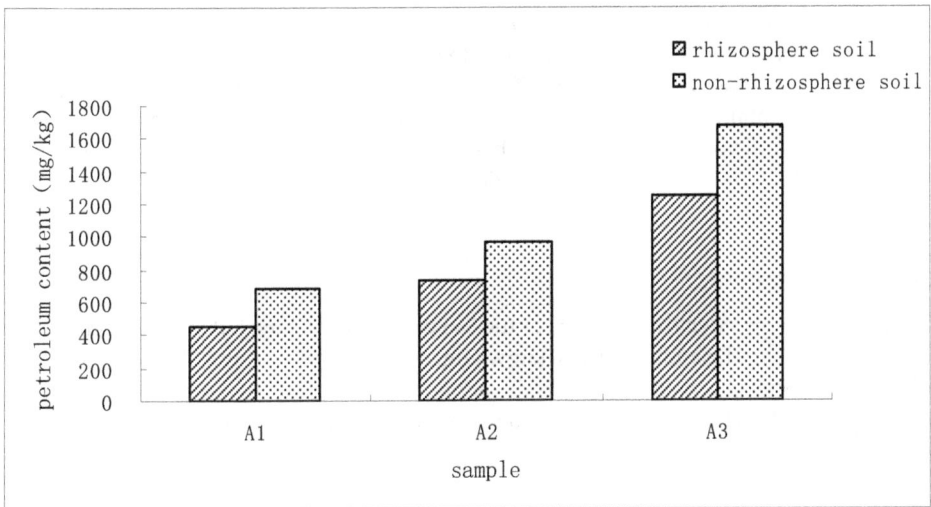

Fig. 9. Petroleum content in non-rhizophere soil and rhizosphere soil of the winter wheat after combined remediation.

The physical and chemical characteristics in rhizosphere soil were significantly different from that in non-rhizosphere soil. After the combined remediation of 126 days, there was a significant difference between the TPH removal in the rhizosphere soils and that in the non-rhizosphere soils, as showed in figure9. The TPH removal in the rhizosphere soil was higher than that in the non-rhizosphere soil.

4. Conclusions

The total bacterial concentration at 10°C was 6.2 × 10^5 CFU/mL and which under the conditions of 25°C was 4×10^8 CFU/mL. It was obvious that the community composition was more diverse under the conditions of 10°C than of 25°C. The microbial community diversity was significantly higher under the conditions of 10 °C. The dominant bacteria were different in different temperature conditions and Rhizobium was the only same genus. 40% of the total community composition was Rhizobium at 25°C and the proportion at 10°C was 3%.As a result, the temperature had a strong impact on the composition of soil microbial communities.

31 strains had been isolated in the low temperature conditions. With preliminary degradation test at 10°C, D17 and D24 had been selected to be used. D17 was identified as *Paracoccus sp.*, and D24 was identified as *Halomonas sp.* The degradation rate was 25.8% by D17, 25.3% by D24 and 21.9% by D17+D24 at 25°C while which was 11.3%, 10.2% and 10.3% respectively. It was suggested that the degrading of petroleum by cold-adapted bacteria was feasible at lower temperature.

The combination of the winter wheat and the cold-adapted bacterial had great potential ability in the remediation of the petroleum contaminated soil. In the rhizosphere soil system, TPH disappeared faster than that in unvegetated pots soil. The main period for the TPH removal was from the turning green stage to the flowering stage. During the period of combined remediation, the TPH removal rate varied in different stages. And a further study was needed in order to study the influencing factors in different stages of the winter wheat, thus we could take different measures in different stages. As incubation proceeded, the catalase activity in rhizosphere soil was improved by planting winter wheat, which indicated that the microbial growth and the metabolic activity were enhanced. And there was positive correlation between the catalase activity and the TPH removal.

5. Acknowledgment

The authors acknowledge the supports of the 863 program of China [2007AA06Z308] during the whole research. The authors also acknowledge the supports of Key Laboratory for Water and Sediment Sciences of Ministry of Education and College of Water Sciences of Beijing Normal University which provided the laboratory and experimental condition.

6. References

[1] U. Welander. Microbial Degradation of Organic Pollutants in Soil in a Cold Climate[J]. Soil and Sediment Contamination, 2005,14(3): 281-291.

[2] Dale R. Van Stempvoorta, James Armstrongb and Bernhard Mayer. Microbial reduction of sulfate injected to gas condensate plumes in cold groundwater[J] . Journal of Contaminant hydrology, 2007, 92(3-4):184-207.

[3] Luigi Michaud, et al.The biodegradation efficiency on diesel oil by two psychrotrophic Antarctic marine bacteria during a two-month-long experiment[J]. Marine Pollution Bulletin, 2004, 49(5-6): 405-409.

[4] D. Juck, T. Charles , L.G. Whyte , C.W. Greer. Polyphasic microbial community analysis of petroleum hydrocarbon-contaminated soils from two northern Canadian communities[J]. FEMS Microbiology Ecology, 2000,33(3):241-249.

[5] Mikael Eriksson, et al.Effects of low temperature and freeze-thaw cycles on hydrocarbon biodegradation in Arctic tundra soil[J]. Appl. Environ. Microbiol, 2001, 67:5107-5112.

[6] Yingfei Ma , Lin Wang and Zongze Shao. Pseudomonas, the dominant polycyclic aromatic hydrocarbon-degrading bacteria isolated from Antarctic soils and the role of large plasmids in horizontal gene transfer[J]. Environ Microbiol. 2006, 8(3):455-65.

[7] Philipp Bergauer, et al. Biodegradation of phenol and phenol-related compounds by psychrophilic and cold-tolerant alpine yeasts[J]. Chemosphere, 2005, 59(7): 909-918.

[8] Ji Xiuling,Wei Yunlin, "Progress in studies on environmental bioremediation technology with cold-adapted microorganism," Techniques and equipment for environmental pollution control, vol.7(10):6-11, 2006.

[9] Li Yaxuan, Zhang Xiaoling, Jiang Anxi,et al, "Application of cold-adapted microorganisms in the treatment of sewage at low temperature," Guang Zhou environmental sciences, vol.21(2):14-18, 2006.

[10] Chen Yan, Li Ganghe, Zhang Xu, et al, "Effect of petroleum biodegradation and rhizosphere microeco-system in phytoremediation of the polluted soil in oilfield," J Tsinghua Univ(sci&Tech), 2005, vol.459(6):784-787, 2005.

[11] Claudia Moreno1, Jaime Romero1 and Romilio T. Espejo. Polymorphism in repeated 16S rRNA genes is a common property of type strains and environmental isolates of the genus Vibrio[J]. Microbiology, 2002, 148, 1233-1239.

[12] Frassinetti S, Setti L, Corti A, et al. Biodegradation of dibenzothiophene by a nodulating isolate of Rhizobium meliloti.Canada Journal of Microbiology, 1998, 44:289-297.

[13] WANG Hui ZHAO Chun-yan LI Bao-ming SUN Jun-de. The Studies on Separation and Selection of Bacteria in Oil Polluted Soil[J].Chinese Journal of Soil Science, 2005, (2):237-239.

[14] Li Chun-Rong Wang Wen-Ke Cao Yu-Qing Wang Li-Juan. Petroleum Pollutions Degraded by Microorganisms[J].Journal of Earch Sciences and Environment, 2007,29(2):214-216.

[15] Zosim Z, Gu Tnick D, Rosenberg E. Properties of hydrocarbon-in-water emulsions stabilized by Acinetobacter RAG-1 emulsan [J] . Biotechnol Bioeng , 1982, 24 :281-292.

[16] Jianfeng Ren,"Effect of Petroleum Pollutants in Typical Soil on Growth of Crop in Jilin Province," Ji Lin:Ji Lin university, 2010.

[17] Wang Dawei,An Lizhe,Wang Xunling, "The effect of crude oil pollution of soil on growth of springwheat and buckwheat," Acta Bot.Boreal.-Occident.Sin.,vol.15(5), pp. 65-70, 1995.

[18] Lu Mang,Zhang Zhong-zhi,Sun Shanshan,et al,"Rhizosphere enhanced remediation of petroleum contaminated soil," Environmental science,vol.30(12):3703-3709,2009.

[19] Chen Haiyan,Hai Reti,Li qianting, "Effect of waste plastics recycling products on activities of soil catalase and urease," Asian Journal of Ecotoxicology, vol.4(6), pp.874-880,2009.

[20] Zhang Songlin,Dong Qingshi,Zhou xibin,et al, "Field Medicago sativa L. phytoremediation on the areificial oil-polluted soil," Journal of Lanzhou University, vol.44, pp.47-50, 2008.

[21] Song X Y,Song Y F,Sun T H,et al, "Limited effect of introduced microbial inoculants in the bioremediation of petroleum-contaminated soils[J]," Acta Scientiae Circumstantiae, vol.27(7), pp.1168-1173, 2007.

[22] Niu Liyuan,Ru Zhen-gang,Shi Ming-wang,Chen Cui-ling, "Studies on the Systemic Changes of Root Character in Winter Wheat and Breeding Meaning," Journal of Henan Vocation-Technical Teachers College, vol.28, pp.1-4, 200.

15

Removal of Effluent from Petrochemical Wastewater by Adsorption Using Organoclay

Jorge V.F.L. Cavalcanti[1], César A.M. Abreu[2],
Marilda N. Carvalho[2], Maurício A. Motta Sobrinho[2],
Mohand Benachour[2] and Osmar S. Baraúna[3]
[1]Universidade Federal Rural de Pernambuco
[2]Universidade Federal de Pernambuco
[3]Instituto de Tecnologia de Pernambuco
Brazil

1. Introduction

Given that environmental policy is increasingly severe, the industries are constantly seeking to establish standards of ever lower concentrations of the pollutants in wastewater, and are gradually adjusting their processes, with new procedures in order to generate less and remove more toxic elements present in their discharges.

The production stages of a petroleum industry, such as extraction and refining, are potentially responsible for generating large volumes of effluent to be discarded in the environment.

Water contaminated with petroleum derivates is produced in large volumes in many stages of refining oil. This mixture should be treated to separate these derivates from water before it can return to the environment. However, treatment with conventional processes is very often not economically feasible, or do not have the appropriate efficiency with regard to separation, or produce large amounts of mud that also need treatment (Almeida Neto et al., 2006).

The waste generated in oil refineries contains many different chemical compositions, depending on the complexity of the refinery, the existing processes and the type of oil used. The effluents are produced mainly by physical separation processes, such as atmospheric distillation and vacuum distillation, deparaffinization, deasphalting and also by processes involving chemical conversions by isomerization, alkylation, etherification, catalytic reform, etc.

The risks of environmental contamination and to human health caused by volatile organic compounds have driven a lot of research designed to eliminate or remedy its deleterious effects. Several of these compounds, such as phenol and BTEX compounds (benzene, toluene, ethyl benzene and the isomers of xylene) are found in effluents from oil refineries, and they are important contaminants due to their high toxicity (Akhtar, 2007).

Volatile organic compounds (VOCs) are usually harmful and carcinogenic and may cause serious environmental problems which affect ecosystems. The VOCs also may cause adverse effects to human health, even in low concentrations (Benmaamar & Bengueddach, 2007). They are known to integrate a group of compounds that most contribute to the formation of photochemical ozone and secondary organic aerosols (SOAs), increasing to global warming (Hu et al., 2008).

Different technologies such as advanced oxidation, biofiltration, separation by membrane, absorption and adsorption have been studied and developed in order to remove these organic compounds. It happens that many of these procedures, due to high operating costs, become unviable.

The phenols appear in the effluents from an oil refinery in the stages of catalytic cracking, production of lubricants and solvents, and in the rinse water of the gasoline. According to a survey, the average concentration of phenol in the effluents was 154 mg L^{-1} (Mariano, 2001). Another study found that the concentrations of phenolic effluents from a refinery were between 0.9 mg L^{-1} and 60.0 mg L^{-1} (Barros Junior, 2004).

Among the organic pollutants present in effluents from oil refineries at higher concentrations, phenol stands out. The phenol is a pollutant that is generated mainly in catalytic cracking and fractioning of crude oil. In addition to these sources, some processes use phenol as a solvent, increasing your concentration in the effluent (Otokunefor & Obiukwu, 2005). This study, based in Nigeria, evaluated the impact caused by the release of effluents from a refinery into a body of water. This study found that the quantity of phenol in the effluent treated released in the river delta showed a level of 1.84 mg L^{-1}, higher than the maximum allowed under Nigerian law which sets it at 0.5 mg L^{-1}. The phenol is one of the most difficult wastes to be removed, usually involving processes that are far from satisfactory and have high operating costs. Moreover, it is a highly toxic substance that can kills fish and other aquatic organisms.

The presence of hydroxyl groups confers to phenols the ability to form hydrogen bonds, which gives the same boiling points above organic compounds of molecular weights close, in addition to presenting high solubility in water. The phenol has solubility 8.3 g per 100 g of water at 25°C, therefore, the phenol may go dissolved in the aqueous industrial waste (Solomons & Fryhle, 2002).

The BTEX compounds are considered organic pollutants with high toxicity. One research found that among organic pollutants common in industrial accidents, leaking storage tanks or organic effluents, the BTEX compounds are present at high frequency, being directly responsible for contamination of waters and soils. These aromatics are extremely harmful to human health, fauna and flora (Sharmasarkar et al., 2000).

Liquid wastes generated by refineries have different chemical compositions including oil and grease, phenols, BTEX, ammonia, suspended solids, cyanide, sulfide, nitrogen compounds and heavy metals such as iron, cadmium, nickel, chromium, copper, molybdenum, selenium, vanadium and zinc. Seeking to evaluate and monitor the environmental impact caused by the discharge of effluents in water bodies, toxicity tests are often used as indicators of damage to the aquatic environment. Tests on fish, invertebrates and seaweeds have revealed that most of effluents from refineries is toxic, causing in these organisms not only lethal effects as well as changes to their growth and reproduction (Wake, 2004).

Many hydrocarbons, particularly the aromatics, have certain solubility in water, such as benzene, 1800 PPMV, toluene, 470 PPMV, ethyl benzene, 150 PPMV and xylenes, 150 PPMV. The process of desalter was very relevant to the contamination of BTEX in aqueous effluents at refineries studied (Worrall & Zuber, 1998).

Experiments have shown that different formulations of gasoline may affect the fate and transport of BTEX compounds. The use of ethanol as an ingredient in the formulation of gasoline has increased worldwide, in order to minimize atmospheric pollution resulting from combustion. The ethanol is completely miscible with water. The ethanol increases the solubility of BTEX compounds in water due to the effect of co-solvent. The International Agency for Research on Cancer (France) and the National Institute for Occupational Safety and Health (USA) include benzene in their lists of products carcinogens. Among the cancers, leukemia is the most frequent. In Brazil, the carcinogenic action of benzene was officially recognized as of March 1994. Your capacity to cause chromosomal damage and to bone marrow was widely demonstrated in humans and animals (Tiburtius & Zamora, 2004).

By understanding the development compatible with the preservation of natural resources, new techniques for wastewater treatment are needed. The clays have a high technical economic feasibility due to the adsorption potential, which associated with its abundant availability make the material more accessible adsorbents (Rodrigues et al., 2004).

The adsorption using clay constitutes one of the techniques recently applicable for the treatment of contaminated effluents. The material is abundant, has low cost and has large surface area. Moreover, it has high rates of mass transfer (Qu et al., 2009).

The increasing use of compounds and products originating in oil has caused serious problems to human health and the environment. The hydrophilic montmorillonite clay is an ineffective adsorbent for aromatic compounds that move often from contaminated sites. The effect of adsorption is suppressed by the competition of water in relation to non-polar compounds to the surface of the adsorbent material. The adsorption of organic compounds on montmorillonite can be enhanced by replacing the inorganic cations presents in the original structure to surfactant cations, such as: hexa-decile-trimethyl-ammonium, HDTMA, tetra-methyl-ammonium (TMA), tetra-ethyl-ammonium (TEA), tetra-butyl- ammonium (TBA), and tri-benzyl-methyl-ammonium (TBMA). Adsorbents surfactants slow the migration of pollutants in subsurface becoming an effective barrier transport of these (Yasser & Jamal, 2004).

Physical and chemical modifications of clays have been used for the production of materials for practical applications. The properties of clay can also be modified by adsorption and intercalation of organic polymers. The pillarization is a modification method that usually involves the intercalation of cationic species acting as pillars to support the mineral layers, separating them, creating a porous material useful for adsorption of organic compounds and other environmental applications (Bergaya et al., 2006).

The pillared clays (PILCs) are microporous materials obtained by intercalation of inorganic species in natural or synthetic expandable clays. Currently the PILCs are considered promising materials in various processes of adsorption and catalysis, and in particular in processes that relate to environmental protection. They result from isomorphic substitution of aluminum and silicon atoms by atoms of lower valence. Their applications can be found in the removal of NO_x, adsorption of toxic organic compounds, among others (Pires et al., 2000).

The process of pillarization consists in intercalation of cationic complexes, followed by calcination. Inorganic complexes can be obtained by hydrolysis of the metals of group 13 and transition metals, by changing the clay basal spacing of 1.2 nm to 1,8 nm, or more (Leite et al., 2000).

The abundance of world reserves of bentonite makes it impossible to estimate the quantity of these resources in a global context. In Brazil the estimated reserves are approximately 47 million tonnes, of which 47.7% are in the State of Paraná, in the municipality of Quatro Barras, 26.6% are in the State of São Paulo, in the municipalities of Taubaté and Tremembé and 25.3% are in the State of Paraíba, in the municipalities of Campina Grande and Boa Vista. The main consumers of these materials are petroleum companies, to employ the clay as thixotropic agent for use in drilling wells, industries that work with iron ore pellets for use as binding agent, paint and varnish industries, among others (Oliveira, 2004).

2. Fundamentals of adsorptive processes

The phenomenon of adsorption is known since the eighteenth century, when it was observed that a certain kind of coal had retained in their pores large amount of water vapor. Nowadays, the adsorption processes are applied in the purification and separation of substances, presenting itself as an important alternative and economically viable in many cases (Fogler, 2002; Ruthven, 1984).

Adsorption phenomena are classified in two types: physical adsorption and chemical adsorption. In the chemical adsorption exists an effective exchange of electrons between the solid and the adsorbed molecule, causing the following: formation of a single layer on the solid surface, irreversibility, and great force of attraction between the adsorbent and adsorbate. For this reason this type of adsorption is favored by increased temperature and pressure. The physical adsorption is a reversible phenomenon, where usually observed the deposition of more than one layer of adsorbate on the adsorbent surface. The forces acting on the physical adsorption are the Van Der Waals forces. The energy released is relatively low and rapidly reaches equilibrium. In this process, the temperature increase is detrimental to the adsorption efficiency.

The adsorption separation processes are widely used in industry, particularly in oil refineries and petrochemical industries (Ruthven, 1984). Knowledge of physical and chemical principles in which they are inserted adsorptive processes is fundamental to the interpretation of adsorption phenomena. The kinetic aspects and the adsorption equilibrium form the theoretical basis for understanding between the fundamental principles and industrial practices. These parameters are therefore essential to the analysis and interpretation of experimental data serving as a support for the dynamic study of the adsorption columns.

Research involving adsorptive processes between organoclay and organic compounds found their interpretations of experimental data supported according to the Langmuir's model, which assumes the existence of adsorption sites, all energetically equivalent, where only one molecule is adsorbed per site, without any interaction with molecules adsorbed on neighboring sites (Burns et al., 2003; Lin et al., 2005). However, there are studies that point to other adsorptive representative models such as Freundlich's model that considers non-uniformity in terms of surface adsorption sites (Boufatit et al., 2007; Cavalcanti et al., 2009; Irene et al., 1998; Kessaïssia et al., 2004; Ko et al., 2007; Richards & Bouazza, 2007; Sameer et al., 2003; Viraraghavan & Alfaro, 1998). Other studies have used the model of Dubinin-Raduskevich, which represents a more comprehensive model. In this model the surface of the adsorbent is not homogeneous and the interaction energy is not constant (Akçay, 2004, 2005).

2.1 Langmuir's isotherm

The kinetic of adsorption starts from the net adsorption rate r_A (mg g^{-1} min^{-1}), represented by Equation 1.

$$r_A = \frac{dq_A}{dt} = k_{AD}C_A(1 - \theta_A) - k_D\theta_A \tag{1}$$

Being C_A the concentration of the adsorbate in the liquid phase (mg L^{-1}), k_{AD} the kinetic constant of adsorption, k_D the kinetic constant of desorption and θ_A the fraction of adsorption sites occupied by component A of the total number of sites occupied at saturation.

At equilibrium, the net rate of adsorption is null, so we can rewrite Equation 1 according to Equation 2.

$$k_{AD}C_A(1 - \theta_A) = k_D\theta_A \tag{2}$$

By introducing the constant of equilibrium K_A, which represents the ratio between the kinetic constants of adsorption and desorption, in Equation 2, we have Equation 3.

$$K_A = \frac{k_{AD}}{k_D} = \frac{\theta_A}{C_A(1 - \theta_A)} \tag{3}$$

Reorganizing the terms and rewriting Equation 3 as a function of the fraction of sites occupied θ_A, we obtain Equation 4.

$$\theta_A = \frac{K_A C_A}{1 + K_A C_A} \tag{4}$$

Considering the fraction of sites occupied θ_A as the ratio of the concentration of the adsorbate on the solid phase q_A (mg g^{-1}) and the concentration of adsorbate in solid phase in saturation q^{max} (mg g^{-1}) also known as maximum adsorption capacity, we have Equations 5 and 6, the latter being known as the Langmuir's Equation.

$$\theta_A = \frac{q_A}{q^{max}} \tag{5}$$

$$\frac{q_A}{q^{\text{max}}} = \frac{K_A C_A}{1 + K_A C_A} \tag{6}$$

The values of C_A can be obtained experimentally and the values of q_A can be calculated according to the material balance representing in the equation 7.

$$q_A = \frac{(C_{A0} - C_A)V}{m_s} \tag{7}$$

The Langmuir's equation can be linearized, yielding experimental values of the equilibrium constant K_A and maximum adsorption capacity q^{max}, according to Equation 8.

$$\frac{1}{q_A} = \frac{1}{K_A q^{\text{max}}}\left(\frac{1}{C_A}\right) + \frac{1}{q^{\text{max}}} \tag{8}$$

The Langmuir's isotherm is one of the most common models used to represent equilibrium behavior in the adsorption process. Its main concept proposes the uniformity in terms of surface adsorption sites with adsorption monolayers. Also considered independent interactions of neighboring sites and possibilities of dynamic equilibrium of adsorption and desorption (Fogler, 2002; Ruthven, 1984).

2.2 Freundlich's isotherm

It is an equilibrium model for indicating a greater degree of generalization of the adsorptive process. Now, it is considered the non-uniformity in terms of surface sites and multilayer adsorption (Ruthven, 1984). Its general expression for a single component system is represented by Equation 9, known as the Freundlich's Equation.

$$q = K C^{\frac{1}{n}} \tag{9}$$

The Equation 9 assumes a linear form as shown in Equation 10.

$$\ln q = \ln K + \frac{1}{n}\ln C \tag{10}$$

The parameters of Equation 10 will be obtained experimentally. Being K the equilibrium constant feature of the Freundlich's isotherm and $1/n$ adsorption intensity.

2.3 Dubinin-Raduskevich's isotherm

Another adsorptive equilibrium model was also used in adsorption of phenol and BTEX in organoclay. This is the Dubinin-Raduskevich's model, represented by Equation 11 (Akçay, 2004, 2005).

$$\frac{q}{q^{máx}} = \exp(-\beta\varepsilon^2) \tag{11}$$

The Dubinin-Raduskevich's isotherm represents a most comprehensive model, considering that the surface of the adsorbent is not homogeneous and that the interaction energy is not constant. Being q the concentration of the adsorbate on the solid phase (mg g^{-1}), β the constant related to adsorption energy (mol^2 kJ^{-2}), q^{max} the maximum adsorption capacity and ε the Polanyi potential, calculated according to Equation 12.

$$\varepsilon = -RT \exp\left(\frac{1}{C_A}\right)_{eq}$$

(12)

2.4 Kinetic models adsorptive

Some adsorptive kinetic models were used in different works of adsorption of organic compounds in organophilic clays, including models of Lagergren, also called models of pseudo-first order and pseudo-second order, represented by Equations 13 and 14 (Akçay, 2004, 2005; Cavalcanti et al., 2008; Yilmaz & Yapar, 2004).

$$\frac{dQ}{dt} = k_{1ads}(Q - Q_{eq})$$

(13)

$$\frac{dQ}{dt} = k_{2ads}(Q - Q_{eq})^2$$

(14)

Which Q_{eq} represents the equilibrium concentration in solid phase and Q the concentration in the solid phase over time, k_{1ads} (min^{-1}) and k_{2ads} (g mg^{-1} min^{-1}) the kinetic constants adsorptive related to models of pseudo-first order and pseudo-second order.

The analytical solution of Equations 13 and 14 are described by Equations 15 and 16 respectively.

$$\ln(Q - Q_{eq}) = \ln Q_{eq} - k_{1ads}t$$

(15)

$$\frac{t}{Q} = \frac{1}{k_{2ads}Q_{eq}^2} + \left(\frac{1}{Q_{eq}}\right)t$$

(16)

According to the model represented by Equation 15, we can determine the kinetic constant of first-order k_{1ads} through the slope of the straight line plotted between ln $(Q - Q_{eq})$ versus t, and we can determine the kinetic constant of second-order k_{2ads} through the interception of Equation 16.

3. Utilization of organophilic smectite clays as adsorbents of phenol and BTEX compounds

Below, we highlight the study of the smectites clays, the main procedures to make them organophilic and its application as an adsorbent for phenol and BTEX.

3.1 General structure of smectite clays and their properties

The classic definition refers to clay as a natural material, earthy, fine-grained that when moistened with water features plasticity (Souza Santos, 2002). The clay minerals are hydrated silicates that have layered structure composed of solid sheets formed by tetrahedral of silicon (or aluminum) and oxygen, and leaves formed by octahedral of aluminum (magnesium, iron), oxygen and hydroxyls. One way of classifying clays is related to the type of cation present in the octahedral sheet. If that is bivalent cation (such as Ca^{2+} or Mg^{2+}) all octahedral are occupied and the clay would be classified as type tri- octahedral. For trivalent ions (such as Al^{3+}), only 2/3 of the sites will be busy, so it is denominated as bi-octahedral type clays. The nomenclature for the types of layers is a simple expression of the ratio between the tetrahedral sheets and octahedral sheets. Therefore, a 1:1 layer clay mineral has a tetrahedral and octahedral sheet, while a 2:1 clay mineral type has two tetrahedral sheets and one internal octahedral. The distance d_{001} is called interlayer distance and is also used to classify the different clays exist (Neumann et al., 2000).

The clay minerals are formed mainly of hydrated silicates of aluminum, iron and magnesium, containing certain amount of alkaline and alkaline earth elements (Rianelli & Pereira, 2001). Clays are minerals that have unique physical and chemical characteristics. They are formed by small crystals, usually in the form of hexagonal platelets, which agglutinates to form conglomerates. These platelets are composed of aluminosilicates organized, composed of the elements silicon, aluminum and oxygen, and others in smaller proportions, such as magnesium and iron. When hydrated, allows the separation of lamellar layers and intercalation of ions and molecules. The understanding of the properties of organized systems leads to interest in the study of systems consisting of the structured clays (Pastre et al., 2003). For being generally of sedimentary origin, the clay can contain a large number of other minerals, represented by grains of varying diameters, soluble salts and organic matter.

The clay minerals are subdivided into crystalline silicates with layer lattices or lamelar, and crystalline silicates with fibrous lattice structure, chain lattices. The silicates of lamellar structure can be divided into three groups or families: dimorphic (layer 1:1), trimorphic (layers 2:1) and tetramorphic (layers 2:2). The nomenclature 1:1 and 2:2 refers to the number of layers of SiO_4 tetrahedral and octahedral of hydroxides, respectively, entering the establishment of the unit cell of the crystalline lattice of clay minerals.

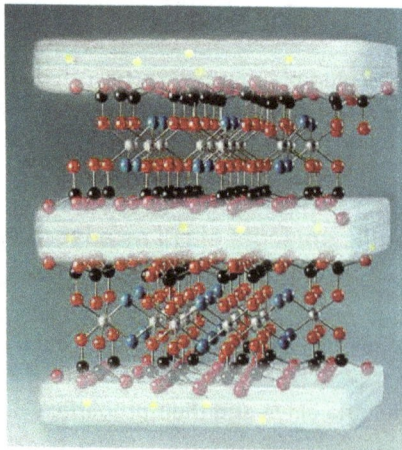

• Exchangeable Cations, • Oxygen, • Hydroxyl, • Silicon, • Aluminum

Fig. 1. Schematic representation of the crystal structure of smectite.

The paligorsquita or attapulgite and sepiolite are the only representatives of the fibrous structure of silicates.

The basal interplanar distance, the degree of substitution in the octahedral layer of the unit cell and the possibility of the basal layers to expand with the introduction of polar molecules are structural properties that allow the subdivision of lamellar structure of clay minerals.

The group of smectite clay minerals (montmorillonite, beidelita, nontronite, volconscoíta, saponite, sauconita, hectorita) is composed of two silicate tetrahedral sheets with a central octahedral sheet, joined by common oxygen atoms to the leaves, as shown in Figure 1. The leaves are continuous in the directions of crystallographic axes.

Sodium is the predominant exchangeable cation to increase the interlayer of smectite. The increase occurs by adsorption of water molecules by sodium, as shown in Figure 2.

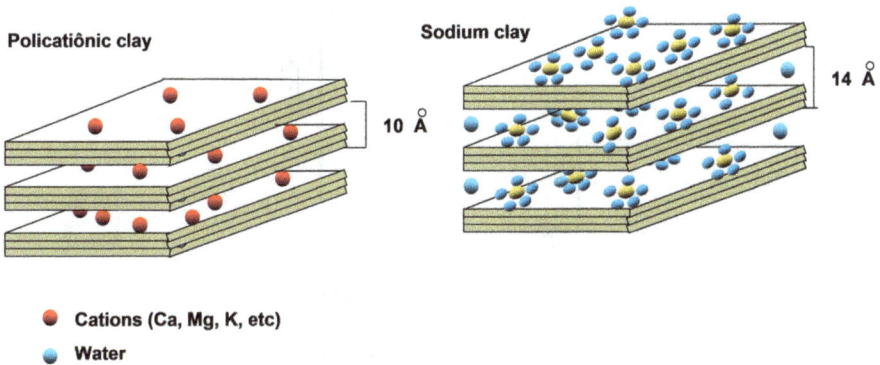

Fig. 2. Polycation not hydrated smectite clays and hydrated with sodium.

The clay treated with quaternary salts develops the ability to swell by adsorption of organic molecules, resulting in the organoclay, as shown in Figure 3.

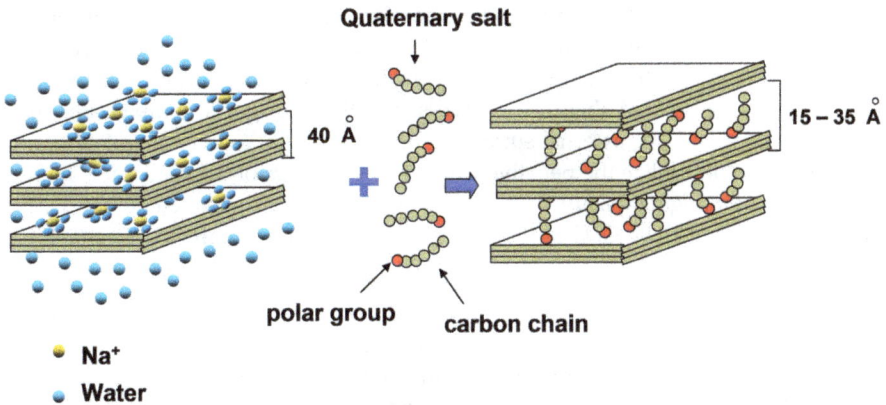

Fig. 3. Development of organoclay treated with quaternary salt (Martins, 2007).

Sodium clays can be used as adsorbents of inorganic compounds, and also as a thixotropic agent for drilling of wells water-based. Organoclay can be used as adsorbents of organic compounds, and also as a thixotropic agent for drilling of wells oil-based. The thixotropic property of gels is extremely important in drilling for oil exploration because they promote increased service life of the drill, torque reduction, increased stability and temperature control in the well. In a general way, any outages in polls, the rocky debris produced by the action of the drill are kept in suspension in the fluid, preventing the loss of the pit of debris by sedimentation and subsequent burial of partial probing tools, as shown in Figure 4.

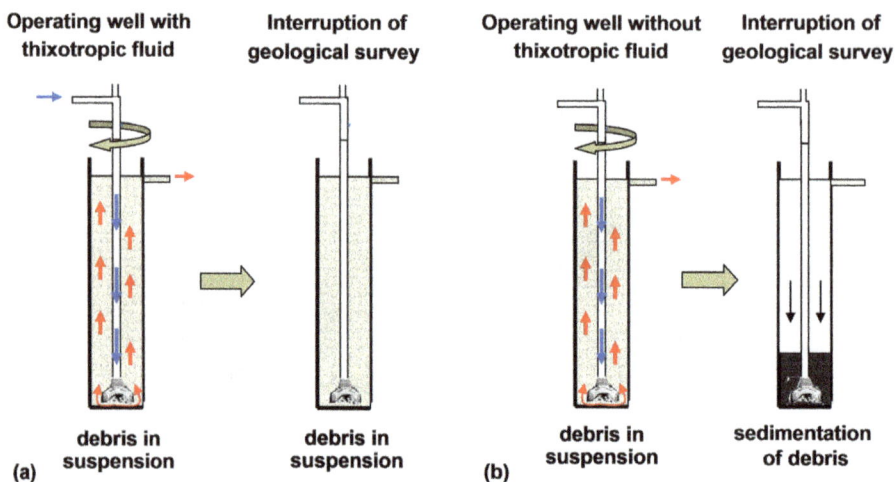

Fig. 4. Oil well drilling - real situation (a) and hypothetical situation (b).

The group of kaolinite is formed by regular stacking 1:1, where each layer consists of a sheet of SiO_4 tetrahedra and a sheet of octahedra $Al_2(OH)_6$ linked together in a single layer through common oxygen.

Replacements of ions Si^{4+} leaves tetrahedral by ions of Al^{3+}, or Al^{3+} by Mg^{2+} or Ca^{2+} in the leaves octahedral, are called substitutions isomorphic, and do not cause distortion in the structure of the lamellae, being all these atoms of similar size. In contrast, these substitutions generate an excess of negative charge in the layers of clay, which is responsible for some of the interesting properties of them. The excess negative charge is compensated by the adsorption of cations on the external surfaces of the layers, where the amount of adsorbed cations required to neutralize the negative charges in the layers of the material is measured by the cationic exchange capacity (CEC). This property is also related to the extent of isomorphic substitution. The CEC of the montmorillonite clays is between 40 and 150 meq (or mmol currently) per 100 g of clay (Gomes, 1988).

Clays are widely used as adsorbents due to its large surface area. On the other hand, its adsorption capacity for organic molecules, water soluble is very low. This is due to the hydrophilic nature of mineral surfaces. Treatment of clays with organic and inorganic reagents increases its adsorption capacity. The modified clays can be used as adsorbents in treatment systems of wastewater, thus preventing contamination of groundwater. The potential of organically modified clays as adsorbents for organic contaminants effective in preventing and reducing pollution has stimulated studies on their adsorptive properties. The adsorption of organic compounds in organoclay occurs through several mechanisms, such as ionic reactions and complexation (Vieira et al., 2004).

The natural smectite clays, whose cation for exchange is sodium, produce gels that can be used as thixotropic agents in drilling fluids for oil wells based on water. Due to the hydrophilic environment on the surface, natural smectite is not an effective sorbent for nonionic organic compounds in water even if the clay has a high specific surface area. Reactions of ion exchange may greatly modify the surface properties of natural smectite. When, for example, organic cations of formula $(CH_3)_3NR^+$, where R is an alkyl hydrocarbon, occupy the local exchange of smectite, the surface properties transforms from hydrophilic to hydrophobic (Chiou et al., 1983). Once such organic cations are fixed on the surface of smectite, an organic phase is formed which is derived from the alkyl hydrocarbon. This effectively removes the nonionic organic compounds from water with hydrophobic interactions.

The modified clays are normally prepared using a quaternary ammonium cation of general formula: $[(CH_3)_3NR]^+$ or $[(CH_3)_2NRR']^+$ where R and R' are hydrocarbons. The adsorptive properties of clays depend on the molecular size of the groups R and R' (Bodocsi et al., 1997). When the sizes R and R 'are small, such as: $-CH_3$ and phenyl, such clays are classified as adsorbents. When R or R' are long-chain, these clays are classified as organophilic clays. Many studies indicate that polar organic compounds such as alcohols, amines and ketones were adsorbed on the clay interlayer spaces, and probably at the edges of clay particles by electrostatic attractions and by ion exchange reactions. Herbicides and insecticides can be also adsorbed by organophilic clays and become inactive.

One research revealed that treatment of smectite clays with quaternary ammonium salts in concentrations higher than their CEC corroborates with the increase in adsorptive capacity due to the setting of quaternary ammonium ions in the interlayer regions and in the external surface of clay minerals, with the possibility of their interlayer regions exceeds 4.0 nm, resulting in delamination of the clay (Menezes et al., 2007).

Researchs confirm the possibility of development of smectite clays in adsorptive properties of the sedimentary watershed of the Araripe-PE, from the establishment of optimal experimental conditions grounded in factorial planning and response surface analysis, as tools of optimization of the process of activation of the clays. These studies proved the existence of calcic smectite clays with potential to develop organophilic properties, present in the sedimentary watershed of various municipalities in the Northeast. These clays have naturally polycations, with a predominance of magnesium and calcium, which, after exchange for sodium, develops thixotropy (Baraúna, 2000). The results of tests conducted on clays of the region showed that: a) these clays are composed of groups of clay minerals smectite and kaolinite; b) the mineral assemblage consists essentially of the smectite group; c) the predominants exchangeable cations are calcium and magnesium; d) clays treated with sodium develops thixotropy, with your plastic viscosity in aqueous suspensions with a concentration equal to 4.86% by mass equal to 3.0 mPas. This result is close to that recommended by Petrobras which recommends, among other analysis, minimum plastic viscosity equal to 4.0 mPas (Petrobras, 1997a, 1997b, 1998a, 1998b).

Another academic work was consisted in the preparation of an organophilic adsorbent material, starting from five different types of clays taken from watershed located in the Northeast of Brazil, as follows: "Verde Lodo", "Bofe" and "Chocolate" taken from the municipality of Campina Grande, State of Paraíba; "Lagoa de Dentro" e "São Jorge", taken from the Plaster Pole of Araripe, Pernambuco's State. The fresh clay were treated with a solution of sodium carbonate in order to exchange the existing polycations in their interlayer regions by sodium in a batch reactor with different concentrations of this cation, contact time and temperature, according experimental design with a 23. As an answer to treatment with sodium carbonate, it was the reading of the viscosities of aqueous suspensions of plastic material with 4.86% by mass. The results showed that the clay Chocolate treated with sodium, in the best condition of the experimental planning, got a plastic viscosity equal to 4 mPas, according to the methodology described in Petrobras 1998a and Petrobras 1998b. Then, there was a treatment with quaternary ammonium salt (chloride hexa-decile-tri-methyl-ammonium), and measured up to its apparent viscosity for dispersions in an oily medium containing 336 mL of 5391 Ultra-lub ester of Oxiteno and 84 mL of saturated solution of sodium chloride, with 2.4 g, 6.0 g, 9.6 g and 13.2 g of organoclay, resulting in 8.5, 10.0, 12.5 and 16.0 mPas, respectively, according to the methodology described in Petrobras 1997a and Petrobras 1997b. These results suggested that sodium clay and organoclay could be indicated as thixotropic fluids for drilling wells. Then, a study was conducted for adsorptive removal of phenol in synthetic wastewater using this organoclay produced. The results indicated a more than 80% removal of organic contaminants (Cavalcanti et al., 2009; Cavalcanti et al., 2010; Portela et al., 2008).

3.2 Adsorption of phenols and BTEX onto organoclays

Several physicochemical methods and biological have been used to remove organic compounds in industrial effluents. Application of membrane filtration systems and adsorption processes in water treatment and effluent was used by a group of researchers. They developed a system for removal of phenol from an aqueous solution through a combined process of ultrafiltration and adsorption using kaolinite and montmorillonite. The adsorption experiments were performed in batch with 0.2 g of clay and 100 mL of water contaminated in the range of variable concentration of the organic compound from 20 to 1000 mg L^{-1}, stirred for 12 h at 25°C. The results showed that the phenol removal efficiency was 80% and a maximum adsorption capacity equal to 40 mg g^{-1} (Lin et al., 2005).

Other survey was developed to increase the range of applicability of the clays in the treatment of organic waste present in wastewater. The retention properties by adsorption of volatile organic compounds such as phenol, chlorobenzene and orthoxylene by a geomaterial composed of bentonite, activated carbon, cement and a water-soluble polymer in different operating conditions were evaluated. The operation was performed in batch at 25°C, using phenol, orthoxylene and chlorobenzene, without further treatment. The solutions were prepared in concentrations ranging from 50 to 400 mg L^{-1} with 0.3 g of the adsorbent material in 50 mL of solution distributed in erlenmeyer flasks. After the equilibrium established, the contaminated samples were centrifuged for 20 min at 2000 rpm. The kinetic results obtained for adsorption of these compounds in geomaterial, an initial concentration of 100 mg L^{-1} indicated residual concentrations of orthoxylene, chlorobenzene and phenol, respectively 40, 30 and 20 mg L^{-1} for a balance established in 12 h (Kessaïssia et al., 2004).

Two researches studied the adsorptive properties of montmorillonite clay modified by tetra-butyl ammonium (Akçay, 2004, 2005). The adsorption of p-chlorophenol in this clay was done in batch with 20 mL of pollutant solution to 0.1 g of clay, at 25°C for 16 h. The adsorption isotherms were adjusted according to the models of Freundlich and Dubinin-Radushkevich. The kinetic and thermodynamic parameters pointed to the application of organoclay as adsorbent effective of phenolic compounds in contaminated effluents.

Essays observed the adsorption, in batch, of phenolic compounds such as phenol and the isomers 2, 3 and 4-chlorophenol in smectite clays treated with chloride of tetra-methyl ammonium (TMA) and bromide of tetra-methyl-phosphonium (TMP). The experiments were performed with 100 mg of clay in solutions of adsorbate equal to 0.02 to 0.15 (mmol/100mL) in erlenmeyer of 250 mL, at 20°C. It was found that the smectite clay treated with TMP was better than the same adsorbent treated with TMA (Lawrence et al., 1998).

Other research confirmed the influence of the modification of bentonite in their adsorptive properties. The adsorption isotherms were determined with solutions of phenol in concentrations 50 mg L^{-1} to 1000 mg L^{-1}, in pH 6.5, at 20°C and 24 h. Structural changes were carried out with tetra-decyl-trimethyl-ammonium bromide (TDTAB) and hexa-decyl-trimethyl-ammonium bromide (HDTAB), with changes in 25%, 50% and 100% of capacity exchange cation. The equilibrium time was approximately 7 h and kinetic results indicated the possible presence of heterogeneous regions on the surfaces of clays modified with 25% and 50% of its cation exchange capacity. The clays modified with TDTAB and HDTAB in 100% obtained the best results for the removal efficiencies (Yilmaz & Yapar, 2004).

Studies evaluating the adsorptive efficiency in modified bentonite to remove aromatic compounds (BTEX) present in aqueous solutions were performed. The results showed that the agent "dimethyl di-hydrogenated" increased the basal spacing of bentonite in 119%. The study demonstrated that the removal efficiency by adsorption to the compounds, benzene, toluene, ethyl benzene and o-xylene was 75%, 87%, 89.5% and 88.5%, respectively (Bodocsi et al., 1997).

In another study were investigated the effects of phenol in adsorptive organoclay with different characteristics. The adsorption experiments were performed in batch with 0.2 g of modified clay with 22 mL of phenol solution at pH 7 for 4 h. The equilibrium time, in preliminary studies pointed 2 h. The results showed the influence of modified clays in adsorptive power. According to the author, a change in the smectite interlayers increased the effectiveness of interactions between the phenol and the organoclay. Therefore, this clay is an effective adsorbent material for removal of nonionic organic compounds (Shen, 2005).

The organoclay dicetyldimethylammonium-bentonite was studied for adsorption of the compounds BTEX (nonionics and nonpolar) and of the compounds 2-chlorophenol, 2,4-dichlorophenol and 2,4,6-trichlorophenol (ionizables and polars). An experiment showed this preference in the adsorptive efficiency order: benzene < toluene < ethyl benzene < ortho-xylene; phenol < 2-chlorophenol < 2,4-dichlorophenol < 2,4,6-trichlorophenol (Irene et al., 2006).

3.3 Factors that can influence the process adsorptive

Some authors investigated physical factors that can interfere with the adsorptive process, such as pH and temperature, interfering on the steady state (Cavalcanti et al., 2009; Irene et al., 1998; Lin et al., 2005; Vianna et al., 2001).

3.3.1 Influence of pH

The rejection of the removal of phenol and o-cresol was related to the increase in pH of the system. For phenol, the rejection at the kaolinite was at pH 8.2, while for montmorillonite at pH 9.1. For o-cresol, the maximum rejection at the kaolinite was at pH 9.2, while for montmorillonite at pH 10.2. This result can be explained by the difference between the pKa of the phenolic compounds and the zeta potential of clays (Lin et al., 2005).

Studying the influence of pH on the adsorptive process, it was observed that changing the pH (5, 7, 9) was not as relevant for changes in the adsorption isotherms of BTEX compounds in "DCDMA-bentonite" (dicetyldimethylamonium-bentonite). The result observed is not due to dependence on pH with the adsorptive process, but the forces of attraction between molecules of BTEX and sites of modified clay. For phenol, this research group concluded that the change of pH (5, 7, 9), leads to a slight change in the adsorption isotherms. The increase in pH decreases the affinity of the clay used with phenol. This is due to the fact that the lower the pH in relation to the pK_a of phenol, there will be smaller fractions of phenol in the ionized form (Irene et al., 1998). This same result was identified in another research that used organophilic clay with chloride-hexadecyl-trimethyl-ammonium. This work used three pH values in the system clay-phenol, and they were 7.0, 8.3 (representing the natural state) and 9.0. The results indicated that at pH 7.0, there was better adsorption efficiency. The pK_a of phenol is 9.95 (Teng & To, 2000), so when the pH of the system is below this value, the phenol is not completely disassociated with negative charges. However, the zeta potential of the organoclay was increasingly negative as the pH increases, so it is expected that there is at alkaline pH favoring the desorption process at equilibrium, due to electrostatic repulsion on the surface of the adsorbent (Cavalcanti et al., 2009).

3.3.2 Influence of temperature

It was found that the increased of temperature in the system clay-phenol, using 30°C and 40°C discourages the adsorptive process. At high temperatures there is a greater solubility of phenol in water, shifting the equilibrium towards of dessorption (Cavalcanti et al., 2009). Using organoclay as adsorbent effluent derived from the stillage, such as phenol, dextran, glucose, fructose, glycerin and glycine, it was observed that the phenol appeared as the more adsorbed substance, in more low temperature, comparing 33°C and 43°C (Vianna et al., 2001).

4. Conclusion

According to several studies cited, it is possible to use effective organophilic clays as adsorbents for the removal of organic substances with high toxicity, such as phenols and BTEX compounds.

The smectite clays are formed by a region called interlayer region, endowed with various inorganic cations, giving it a hydrophilic nature. The replacement of inorganic cations by organic cations, through ion exchange, creates the organoclays, which are effective adsorbents of organic compounds, having many environmental and technological applications.

In addition to the structure and size of the quaternary ammonium ion, density and orientation of alkyl groups on the surface of the clays are determining factors in the interactions between modified clays and organic compounds. Consequently, the mechanisms that control these interactions depend on the type of cation that forms. The observations suggest the use of organoclays as adsorbents of organic compounds in contaminated water.

To obtain organophilic clays, the natural clays can be initially treated by replacement of your different cations, by sodium cation. And after, the organoclays can be treated by organic cations, such as quaternary ammonium.

Alterations in physicochemical adsorptive system, such as temperature and pH can to influence the efficiency of the process. It was seen that in some citations the decreased of pH favored the adsorptive process of phenol but did not changed significantly the adsorption of BTEX. In addition to pH, the increased of temperature also seemed to be relevant, shifting the equilibrium towards desorption, decreasing the adsorptive efficiency.

Among the isotherms that portray the adsorptive equilibrium, the Langmuir isotherm, Freundlich and Dubinin-Raduskevich were cited. For the adsorptive kinetic models, stand out the Lagergren, in models of pseudo-first order and pseudo-second order.

Alternative methods for removing harmful substances to the environment should be diffused and increasingly applied, decreasing the concentration of industrial effluents and decreasing costs.

5. References

Akçay, M. (2004). Characterization and determination of the thermodynamic and kinetic properties of p-CP adsorption onto organophilic bentonite from aqueous solution. *Journal of Colloid and Interface Science*, vol. 280(2), pp. 299-304. ISSN 0021-9797.

Akçay, M. (2005). Characterization and adsorption properties of tetrabutylammonium montmorillonite (TBAM) clay: Thermodynamic and kinetic calculations. *Journal of Colloid and Interface Science*, v.294, pp. 1-6. ISSN 0021-9797.

Akhtar, M., Hasany, S. N., Bhanger, M. I. & Iqbal, S. (2007). Sorption potential of Moringa oleifera pods for the removal of organic pollutants from aqueous solutions. *Journal of Hazardous Materials*, vol. 141. pp. 546–556. ISSN 0304-3894.

Almeida Neto, A. F., Silva, A. A., Valenzuela-Díaz, F. R. & Rodrigues, M. G. F. (2006). Study of organophilic clays in the adsorption of xylene and toluene. *Proceedings of VI Encontro Latino Americano de Pós-Graduação*, ISBN 1517-3275, São José dos Campos, SP, Brazil, 2006.

Baraúna, O. S., Meneses Junior, V.P. & Souza Santos, P. (2000). Development of thixotropic properties of smectite clays from Santana - St. Jorge Mine - Ouricuri - PE, Brazil. *Proceedings of 55° Congresso da Associação Brasileira de Materiais*, Rio de Janeiro, RJ, Brazil, 2000.

Barros Junior, L. M. (2004). Removal of Phenol in Wastewater Oil Refineries. *Revista Petroquímica, Petróleo, Gás & Química*, vol. 266, pp. 58-62. ISSN 0329-5001.

Benmaamar, Z. & Bengueddach, A. (2007). Correlation with different models for adsorption isotherms of m-xylene and touene on zeolites. *Journal of Applied Sciences in Enviromental Sanitation*, vol. 2, pp. 43-53. ISSN 0126-2807.

Bergaya, F., Theng, B. K. G. & Lagaly. (2006). Modified Clay and Clay Minerals, In: *Handbook of Clay Science*, pp. 261-422, Elsevier, ISSN 1572-4352, Oxford, United Kingdom.

Bodocsi, A., Huff, W., Bowers, M. T. & Gitipour, S. (1997). The efficiency of modified bentonite clays for removal of aromatic organics from oily liquid wastes. *Spill Science & Technology Bulletin*, vol. 4, pp. 155-164. ISSN 1353-2561.

Boufatit, M., Ait-Amar, H. & McWhinnie, W. R. (2007). Development of an Algerian material montmorillonite clay. Adsorption of phenol, 2-dichlorophenol and 2,4,6-trichlorophenol from aqueous solutions onto montmorillonite exchanged with transition metal complexes. *Desalination*, vol. 206, pp. 394-406. ISSN 0011-9164.

Burns, S. E., Bartelt-Hunt, S. L. & Smith, J. (2003). Sorption and permeability of gasoline hydrocarbons in organobentonite porous media. *Journal of Hazardous Materials*, vol. 96, pp. 91-97. ISSN 0304-3894.

Cavalcanti, J. V. F. L., Da Motta, M. A., Abreu, C. A. M., Portela, L. A. P. & Baraúna, O. S. (2008). Equilibrium and kinetic study on the use of organoclay as adsorbent of phenol. *Proceedings of XVII Congresso Brasileiro de Engenharia Química*, Recife, PE, Brazil, 2008.

Cavalcanti, J. V. F. L., Da Motta, M. A., Abreu, C. A. M., Portela, L. A. P. & Baraúna, O. S. (2009). Preparation and use of a organophilic smectitic clay for adsorption of phenol. *Química Nova*, vol. 32, pp. 2051-2057. ISSN 0100-4042.

Cavalcanti, J. V. F. L., Da Motta, M. A., Abreu, C. A. M., Portela, L. A. P. & Baraúna, O. S. (2010). Utilization of smectit clays from Brazil northeast for preparing an organophilic adsorber. *Cerâmica*, vol. 56, pp. 168-178. ISSN 0366-6913.

Chiou, C.T., Porter, P.E. & Schmedding, D.W. (1983). Partition equilibriums of nonionic organic compounds between soil organic matter and water. *Environment Science Technology*, vol. 17, pp. 227-231. ISSN 0013-936X.

Menezes, R. R., Ávila Junior, M. M., Santana, L. N. L., Neves, G. A. & Ferreira, H. C. (2008). Expansion behavior of organophilic bentonite clays from the State of Paraíba. *Cerâmica*, vol. 54, pp. 152-159. ISSN 0366-6913.

Fogler, H. S. (2002). *Chemical Reaction Engineering*. 1 ed. São Paulo, LTC, Brasil, 2002.

Gomes, C.F. (1988). *Clays - What they are and what you use*. Editora Fundação Caloust Gulbenkian, ISBN 9723100274, Lisbon, Portugal.

Hu, Q., Li, J. J., Hao, Z. P., Li, L. D. & Qiao, S. Z. (2008). Dynamic adsorption of volatile organic compounds on organofunctionalized SBA-15 materials. *Chemical Engineering Journal*, vol.149, p.281–288. ISSN 1385-8947.

Irene, M. C., Samuel, C. H. & Raymond, K. M. (1998). Sorption of nonpolar and polar organics on dicetyldimethylammonium-bentonite. *Waste Management Research*, vol. 16, No 2, pp. 129-138. ISSN 0734-242X.

Kessaïssia, Z., Ait Hamoundi, S., Houari, M. & Hamdi, B. (2004). Adsorption of some volatile organic compounds on geomaterials. *Desalination*, vol. 166, pp. 449-455. ISSN 0011-9164.

Ko, C. H., Fan, C., Chiag, P. N., Wang, M. K. & Lin, K. C. (2007). p-Nitrophenol, phenol and aniline sorption by organoclays. *Journal of Hazardous Materials*, vol. 149, pp. 275-283. ISSN 0304-3894.

Lawrence, M. A. M., Kukkadapu, R. K. & Boyd, S. A. (1998). Adsorption of phenol and chlorinated phenols from aqueous solution by tetramethylammonium and tetramethylphosphonium exchanged montmorillonite. *Applied Clay Science*, vol. 13, pp. 13-20. ISSN 0169-1317.

Leite, S. Q. M., Dieguez, L. C., San Gil, R. A. S. & Menezes, S. M. C. (2000). Employment pillared clay in alqulation of benzene with 1-dodecene. *Química Nova*, vol. 23, No 2, pp. 149-154. ISSN 0100-4042.

Lin, S. H., Hsiao, R. C. & Juang, R. S. (2005). Removal of soluble organics from water by a hybrid process of clay adsorption and membrane filtration. *Journal of Hazardous Materials*, vol. 135, pp. 134-140. ISSN 0304-3894.

Mariano, J.B. (2001). Environmental Impacts of Petroleum Refining. *Dissertation to degree of Master of Science*, COPPE/UFRJ, Rio de Janeiro, RJ, Brazil.

Martins, A. B. (2007). Development of organoclay for use in non-aqueous fluids with low aromatic content. *Proceedings of 4° PDPETRO*, Campinas, SP, Brazil, 2007.

Neumann, M. G., Gessner, F., Cione, A. P. P., Sartori, R. A. & Schmitt Cavalheiro, C. C. (2000). Interaction between dyes and clays in aqueous suspension. *Química Nova*, vol. 23, No 6, pp. 818-824. ISSN 0100-4042.

Oliveira, M. L. (2003). Bentonite, In: *National Department of Mineral Research*, www.dnpm.gov.br/assets/galeriadocumento/sumariomineral2004/BENTONITA%202004.pdf, accessed in July of 2011.

Otokunefor, T. V. & Obiukwu, C. (2005). Impact of Refinery Effluent on the Physicochemical Properties of a Water Body in the Niger Delta. *Applied Ecology and Environmental Research*, vol. 3, pp. 61-72. ISSN 1589-1623.

Pastre, I.A., Fertonani, F. L. & Souza, G. R. (2003). Spectro-eletrochemical study of clay-dye structured systems, *Eclética Química*, vol. 28, No. 1, pp. 77-78. ISSN 0100-4670.

Petrobras. 1997 (a). Organoclay for oil-based fluids in oil exploration. N-2258.

Petrobras. 1997(b). Organoclay tests for oil-based fluids in oil exploration. N-2259.

Petrobras. 1998(a). Viscosity for water-based fluids in oil exploration. N-2604.

Petrobras. 1998(b). Viscosity tests for water-based fluids in oil exploration. N-2605.

Pires, J., Carvalho, A. P. & Carvalho, M. B. (2000). Potential of clays with pillars and microporous adsorbents, In: *Department of Chemistry and Biochemistry, Lisbon Faculty of Sciences*, www.icp.csic.es/cyted/Monografias/Monografias1998/A5-169.pdf, accessed in August of 2011.

Portela, L.A.P., Baraúna, O. S., Cavalcanti, J.V.F.L., Da Motta, M. A. & Abreu, C. A. M. (2008). Development of an organoclay and maximizing their adsorptive properties. *Proceedings of 7° Encontro Brasileiro sobre Adsorção*, Campina Grande, PB, Brazil, 2008.

Qu, F., Zhu, L. & Yang, K. (2009). Adsorption behaviors of volatile organic compounds (VOCs) on porous clays heterostructures (PCH). *Journal of Hazardous Materials*, vol.170, pp. 7-12. ISSN 0304-3894.

Rianelli, R. S., Pereira, W. C. (2001). Clays as adsorbents of elements potentially contaminating. In: *Department of Industrial Chemical and Environmental Geochemical*, www.cetem.gov.br/publicacao/serie_anais_IX_jic_2001/Renatari.pdf, accessed in August of 2011.

Richards, S. & Bouazza, A. (2007). Phenol adsorption in organo-modified basaltic clay and bentonite. *Applied Clay Science*, vol. 37, pp. 133-142. ISSN 0169-1317.

Rodrigues, M. G. F., Silva, M. L. & Da Silva, M. G. C. (2004). Characterization of smectite clays for application in the removal of lead from synthetic effluents. *Cerâmica*, vol. 50, pp. 190-196. ISSN 0366-6913.

Ruthven, D. M. (1984). Principles of Adsorption and Adsorption Processes. 2 ed., New York, John Wiley & Sons, Inc., USA, 1984.

Sameer, A. A., Banat, F. & Leena, A. A. (2003). Adsorption of phenol using different types of activated bentonites. *Separation and purification technology*, v. 33, pp. 1-10. ISSN 1383-5866.

Sharmasarkar, S., Jaynes, W. F. & Vance, G. F. (2000). BTEX sorption by montmorillonite organo-clays: TMPA, ADAM, HDTMA. *Water, air and soil pollution*, vol. 119, pp. 257-273. ISSN 0049-6979.

Shen, Yun-Hwei. (2005). Phenol sorption by organoclays having different charge characteristics. *Colloids and Surfaces A: Physicochemical and Engineering Aspects*, vol. 232, pp. 143-149. ISSN 0927-7757.

Solomons, G. & FRYHLE, C. (2002). Organic Chemistry, 7 ed., São Paulo, LTC, 2002.

Souza Santos, P. (2002). Science and technology of clays. 2 ed., São Paulo, Edgard Blucher, 2002.

Teng, H. & To, C. (2000). Liquid-phase adsorption of phenol onto activated Carbons Prepared with Different Activation Levels. *Journal of Colloid Interface Science*, vol. 230, pp. 171-175. ISSN 0021-9797.

Tiburtius, E. R. L., Peralta-Zamora, P. & Leal, E. S. (2004). Contamination of waters by BTXs and processes used in the remediation of contaminated sites. *Química Nova*, vol. 27, pp. 441-446. ISSN 0100-4042.

Vianna, M. M. G. R., Valenzuela Díaz, F. R. & Büchler, P. M. (2001). Study of adsorption of organic components of vinasse in organophilic smectites. *Proceedings of 21° Congresso Brasileiro de Engenharia Sanitária e Ambiental*, João Pessoa, PB, Brazil, 2001.

Vieira, A. L., Hanna, R. A., Vieira Coelho, A. C., Valenzuela-Díaz, F. R. & Souza Santos, P., Franca's smectites as organophilic clays. *Proceedings of 8° International Congress on Applied Mineralogy*, Águas de Lindóia, SP, Brazil, 2004.

Viraraghavan, T., Alfaro, Flor de Maria. (1998). Adsorption of phenol from wastewater by peat, fly ash and bentonite. *Journal of Hazardous Materials*, vol. 57, pp. 59-70. ISSN 0304-3894.

Wake, H. (2004). Oil refineries: a review of their ecological impacts on the aquatic environment. *Estuarine, Coastal and Shelf Science*, vol. 62, pp. 131-140. ISSN 0272-7714.

Worrall, M. & Zuber, I. (1998). Control VOCs in Refinery Wastewater. In: *Process Optimization Conference*, www.amcec.com/case3.html, accessed in August of 2011.

Yasser E. N. & Jamal S. (2004). Adsorption of benzene and naphthalene to modified montmorillonite. *Journal of food, agriculture & environment*, vol. 3, pp. 295-298. ISSN 1459-0255.

Yilmaz, N. & YAPAR S. (2004). Adsorption properties of tetradecyl and hexadecyl trimethylammonium bentonites. *Applied Clay Science*, vol 27, pp. 223-228. ISSN 0169-1317.

16

Safety Cognitive Concept and Countermeasure Analysis for Petrochemical Industry

Mu Shan-Jun, Shi Hong-Xun, Wang Xiu-Xiang,
Zhang Xiao-Hua and Bian Min
China Petroleum & Chemical Corporation
Qingdao Safety Engineering Institute; Qingdao Shandong
China

1. Introduction

In recent years, we collect a large number of accidents about the petroleum and petrochemical enterprise, summarized the data analysis shows that, along with the progress of science and technology, the improvement of technical equipment, petroleum and petrochemical enterprise in essence safety equipment has made rapid progress, but also see the man-made accidents still occur frequently, makes the enterprise safety production situation facing new challenges. In daily work, employees why always there will be some unsafe behavior, the main reason is enterprise managers in the safety cognition concept is still not aware of the employee behavior of the importance of safety, thus let employees formed "see be used to, dry to, used to" idea, the idea for a good safe for behavior is an advantage, but for unsafe behavior as it may cause an accident. So from safety cognition concept to looking for reasons of the occurrence unsafe behavior, thus looking for breach in behavior safety management, through the application of behavior safety management tools, regulate employee behavior of work.

2. Widespread safety cognitive concept

Thought Analysis of petroleum and petrochemical enterprise field work activities, observe homework personnel work behavior and interview with the staff, using the research methods of system theory, research and get enterprise employees and managers widespread security cognitive concept.

Through the analysis of the safety of the employees and managers cognitive concept, we find inconsistency in the enterprise employees and managers. From the two aspects to this paper discusses enterprise employees and managers of the safety cognitive concept.

1. Employees of the cognitive concept generally
   In the enterprise, the staffs exist in the following cognition about unsafe behavior :
   - Unsafe behavior can improve the work efficiency.
   - Unsafe behavior may not cause an accident.
   - If know accident, or should be safety work.

Due to the shortcut and less effort to finish the job is the instinct of a man. In the work it will often be not according to procedures system operating practices, this operations mostly are unsafe behavior. In the process of operation, the unsafe behavior of the employees on one hand is that the employees don't by the operation regulation of generation for the saves time and effort. On the other hand may be produce the unsafe behavior that due to the managers blind pursue assignments, ignore security operations etc.Once the accident happened, safety awareness of employees will have been strengthened. For the management of the unreasonable requirement they will also put forward some suggestions, but improved safety awareness will fade with time passing by (Fig 1).

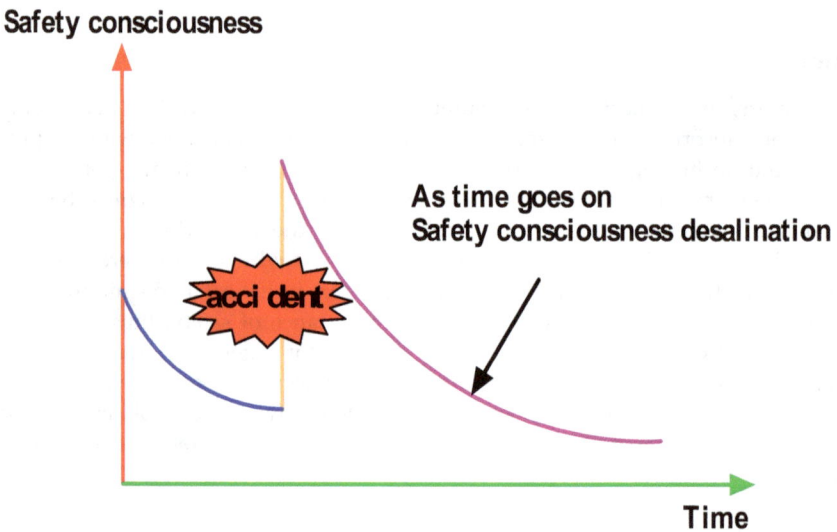

Fig. 1. Employee of safety consciousness and time relationship.

At the same time, due to the small scope of serious accident frequency will not be too high, It will make the staff produce the idea that is "unsafe behavior may not cause the accidents, unsafe behavior can improve the work efficiency". In addition, there is another important factor that is employees of risk cognition. In conventional practice, employees of risk in cognitive on one hand is to rely on their own in the process of operation risk know common sense, On the other hand is enterprise managers for enterprise of various kinds of activities of the risk of system analysis and understanding, and through different means told operators, such as shift-overlap before operation, safety identification tips, Into the factory education, etc. But because the employee safety consciousness, they lack the necessary risk cognitive common sense. This is the main causes of the unsafe behavior. Thus, it can come to the employees of safety cognitive mode (see fig 2).

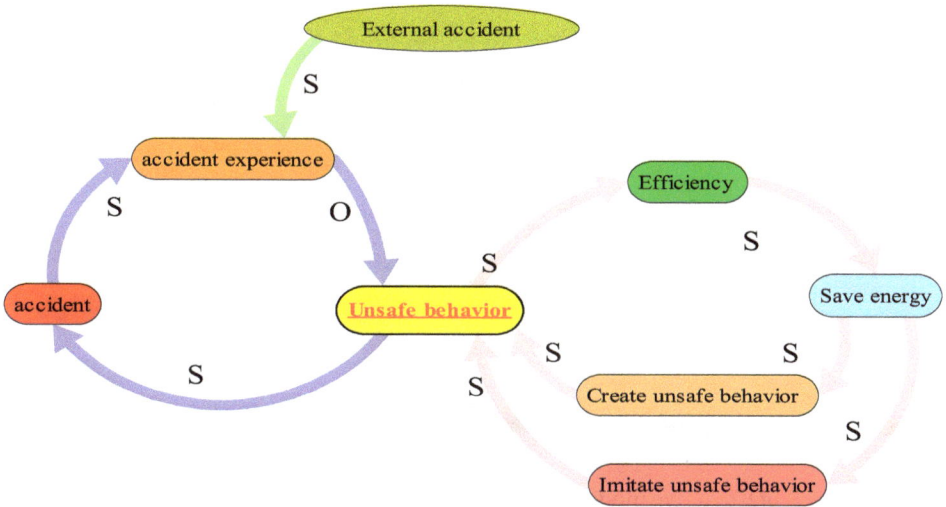

Note: S is Positive correlation; O is Negative correlation

Fig. 2. Employee of Cognitive model.

2. Managers of the cognitive concept generally
    In the enterprise, the managers exist in the following cognition about unsafe behavior：
    - Unsafe behavior is the key link, mechanical intergrity is the basis of production safety
    - If unsafe behavior gradually spread, it will make the vicious cycle of production safety. If unsafe behavior gradually reduced, it will push the virtuous cycle of production safety.

From the view of the managers, unsafe behavior of accidents will increase enterprise's safety management of pressure and affect the normal operation of the enterprise, further influence enterprise overall efficiency. At the same time, under pressure in the business, some managers will have to take the measures put into lower safety, but if the safety input is reduced, it will affect the equipment of intrinsically safe and safety training, this will cause unsafe behavior and the unsafe condition of the increase. It will lead to the increase of the accidents and form the vicious circle.

On the contrary, if the managers take measures to reduce unsafe behavior and unsafe condition, it will lower accident rates. The managers can put the energy and resources into equipment of intrinsically safe and employee safety quality improvement. It will further enhance the enterprise the overall level of safety management. Thus, it can come to the managers of safety cognitive mode (see fig 3).

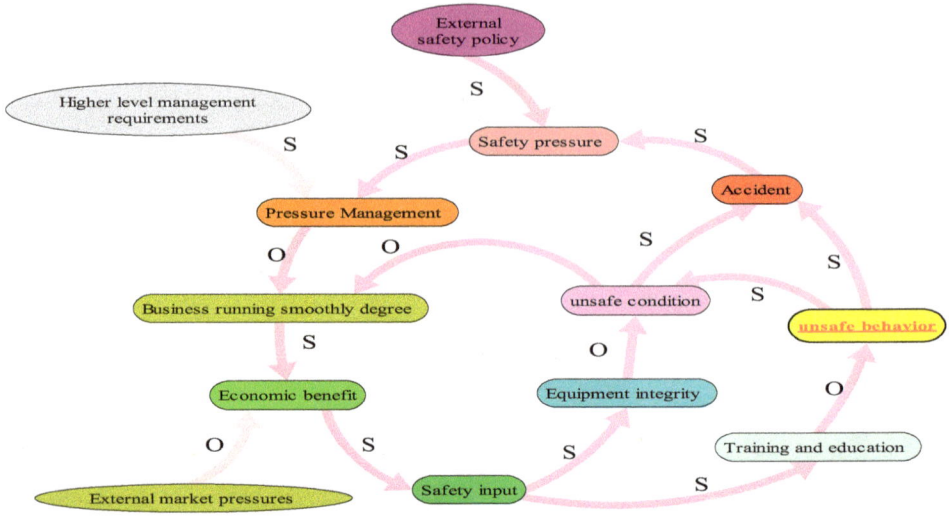

Fig. 3. Managers of Cognitive model.

3. Recommend common safety cognitive concept

Through the analysis of enterprise employees and managers of unsafe behavior cognition, and combined with the employees and managers of the safety cognitive model, it can come to the employees and managers of common safety cognitive concept. That is the uniform safety cognitive concept, see fig 4.

The common safety the cognitive concept of idea lies in:

- Prevent accidents and avoid damage are our common goals and commitments.
- Strengthen communication, promote safety behavior and develop HSE cultural are to promote safe production of the benign circulation pry points.
- Pay attention to the draw lessons from the accident outside enterprises is the best way.
- Eliminate equipment safety problems and Improve device integrity level are the basis of production safety.

Safety needs the managers and employees involved in. The managers and employees follow the principle of uniform management and form the common perception mode. Eliminate unsafe behavior is the key to prevent accidents. Eliminate unsafe behavior is the enterprise safety development necessary means. Eliminate unsafe behavior is everyone's responsibility. Only in form the correct on the basis of common cognitive mode, the system of safety management can be enforced, the safety management measures can be truly carried out. Through the model the virtuous cycle of their own, it can improve the device of essence safety, it can strengthen and standardize the safety of the employees behavior. It will radically reduce or eliminate unsafe condition of object and the unsafe behavior of the people, so as to achieve the purpose of reducing and put an end to the accidents.

We analysis the concept of the common safety cognitive purpose that is to through the HSE idea infusion, safety culture atmosphere build, formation and strengthen the correct cognition model, standard work behavior, find and eliminate unsafe behavior and unsafe condition.At the same time, it makes enterprise employees and managers know to eliminate unsafe behavior and unsafe condition. It not only need to the employees own safety consciousness and homework habit of ascension,but also need to enterprise managers in the construction of safety culture, safety input, etc to ascend, and take effective measures and means of management, to reduce the man-made accidents. It will promote the enterprise the safety operation level from the whole.

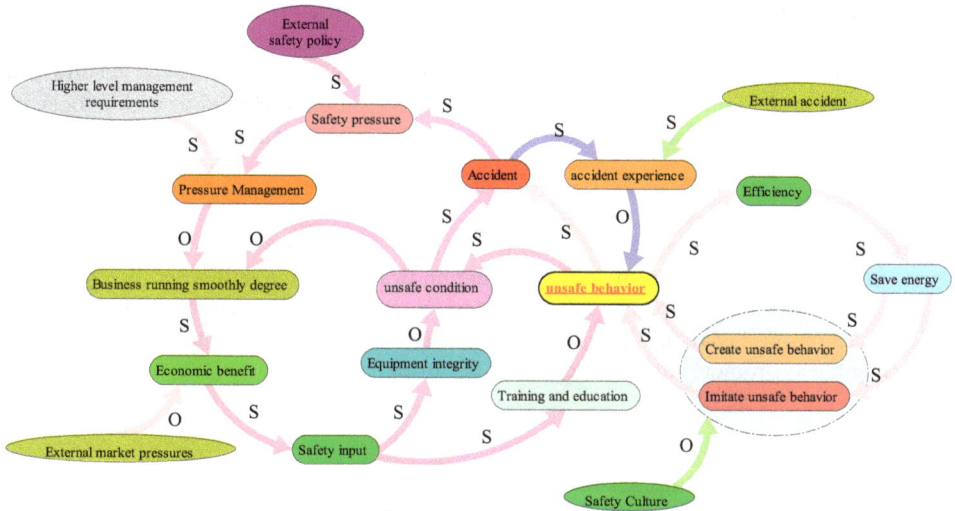

Fig. 4. Common cognitive model.

4. Effective solution

Through the common cognitive model that can be seen, in addition to model their internal circulation outside, the whole cycle graph of the external input have external safety policy, superior management requirements, external market pressure, external accidents and safety culture. In the five aspects, for enterprise in solving unsafe behavior of the effective input is only external accidents and safety culture. For external accidents, it's just a short accident education process, it can't form long-term input to reduce the enterprise employees's unsafebehavior. But safety culture as a stable and effective long-term input, the enterprise through the safety culture construction and propaganda, it can gradually reduce the incidence of enterprise unsafe behavior and effectively reduce the accident rates.

The construction of safety culture of the enterprise is a long process,in the process of how to find and correct the employees of unsafe behavior, to gradually regulate the employees of safety behavior is particularly important. In this respect the effective method is to implement behavior safety management tools. This paper introduced at present in SINOPEC widely applied behavior safety management tool - HSE observation card.

HSE observation card is a method that is to encourage employees voluntarily observed the operating personnel behavior. Its purpose is to encourage safety behavior, stop and correct unsafe behavior. Through the effective communication, it can be to strengthen operation personnel to safety consciousness and standardize safety behavior. At the same time, it will see that the work environment and facilities and the use of tools, and found that and deal with unsafe condition of object, effectively avoid and reduce accident and the occurrence of injury.

5. Conclusion

If we will enhance safety culture and standardize employee safety behavior, it cannot leave the application of behavior-based safety management tools. The behavior-based safety management tools are an important means that is to effectively promote the safety culture of enterprise. At present, in SINOPEC comprehensive application of HSE observation tools, its core is to conduct field observation and analysis and communication, to interfere with the way or intervention, make people know the dangers of unsafe behavior, prevent and eliminate unsafe behavior.

Implementation of behavior safety management and correction employee in the work unsafe behavior, it is at home and abroad common practice in the industry. Through these tools and methods are to implement and promote, it can reduce the frequency of occurrence of employee unsafe behavior, thus it can avoid employees of unsafe behavior which led to the accident, and promote enterprise of safety performance.

6. References

Sun Ai-jun, LIU Mao.The Implementing Predicament of Behavior-based Safety Management Theory and Its Solution[J]. China Safety Science Journal, 2009,(09).

Xiu Jing-tao. Exploration and practice of behavior safety management[J]. Journal of Safety Science and Technology, 2008,(04).

Zhang Ya-wei; Wang Bing-jian.Theoretical foundation of three violation psychology in coal mine [J]. Shandong Coal Science and Technology, 2009,(03).

Sun Shu-ying. Unsafe behavior in furniture enterprise[J]. Journal of Safety Science and Technology, 2009,(01).

Cao Qing-ren. Study on the Cognitive Diversities between Managers and Workers in Controlling Unsafe Behavior[J]. China Safety Science Journal, 2007,(01).

Zhao En-qian. Case Analysis and Psychological Discussion on Behavior Safety [J].Petrochemical Safety Technology, 2006,(04).

Liang Li. An Analysis of Violation Behavior and Psychological Factors[J]. China Safety Science Journal, 2006,(08).

Lu Bao-he. Analysis and Control of Rule-Breaking Behavior[J]. China Safety Science Journal, 1997,(05).

Liu Yi-song. The Discuss of Unsafe Behavior in Safety Management[J]. West-china Exploration Engineering, 2005,(06).

Dai Licao; Engineer Zhang Li Huang Shudong. Analysis of Fundamental Causes of Man-made Errors in Complex Industrial System[J]. China Safety Science Journal, 2003,(11).

Permissions

The contributors of this book come from diverse backgrounds, making this book a truly international effort. This book will bring forth new frontiers with its revolutionizing research information and detailed analysis of the nascent developments around the world.

We would like to thank Dr. Vivek Patel, for lending his expertise to make the book truly unique. He has played a crucial role in the development of this book. Without his invaluable contribution this book wouldn't have been possible. He has made vital efforts to compile up to date information on the varied aspects of this subject to make this book a valuable addition to the collection of many professionals and students.

This book was conceptualized with the vision of imparting up-to-date information and advanced data in this field. To ensure the same, a matchless editorial board was set up. Every individual on the board went through rigorous rounds of assessment to prove their worth. After which they invested a large part of their time researching and compiling the most relevant data for our readers. Conferences and sessions were held from time to time between the editorial board and the contributing authors to present the data in the most comprehensible form. The editorial team has worked tirelessly to provide valuable and valid information to help people across the globe.

Every chapter published in this book has been scrutinized by our experts. Their significance has been extensively debated. The topics covered herein carry significant findings which will fuel the growth of the discipline. They may even be implemented as practical applications or may be referred to as a beginning point for another development. Chapters in this book were first published by InTech; hereby published with permission under the Creative Commons Attribution License or equivalent.

The editorial board has been involved in producing this book since its inception. They have spent rigorous hours researching and exploring the diverse topics which have resulted in the successful publishing of this book. They have passed on their knowledge of decades through this book. To expedite this challenging task, the publisher supported the team at every step. A small team of assistant editors was also appointed to further simplify the editing procedure and attain best results for the readers.

Our editorial team has been hand-picked from every corner of the world. Their multi-ethnicity adds dynamic inputs to the discussions which result in innovative outcomes. These outcomes are then further discussed with the researchers and contributors who give their valuable feedback and opinion regarding the same. The feedback is then collaborated with the researches and they are edited in a comprehensive manner to aid the understanding of the subject.

Apart from the editorial board, the designing team has also invested a significant amount of their time in understanding the subject and creating the most relevant covers. They scrutinized every image to scout for the most suitable representation of the subject and create an appropriate cover for the book.

The publishing team has been involved in this book since its early stages. They were actively engaged in every process, be it collecting the data, connecting with the contributors or procuring relevant information. The team has been an ardent support to the editorial, designing and production team. Their endless efforts to recruit the best for this project, has resulted in the accomplishment of this book. They are a veteran in the field of academics and their pool of knowledge is as vast as their experience in printing. Their expertise and guidance has proved useful at every step. Their uncompromising quality standards have made this book an exceptional effort. Their encouragement from time to time has been an inspiration for everyone.

The publisher and the editorial board hope that this book will prove to be a valuable piece of knowledge for researchers, students, practitioners and scholars across the globe.

List of Contributors

Azzouzi Messaouda
Ziane Achour University of Djelfa, Algeria

Popescu Dumitru
Politehnica University of Bucharest, Romania

Milena Dimitrova and Yordanka Tasheva
University "Prof. Assen Zlatarov"- Burgas, Bulgaria

Solomon Gabche Anagho
University of Dschang, Cameroon

Horace Manga Ngomo
University of Yaounde I, Cameroon

Douglas Falleiros Barbosa Lima and Fernando Ademar Zanella
Refinaria Presidente Getúlio Vargas – REPAR / PETROBRAS, Brazil

Marcelo Kaminski Lenzi and Papa Matar Ndiaye
Universidade Federal do Paraná – UFPR, Brazil

Anne-Claire Texier, Jorge Gómez and Flor Cuervo-López
Universidad Autónoma Metropolitana-Iztapalapa, Mexico

Alejandro Zepeda
Universidad Autónoma de Yucatán, Mexico

José Luis Velázquez Ortega and Suemi Rodríguez Romo
Facultad de Estudios Superiores Cuautitlán/UNAM, Cuautitlán Izcalli, Edo. de Méx., Mexico

Rachad Alami and Abdeslam Bensitel
Centre National de l'Energie, des Sciences et des Techniques Nucléaires, Morocco

Job García and Gabriel García
Instituto de Investigaciones Eléctricas, Mexico

Jyin-Wen Cheng
General Manager, Cepstrum Technology Corp., Ling-Ya Dist. Kaohsiung, Taiwan, ROC

Shiuh-Kuang Yang
Dept. Mechanical and Electro-Mechanical Engr., Natl. Sun Yat-Sen Univ., Kaohsiung, Taiwan, ROC

Ping-Hung Lee and Chi-Jen Huang
Taiwan Metal Quality Control Co., Ltd., Taiwan, ROC

Tao Huang
College of Resource and Environmental Science, Lanzhou University, Lanzhou, P.R, China

Hong Gao
Key Laboratory of Western China's Environmental Systems (Ministry of Education), College of Resource and Environmental Science, Lanzhou University, Lanzhou, P.R. China

Xingguo Cheng
Lanzhou Lubricating Oil R&D Institute of Petro China, Lanzhou, P. R. China

Hoda S. Ahmed and Mohammed F. Menoufy
Egyptian Petroleum Research Institute, Egypt

Yana Koleva and Yordanka Tasheva
University "Prof. Assen Zlatarov"- Burgas, Bulgaria

Hong-Qi Wang, Ying Xiong, Qian Wang, Xu-Guang Hao and Yu-Jiao Sun
College of Water Sciences Key Laboratory for Water and Sediment Sciences of Ministry of Education Beijing Normal University, Beijing, China

Jorge V.F.L. Cavalcanti
Universidade Federal Rural de Pernambuco, Brazil

César A.M. Abreu, Marilda N. Carvalho, Maurício A. Motta Sobrinho and Mohand Benachour
Universidade Federal de Pernambuco, Brazil

Osmar S. Baraúna
Instituto de Tecnologia de Pernambuco, Brazil

Mu Shan-Jun, Shi Hong-Xun, Wang Xiu-Xiang, Zhang Xiao-Hua and Bian Min
China Petroleum & Chemical Corporation, Qingdao Safety Engineering Institute, Qingdao Shandong, China